METHODS IN
ENZYMOLOGY

RNA Modification

METHODS IN ENZYMOLOGY

Editors-in-Chief

JOHN N. ABELSON AND MELVIN I. SIMON

Division of Biology
California Institute of Technology
Pasadena, California

Founding Editors

SIDNEY P. COLOWICK AND NATHAN O. KAPLAN

VOLUME FOUR TWENTY FIVE

METHODS IN
ENZYMOLOGY

RNA Modification

EDITED BY

JONATHA M. GOTT
Center for RNA Molecular Biology
Case Western Reserve University
Cleveland, Ohio

AMSTERDAM • BOSTON • HEIDELBERG • LONDON
NEW YORK • OXFORD • PARIS • SAN DIEGO
SAN FRANCISCO • SINGAPORE • SYDNEY • TOKYO
Academic Press is an imprint of Elsevier

ELSEVIER

Contents

Section II. tRNA Modifications   55

9. Mass Spectrometric Identification and Characterization of RNA-Modifying Enzymes   211

Tsutomu Suzuki, Yoshiho Ikeuchi, Akiko Noma, Takeo Suzuki, and Yuriko Sakaguchi

10. Chaplet Column Chromatography: Isolation of a Large Set of Individual RNAs in a Single Step   231

Takeo Suzuki and Tsutomu Suzuki

Section III. Sno-Mediated Modifications   241

Contributors

Isabelle Behm-Ansmant
Laboratoire de Maturation des ARN et Enzymologie Moléculaire, Nancy Université, Faculté des Sciences et Techniques,Vandoeuvre-les-Nancy, France

Christiane Branlant
Laboratoire de Maturation des ARN et Enzymologie Moléculaire, Nancy Université, Faculté des Sciences et Techniques,Vandoeuvre-les-Nancy, France

Céline Brochier-Armanet
Laboratoire de Chimie Bactérienne, Marseille, and Université de Provence, Aix-Marseille I, France

Bruno Charpentier
Laboratoire de Maturation des ARN et Enzymologie Moléculaire, Nancy Université, Faculté des Sciences et Techniques, Vandoeuvre-les-Nancy, France

Stephanie M. Chervin
Department of Medicinal Chemistry, College of Pharmacy, University of Michigan, Ann Arbor, Michigan

Valérie de Crécy-Lagard
Department of Microbiology and Cell Science, University of Florida, Gainesville, Florida

Wayne A. Decatur
Department of Biochemistry and Molecular Biology, University of Massachusetts, Amherst, Massachusetts

Stephen Douthwaite
Department of Biochemistry and Molecular Biology, University of Southern Denmark, Odense M, Denmark

Louis Droogmans
Laboratoire de Microbiologie et Institut de Recherches, Université Libre de Bruxelles, Bruxelles, Belgique

Jean-Baptiste Fourmann
Laboratoire de Maturation des ARN et Enzymologie Moléculaire, Nancy Université, Faculté des Sciences et Techniques, Vandoeuvre-les-Nancy, France

Maurille J. Fournier
Department of Biochemistry and Molecular Biology, University of Massachusetts, Amherst, Massachusetts

Keith Gagnon
Department of Molecular and Structural Biochemistry, North Carolina State University, Raleigh, North Carolina

George A. Garcia
Department of Medicinal Chemistry, College of Pharmacy, University of Michigan, Ann Arbor, Michigan

David E. Graham
Department of Chemistry and Biochemistry, and Institute for Cellular and Molecular Biology, The University of Texas at Austin, Austin, Texas

Elizabeth J. Grayhack
Department of Biochemistry and Biophysics, University of Rochester School of Medicine, Rochester, New York

Henri Grosjean
Institut de Génétique et Microbiologie, Université Paris–Sud, Orsay, France

Yoshiho Ikeuchi
Department of Chemistry and Biotechnology, Graduate School of Engineering, The University of Tokyo, Bunkyo-ku, Tokyo, Japan

Jane E. Jackman
Department of Biochemistry and Biophysics, University of Rochester School of Medicine, Rochester, New York

John Karijolich
Department of Biochemistry and Biophysics, University of Rochester Medical Center, Rochester, New York

Gérard Keith
Institut de Biologie Moléculaire et Cellulaire, Strasbourg, France

Finn Kirpekar
Department of Biochemistry and Molecular Biology, University of Southern Denmark, Odense M, Denmark

Jeffrey D. Kittendorf
Department of Medicinal Chemistry, College of Pharmacy, and Life Sciences Institute, University of Michigan, Ann Arbor, Michigan

Lakmal Kotelawala
Department of Biochemistry and Biophysics, University of Rochester School of Medicine, Rochester, New York

Gisela Kramer
Department of Chemistry and Biochemistry, The University of Texas at Austin, Austin, Texas

Fabrice Leclerc
Laboratoire de Maturation des ARN et Enzymologie Moléculaire, Nancy Université, Faculté des Sciences et Techniques, Vandoeuvre-les-Nancy, France

Xue-hai Liang
Department of Biochemistry and Molecular Biology, University of Massachusetts, Amherst, Massachusetts

Ben Liu
Dana Farber Cancer Institute, Harvard Medical School, Boston, Massachusetts

E. Stuart Maxwell
Department of Molecular and Structural Biochemistry, North Carolina State University, Raleigh, North Carolina

Yuri Motorin
Laboratoire de Maturation des ARN et Enzymologie Moléculaire, Nancy Université, Faculté des Sciences et Techniques, Vandoeuvre-les-Nancy, France

Sébastien Muller
Laboratoire de Maturation des ARN et Enzymologie Moléculaire, Nancy Université, Faculté des Sciences et Techniques, Vandoeuvre-les-Nancy, France

Hannu Myllykallio
Laboratoire de Génomique et Physiologie Microbienne, Université Paris-Sud, Orsay, France

Akiko Noma
Department of Chemistry and Biotechnology, Graduate School of Engineering, The University of Tokyo, Bunkyo-ku, Tokyo, Japan

Eric M. Phizicky
Department of Biochemistry and Biophysics, University of Rochester School of Medicine, Rochester, New York

Dorota Piekna-Przybylska
Department of Biochemistry and Molecular Biology, University of Massachusetts, Amherst, Massachusetts

Martine Roovers
Laboratoire de Microbiologie et Institut de Recherches, Université Libre de Bruxelles, Bruxelles, Belgique

Yuriko Sakaguchi
Department of Chemistry and Biotechnology, Graduate School of Engineering, The University of Tokyo, Bunkyo-ku, Tokyo, Japan

Stéphane Skouloubris
Laboratoire de Génomique et Physiologie Microbienne, Université Paris-Sud, Orsay, France

David Stephenson
Department of Biochemistry and Biophysics, University of Rochester Medical Center, Rochester, New York

Takeo Suzuki
Department of Chemistry and Biotechnology, Graduate School of Engineering, The University of Tokyo, Bunkyo-ku, Tokyo, Japan

Tsutomu Suzuki
Department of Chemistry and Biotechnology, Graduate School of Engineering, The University of Tokyo, Bunkyo-ku, Tokyo, Japan

Jaunius Urbonavicius
Laboratoire d'Enzymologie et Biochimie Structurales, Gif-sur-Yvette, France

Yi-Tao Yu
Department of Biochemistry and Biophysics, University of Rochester Medical Center, Rochester, New York

Xinxin Zhang
Department of Molecular and Structural Biochemistry, North Carolina State University, Raleigh, North Carolina

PREFACE

The presence of modified nucleotides in cellular RNAs has been known for over 40 years, and over 100 distinct RNA modifications have been characterized to date. While these modifications are often not essential, most contribute to the overall fitness of the organism. Because a significant proportion of genomic information is devoted to RNA modifications, this volume contains computational, bioinformatic, and genome-wide strategies for identifying and characterizing the activities involved, as well as the biochemical approaches that are more typically included in a methods volume.

This volume and its companion (Volume 424: *RNA Editing*) are meant to complement one another. The fields of RNA modification and editing overlap extensively, and it is often impossible to make clear distinctions between changes that are classified as RNA modifications and those that are referred to as RNA editing events (e.g., A to I changes in tRNAs vs. mRNAs). For the purpose of this series, I have designated chapters for the modification or editing volumes based on common conventions, which are largely historical rather than scientific in origin.

I wish to thank the contributors to this volume for their efforts and professionalism; it has been a pleasure reading their chapters. This volume is dedicated to my parents in appreciation for their unfailing love and support throughout my life.

JONATHA M. GOTT

METHODS IN ENZYMOLOGY

MODIFIED NUCLEOTIDES

CHAPTER ONE

IDENTIFYING MODIFICATIONS IN RNA BY MALDI MASS SPECTROMETRY

Stephen Douthwaite *and* Finn Kirpekar

Contents

Abstract

Posttranscriptional modifications on the base or sugar of ribonucleosides generally result in mass increases that can be measured by mass spectrometry. Matrix-assisted laser desorption/ionization mass spectrometry (MALDI-MS) is a direct and accurate means of determining the masses of RNAs. Mass spectra produced by MALDI are relatively straightforward to interpret, because they are dominated by singly charged ions, making it possible to analyze complex mixtures of RNA oligonucleotides ranging from trinucleotides up to 20-mers. Analysis of modifications within much longer RNAs, such as ribosomal RNAs, can be achieved by digesting the RNA with nucleotide-specific enzymes. In some cases, it may be desirable to isolate specific sequence regions before MALDI-MS analysis, and this requires a few additional steps. The method is applicable to the study of modified RNAs from cell extracts as well as RNA modifications added in cell-free *in vitro* systems. MALDI-MS is particularly useful in cases in which other techniques such as those involving primer extension or chromatographic analyses are not practicable. To date, MALDI-MS has been used to localize rRNA modifications that are involved in fundamental processes

Department of Biochemistry and Molecular Biology, University of Southern Denmark, Odense M, Denmark

Methods in Enzymology, Volume 425
ISSN 0076-6879, DOI: 10.1016/S0076-6879(07)25001-3

3

in protein synthesis as well as methylations that confer resistance to antibiotics. For several rRNA sites, MALDI-MS has served an essential role in the
identification of the enzymes that catalyze the modifications.

1. INTRODUCTION

Mass spectrometry (MS) is an extremely sensitive and accurate tool for
measuring the masses of biological macromolecules. In essence, application
of MS involves generation of gas phase ions of the molecules under study,
followed by partitioning the ions to measure their mass-to-charge (m/z)
ratios. The two most commonly used ionization methods are matrix-
assisted laser desorption/ionization (MALDI) and electrospray ionization
(ESI), and these are often coupled with time-of-flight (ToF) analysis to
measure the ion m/z values. MALDI-ToF (Karas and Hillenkamp, 1988)
has for many years been among the standard inventory of techniques used to
study proteins and their posttranslational modifications. When extended to
include ion selection and fragmentation by tandem MS, the technique
allows peptide sequencing in addition to providing positional and structural
information on posttranslational modifications. This chapter describes the
recent expansion of this instrumentation to include the analysis of RNA
sequences and modifications (Kirpekar et al., 2000). Much of what we
describe here can also be achieved using ESI-MS, and we discuss the relative
merits of the two MS techniques later in this chapter.

MALDI mass spectrometry can in principle be used to study posttranscriptional modifications in any type of RNA. Here we concentrate on the
localization of modifications in ribosomal RNAs (rRNAs) and deal with the
specific challenges that are inherent in analyzing RNA molecules that can
be larger than 1 million Daltons. A range of techniques, including MS, have
been used over the years to obtain a comprehensive map of the rRNA
modifications in Escherichia coli and have revealed 11 modified nucleotides
in 16S rRNA and 25 in 23S rRNA, consisting mainly of base and sugar
methylations and pseudouridylations (Andersen et al., 2004; Rozenski et al.,
1999). The posttranscriptional addition of these modifications is carried out
by chromosome-encoded enzymes, and the modifications are thought to
fine-tune various rRNA interactions involved in protein synthesis. We
term these "housekeeping" modifications to distinguish them from antibiotic resistance modifications that tend to be encoded by plasmids or trans-
posons and that have no observable beneficial effect on protein synthesis in
the absence of an antibiotic. More recently, the housekeeping modifications
in Thermus thermophilus 16S rRNA and 23S rRNAs have been characterized
by ESI-MS (Guymon et al., 2006) and MALDI approaches (Mengel-
Jorgensen et al., 2006), respectively; the former data join the growing
database of 16S rRNA modifications (McCloskey and Rozenski, 2005).

Many of the rRNA housekeeping modifications are conserved between different organisms, whereas others are specific to certain bacterial or archaeal groups. A major point of similarity is the spatial location of the modifications, and when superimposed on the ribosome crystal structures (Ban *et al.*, 2000; Harms *et al.*, 2001; Korostelev *et al.*, 2006; Schlünzen *et al.*, 2000; Schuwirth *et al.*, 2005; Selmer *et al.*, 2006), the modifications can be seen to cluster within several discrete regions concerned with essential ribosomal functions (Brimacombe *et al.*, 1993; Decatur and Fournier, 2002; Hansen *et al.*, 2002; Ofengand and Del Campo, 2004). It is also noteworthy that most of these modified rRNA regions participate in intermolecular RNA–RNA contacts (Korostelev *et al.*, 2006; Mengel-Jorgensen *et al.*, 2006). Although the specific functions of individual housekeeping modifications remain difficult to define, their collective importance for efficient protein synthesis has been demonstrated by the superior performance of authentic rRNAs compared with unmodified 16S (Krzyzosiak *et al.*, 1987) and 23S counterparts (Green and Noller, 1999; Khaitovich *et al.*, 1999).

Resistance modifications on the rRNAs are almost invariably methylations. The benefit of resistance methylation first becomes evident when the bacterium is challenged by an antibiotic, and growth can be facilitated at antibiotic concentrations several orders of magnitude higher than required to inhibit strains lacking the rRNA methylation (Gale *et al.*, 1981). Although many resistance methylations were mapped more than two decades ago (Cundliffe, 1989), others remain to be determined. Most recently, several previously unidentified sites of rRNA methylation have been defined by use of MALDI-MS, including sites associated with resistance to the ribosome-targeting antibiotics avilamycin (Treede *et al.*, 2003), capreomycin (Johansen *et al.*, 2006), chloramphenicol (Kehrenberg *et al.*, 2005), and telithromycin (Madsen *et al.*, 2005). In the following section, we outline general protocols for identifying both housekeeping and resistance modification sites in rRNAs.

2. EXPERIMENTAL STRATEGY

An overview of the strategy we use is depicted in Fig. 1.1. RNAs for MALDI-MS analysis can originate from *in vitro* transcription or from living cells. In the latter case, RNAs are normally harvested from mid-log phase cells growing exponentially under optimal conditions of nutrition, aeration, and temperature to ensure the best rRNA yield and quality. When modification enzymes of interest are expressed under particular growth conditions (such being induced by antibiotic), the procedure should be adjusted accordingly. For screening of an entire rRNA sequence, the small and large ribosomal subunits are separated by sucrose gradient centrifugation and are

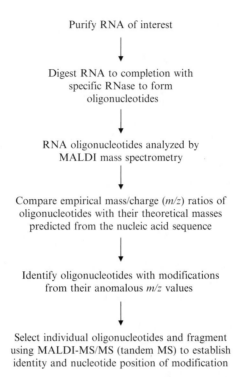

Figure 1.1   Outline of the strategy used to locate and characterize RNA modifications.

extracted to obtain purified 16S and 23S rRNAs, respectively. In cases in which a specific rRNA region is to be studied, no prior separation of subunits is required, and sequences of approximately 50 nucleotides in length can be isolated directly from a mixture of total RNAs by use of a site-directed hybridization procedure.

The purified RNA is digested with ribonucleases such as the guanosine-specific RNase T1 or the pyrimidine-specific RNase A to provide oligonucleotides of suitable length for MS analysis. A denaturing agent is used to ensure complete digestion of the RNA; 3-hydroxypicolinic acid (3-HPA) combines the fortuitous properties of being a good denaturant and also an excellent matrix for the MS ionization step that follows. The digested RNA is usually already pure enough for mass analysis, although, if required, it can be purified further on a small reverse-phase column. The RNA oligonucleotides are ionized by MALDI by use of 3-HPA as the matrix; spectra are recorded in the positive ion mode on an MS instrument with delayed ion extraction and a reflector ToF mass analyzer. Analysis of RNA oligonucleotides from 3-mers up to approximately 20-mers works well with this instrumental setup.

For RNAs of known sequence, the expected masses of the oligonucleotide are then compared with the experimentally derived m/z values to reveal any discrepancies that could be the result of base substitutions or modifications. Oligonucleotides of particular interest (identified by their anomalous mass) are selected for further analysis by tandem MS (also referred to as MS/MS). Tandem MS analysis is carried out in essentially three stages: first, an analyzer selects an oligonucleotide ion of interest (usually within a 2-Da size window) to the exclusion of ions of all other masses; second, the selected oligonucleotide is subjected to fragmentation by collision with an inert gas (here, argon is used); and last, the masses of the resultant fragments are recorded, and these will include a range of progressively shortened oligonucleotides, as well as single nucleotides, nucleosides, and bases. The sequence and modification sites in an RNA oligonucleotide can generally be deduced by piecing together the observed fragment masses.

3. EXPERIMENTAL PROCEDURES

3.1. Isolation of RNA sequences

RNA transcripts used for *in vitro* modification assays are generally between 30 and 150 nucleotides in length and are synthesized from DNA oligodeoxynucleotides or plasmid templates encoding the T7 RNA polymerase promoter (Hansen *et al.*, 2001; Krupp and Soll, 1987). RNAs are recovered from the aqueous phase by ethanol precipitation and are redissolved in 2.5 μl H_2O (double-distilled in all cases). After incubation of RNA (usually 4 pmol per sample) with a modification enzyme, reactions are stopped by extraction with phenol and chloroform. The modified RNA transcripts are purified on reverse-phase Poros 50R2 columns (PerSeptive Biosystems) made in-house (Kirpekar *et al.*, 1998).

In studies on authentic RNAs, these are extracted from mid-log phase cells harvested by centrifugation. This procedure works for most Gram-positive and Gram-negative bacteria: cells are washed twice with 10 mM Tris-HCl, pH 7.5, 10 mM $MgCl_2$, 100 mM NH_4Cl at 4°, before lysis by sonication in the same buffer. Cell debris is removed by centrifugation at 30,000g for 10 min. Ribosomes are then pelleted from the supernatant by centrifugation at 30,000g for 18 h at 4°. The ribosomes are gently resuspended into 10 mM Tris-HCl, pH 7.5, 2 mM $MgCl_2$, 100 mM NH_4Cl, 1 mM dithiothreitol, and ribosomal subunits are fractionated on sucrose gradients in the same buffer. The large and small subunits are collected separately; the rRNA is extracted from the ribosomal proteins by use of phenol and chloroform and is recovered by ethanol precipitation (Hansen *et al.*, 1999).

If specific regions of RNAs are to be investigated, sequences can be isolated by hybridization to complementary oligodeoxynucleotides of approximately 50 residues (Andersen *et al.*, 2004). For each region to be isolated, rRNA at 30 pmol is heated with a 10-fold molar excess of oligodeoxynucleotide for 1 min at 80–90° in 100 μl of 60 mM HEPES (pH 7.0), 125 mM KCl, and cooled slowly during 2 h to 45°. The resultant DNA–rRNA hybrid is digested at 37° with 30 units Mung bean nuclease (New England Biolabs) and 0.5 μg of RNase A to remove the unprotected regions of rRNA. The 50-base-pair hybrid is phenol extracted and recovered by ethanol precipitation; the protected rRNA fragment is released on a denaturing polyacrylamide gel containing 7 M urea and visualized by ethidium bromide staining. The RNA band is excised and extracted by incubation overnight in 100–250 μl 2 M NH$_4$OAc (pH 5.3) at 4°. The RNA is precipitated with two vol. of a 1:1 mixture of ethanol and isopropanol (Andersen *et al.*, 2004). Alternately, rRNA fragments can be isolated by annealing oligodeoxynucleotides at sites flanking the sequence of interest; the rRNA is then cleaved with RNase H at the hybridization sites, and the rRNA fragments are separated on denaturing polyacrylamide gels (Hansen *et al.*, 2002).

One point worthy of note is that precipitation with ethanol/ammonium acetate serves not only to recover the RNA but also displaces metal cations. Metal cations, in particular sodium and potassium ions, form RNA salts that distribute the signal of gas phase RNA ions over numerous masses and can severely complicate the interpretation of mass spectra.

3.2. Digestion of RNA to oligonucleotides

The RNA samples are further reduced in size for optimal analysis by MALDI-MS. Digestion of RNA with the nucleotide-specific RNases A and T1 yields fragments of predictable and suitable sizes. The denaturant 3-hydroxypicolinic acid (3-HPA) is used to ensure that RNAs are digested to completion; some denaturants that are commonly used in other techniques (e.g., urea) interfere with MS analysis and should be avoided. RNA (0.1 μg) is digested at 37° for 4 h in 1–2 μl of 50 mM 3-HPA containing either 0.2 μg RNase A or 10–20 units of RNase T1 (USB). Digestion with RNase T1 can leave a substantial proportion of 2′–3′ cyclic phosphate intermediates, giving rise to oligonucleotides with masses 18 Da less than the 3′-linear phosphate end products. A mixture of these two phosphate forms gives a more complex mass spectrum. The cyclic intermediates can be removed by hydrolyzing to the linear phosphate form: HCl is added to the samples to a concentration of 0.2 M; after 30 min at room temperature, 20 μl of water is added, the samples are dried under vacuum, and are then redissolved in 2 μl H$_2$O.

The RNase digestion products are normally taken directly for mass spectrometric analysis, although their purity can be improved by ion pairing reverse-phase chromatography. Triethyl ammonium acetate (TEAA), pH 7.0 is used as the ion pairing reagent with Poros 50R3 (Applied Biosystems, Toronto, Canada) as the chromatography material; short oligonucleotides are eluted with 10 mM TEAA/6% acetonitrile and larger oligonucleotides with 10 mM TEAA/25% acetonitrile (Kirpekar *et al.*, 1998). Eluants are dried and redissolved in 1–2 μl H$_2$O.

If the ribonuclease digestion is performed directly on a full-length rRNA, many of the resultant oligonucleotides will have similar or identical masses and produce a complex spectrum. The interpretation of such data can be challenging (Fig. 1.2). The use of smaller *in vitro* transcripts or preselection of sequences (as described previously) gives rise to less ambiguous spectra (Fig. 1.3).

3.3. MALDI mass spectrometry analysis

Mass spectra are recorded on a PerSeptive Voyager-DE STR mass spectrometer (Applied Biosystems) with a reflector ToF mass analyzer in positive ion detection mode. Samples for MALDI-MS are prepared by mixing the RNA digestion products (1–2 μl) on the target with 0.7 μl of 0.5 M 3-HPA matrix and a small volume (~0.1 μl) of ammonium-loaded cation exchange material (Nordhoff *et al.*, 1993). The sample is left to air dry, and as much cation exchange material as possible is removed under a microscope by use of a micropipette tip. Although some other matrices yield higher sensitivity and/or resolution for oligonucleotide analysis by MALDI-MS (Asara and Allison, 1999; Zhu *et al.*, 1996), we prefer the 3-HPA matrix. In addition to its favorable denaturing properties, 3-HPA shows less bias between RNAs of different composition or sequence, and 3-HPA can thus be used for the simultaneous detection of most oligonucleotides within a complex mixture.

MALDI mass spectra of RNA are dominated by singly charged ions. These ions will be protonated when viewed in the positive ion detection mode, giving an *m/z* value equal to the oligonucleotide mass in Daltons plus one hydrogen (1.01 Da). Correspondingly, the *m/z* values of deprotonated ions viewed in the negative ion detection mode will be the oligonucleotide mass minus one hydrogen. Contrary to intuitive expectations for nucleic acids, MALDI yields signals of nearly equal intensity for oligonucleotides in both ion detection modes (Wu *et al.*, 1994). In our experience, negative ion detection gives a modest increase in sensitivity, although the positive ion detection provides better instrument stability and is thus preferable in most cases. Spectra can be smoothed and calibrated by use of the manufacturer-supplied "*Data Explore*" software or "*m/z*" freeware (Proteometrics Inc). The masses and *m/z* values used throughout are calculated from

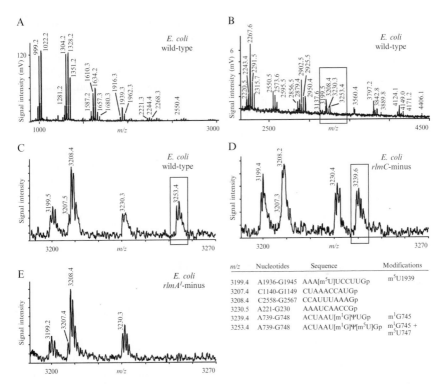

Figure 1.2 Examples of complex MALDI mass spectra resulting from RNase T1 digestion of the *E. coli* large ribosomal subunit RNAs. The resultant oligonucleotides are a mixture of products from 23S rRNA and 5S rRNA, although products from the larger 23S rRNA dominate the spectrum. All fragments of trinucleotides and larger are depicted here and were eluted in two steps: (A) The smaller digestion products eluted with 6% acetonitrile; (B) larger oligonucleotides eluted with 25% acetonitrile. The boxed region is enlarged in (C) to show the m/z 3190 to m/z 3270 region. In addition to the wild-type CP79 strain, rRNAs from two other *E. coli* strains were analyzed. (D) The signal in the *rlmC* strain at m/z 3239.6 is lighter by 14 Da than the corresponding wild-type signal (oligonucleotide 5'-ACUAAU[m^{1}G]Ψ[m^{5}U]Gp at m/z 3253.4). (E) In the IB10 rRNA, this oligonucleotide is cleaved by RNase T1 at the unmethylated (and reactive) nucleotide G745 to forming smaller oligonucleotides; IB10 lacks the *rlmA^I* gene that encodes the specific m^{1}G methyltransferase. The theoretical monoisotopic masses are given in the last panel for the singly protonated RNase T1 fragments of 23S rRNA in the region of interest; these match the measured m/z values to within 0.2 Da. These analyses determined that the *rlmC* gene (also termed *ybjF* or *rumB*) encodes the m^{5}U methyltransferase specific for position 747 (Madsen *et al.*, 2003). The peak at m/z 3199.4 is a decanucleotide from 23S rRNA positions 1936 to 1945, and includes a methyl group at m^{5}U1939 that is added by the *rlmD* gene product (previously termed *ygcA* or *rumA*) (Agarwalla *et al.*, 2002; Andersen and Douthwaite, 2006). The spectra have been electronically smoothed using the Proteometrics Inc "*m/z*" software program. The jagged nature of the peaks visible in the enlargements reflects the natural isotopic distribution of ^{12}C and ^{13}C in the RNA (see text).

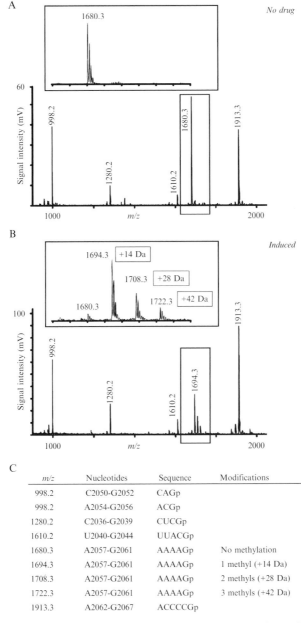

Figure 1.3 Example of a MALDI mass spectrum resulting from digestion of a 53-nucleotide rRNA sequence isolated by use of the hybridization technique (Andersen *et al.*, 2004). Spectra were obtained after RNase T1 digestion of the *Mycobacterium bovis* 23S rRNA region from nucleotides G2035 to G2087 (Madsen *et al.*, 2005). (A) Control rRNA from cells grown without induction. (B) Sample from cells grown in presence

the monoisotopic masses of the elements (i.e., the mass of the most abundant isotope of each element).

The mass spectrometer resolving power and accuracy required for these studies is dictated by the mass difference between uridine and cytidine (the former nucleoside is 0.98 Da heavier). Suitable performance is provided by MALDI instruments equipped with delayed ion extraction and a reflector ToF mass analyzer, and these are now standard features on most commercially available instruments. Sufficient resolution is provided up to approximately m/z 6000 (approximately 20 nucleotides), and oligonucleotides resulting from RNase A or T1 digestion only rarely exceed this size (Zhang et al., 2006). The delayed ion extraction parameters can be varied for optimal resolution at different m/z ranges, typically using one set of parameters for ions below m/z 4000 and another set for ions above this value. Mass accuracy of the RNA signals is generally better than 0.1 Da at the lower end and approximately 0.3 Da at the top end of the m/z range, and therefore the mass difference between pyrimidines remains distinguishable. At the lower end of the mass spectrum, mononucleotides and dinucleotides are not normally included in the analyses, because they become obscured by intense signals from matrix and buffer components.

Above m/z 6000, mass determination is hindered by two main factors. First, the sensitivity of detection decreases with larger mass (mainly because of detector design, but large ions also have a greater tendency to fragment before reaching the detector). The second factor stems from the natural distribution of carbon isotopes (the ratio of $^{12}C/^{13}C$ is approximate 99:1), which leads to a progressive broadening of the isotopic distribution with increasing mass. This is illustrated by the data shown here. For example, in the relatively low mass region showing the pentanucleotide AAAAGp (Fig. 1.3A), the predominant signal arises from the monoisotopic component (containing only ^{12}C) at m/z 1680.3; smaller peaks are evident at m/z 1681.3, m/z 1682.3, and m/z 1683.3 (containing one, two, and three ^{13}C atoms), but these do not hamper interpretation. In comparison, the peaks arising from the four decanucleotides in Fig. 1.2E are appreciably broader, and the monoisotopic signal is no longer the largest component. Nonetheless, interpretation of the m/z 3199.2 and m/z 3230.3 signals remains unambiguous, whereas the other two decanucleotides that differ in composition by only a cytidine or uridine (monoisotopic m/z values of 3207.4 and

---

of the macrolide antibiotic erythromycin, which induces expression of the *erm(37)* methyltransferase gene. The spectral region around the pentanucleotide containing A2058, AAAAGp (m/z 1680.3), is enlarged (boxes) to show ions with one, two, and three methyl groups at m/z 1694.3, 1708.3 and 1722.3, respectively. (C) The calculated singly protonated masses of the RNase T1 fragments from the G2035–G2087 sequence match the experimentally measured m/z values in (A) and (B) to within 0.1 Da.

3208.4, respectively) presents a challenging, but not impossible task, to unravel. Eventually, in appreciably longer RNA fragments, the binomial distribution of the carbon isotopes results in broad polymeric spectral signals with an indiscernible monoisotopic component. As can be imagined, signals from two 20-mer RNA sequences differing by only a single cytidine or uridine would be impossible to deconvolute.

3.4. Tandem mass spectrometry

In tandem mass spectrometry analysis, individual oligonucleotides of interest are isolated and subjected to controlled fragmentation. Oligonucleotides of interest are identified by use of regular MALDI-MS, as illustrated in Fig. 1.3, where the *M. tuberculosis* 23S rRNA oligonucleotide AAAAGp was shown to contain one, two, or three methyl groups. RNA samples were prepared and digested in an identical manner to that described previously for regular MALDI-MS; the oligonucleotide ions at m/z 1680.3, 1694.3, 1708.3, and 1722.3 (Fig. 1.3) were selected individually, fragmented, and the masses of the resultant fragments were recorded on a MicroMass MALDI Q-TOF Ultima instrument (Waters, Manchester, UK) in positive ion mode (Madsen *et al.*, 2005; Mengel-Jorgensen and Kirpekar, 2002) (Fig. 1.4).

The degree of fragmentation is proportional to the energy applied through collision with argon gas, and this generally varies between 30 and 110 eV. Varying the collision energy results in a wider range of fragments and a final spectrum with sequence-informative ions, as well as nucleosides/nucleotides, bases, and ribose derivatives that collectively can be used to pinpoint the location of a modification. The sequence-informative ions are categorized by use of the nomenclature of McLuckey *et al.* (1992) according to the position of the break in the phosphodiester-backbone (Fig. 1.4D). The RNA sequence can be read from the m/z difference of the y-ions, which are the predominant ions seen with the approach described here (Fig. 1.4). Masses of the G, A, U, and C nucleotide residues are 345.05 Da, 329.05 Da, 306,03 Da, and 305.04 Da, respectively, and thus a methyl group will reveal its position by addition of 14.02 Da mass difference (Fig. 1.4A). For a detailed discussion on fragmentation of singly protonated oligonucleotide ions, see Andersen *et al.* (2006).

Tandem mass spectra are calibrated from the mass of the parent ion and are smoothed by use of the MassLynx software supplied by the instrument manufacturer. Various software packages are available to aid calculation of RNase digestion fragments and interpretation of the tandem mass spectra. We have used Mongo Oligo Mass Calculator (http://www.medlib.med. utah.edu/masspec/mongo.htm), Nuke (http://www.zebra-crossing.de/ software/index.html), and an in-house modified version of GPMAW (Lighthouse Data, Odense, DK).

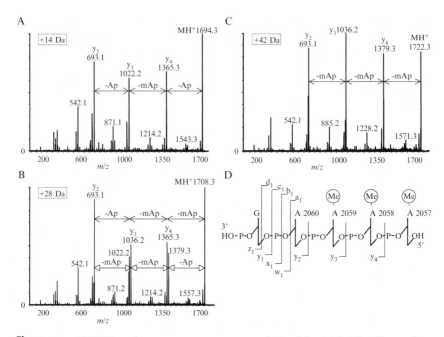

Figure 1.4   MALDI tandem mass spectrometry analysis of the AAAAGp oligonucleotides. (A) The location of the methyl groups can be seen from the shift in the y–ions peaks: loss of an unmodified adenosine nucleotide corresponds to a difference of 329.1 Da, whereas a singly methylated adenosine is 343.1 Da. The mass differences between the monomethylated parent ion (1694.3 m/z), the y_4 ion (1365.3 m/z), the y_3 ion (1022.2 m/z), and the y_2 ion (693.1 m/z) clearly show the methyl group is located on the second adenosine (A2058). (B) The 1708.3 m/z peak contains two methyl groups and is composed of two ion species, one with single methyl groups at A2057 and A2058 (open arrowheads) and the other with methyl groups at A2058 and A2059 (closed arrowheads). (C) The positions of the three methyl groups in the 1722.3-m/z ion are unambiguous, and a single methyl group resides on each of the nucleotides A2057, A2058, and A2059. The structures were verified by other ions present in lower abundance; one such series of ions from m/z 542.1–1571.3 (y–ions missing the guanine base) is indicated. (D) The fragmentation of the RNA backbone results in a, b, c and d ions from the 5′-end, and w, x, y and z ions from the 3′-end (McLuckey *et al.*, 1992), as shown on the phosphate bridging G2061 and A2060. The y–ions predominate in the instrumentation setup used here (Kirpekar and Krogh, 2001). The nucleotide sequence is shown in the reverse (3′–5′) orientation to align the schematic positions of y_2, y_3, and y_4 ions with their corresponding spectral signals. Also indicated are the three N^6-adenosine methylations (Me) added by the Erm(37) methyltransferase to confer antibiotic resistance (Madsen *et al.*, 2005).

3.5. Comparison of MALDI and ESI techniques

The main difference that distinguishes MALDI and ESI for the type of analysis described here is that MALDI mass spectra are dominated by singly charged ions, whereas ESI generates multiply charged ions. MALDI is directly applicable to determining the masses within a complex mixture, because each mass is represented by one signal. If ESI is used, the sample complexity has to be reduced to avoid problems that arise when interpreting mass spectra containing ions with overlapping charge distribution. The state-of-the-art technique for reducing the complexity of RNA digestion samples is ion pairing reverse-phase chromatography coupled online to the ESI mass spectrometer (Felden *et al.*, 1998; Kowalak *et al.*, 1995). For the types of analyses described here, where we wish to determine the component masses of a complex mixture, it can be concluded that ESI is a technically complicated means of providing essentially the same information as MALDI. Furthermore, because no chromatographic separation is necessary with the MALDI approach, less starting material is required.

For more advanced tasks, however, the analytical power of ESI has been enhanced in a manner that would be difficult to achieve with MALDI. The groups of Crain and McCloskey have devised a sophisticated setup for the identification of all nucleosides in an RNA by use of online liquid chromatography together with ESI-tandem MS, followed up with analysis of the RNase T1 fragments from the same RNA by use of online ion pairing reverse-phase chromatography combined with ESI-tandem MS (Felden *et al.*, 1998; Kowalak *et al.*, 1993). This approach, when pushed to its limits, can map sequences as large as 16S rRNA to near completion, localizing modifications to their positions within nucleotides (Guymon *et al.*, 2006). By use of standard and tandem MALDI approaches, we can determine the nucleotide location of a modification, the mass of the modification, and whether it is located on the base or the ribose. However, by itself, our MALDI setup does not tell us at which atom of the base a modification is located, nor is our setup as effective in catching all modifications. The coverage of modification positions can be improved by combining standard and tandem MALDI-MS analyses with biochemical techniques such as primer extension (Hansen *et al.*, 2002; Kirpekar *et al.*, 2005; Mengel-Jorgensen *et al.*, 2006).

In summary, the elaborated ESI-MS approach yields nearly complete information on modifications in a given RNA but is labor intensive, technically challenging, and requires larger amounts of starting material. The MALDI mass spectrometry approaches described here are relatively simple and need less sample, but the information acquired can be less comprehensive.

4. PERSPECTIVES AND CONCLUSION

Mass spectrometry is a valuable tool for identifying RNA modifications, particularly in cases where reverse transcriptase primer extension or chromatographic approaches are not applicable. We demonstrate here the use of MALDI-MS to localize a 5-methyl uridine in *E. coli* 23S rRNA

(Fig. 1.2). Modifications such as m^5U, m^5C, m^7A, and m^6A do not halt the progress of reverse transcriptase and, therefore, cannot be detected by primer extension. Primer extension remains a useful tool for locating m^3U, $m^3\Psi$, m^3C, m^1G, m^2G, m^1A, and m_2^6A and, with a couple of extra steps, can also detect Ψ (Bakin and Ofengand, 1993), $2'$-O-methylribose (Maden et al., 1995), and m^7G modifications (Stern et al., 1988). Thin-layer chromatography is an alternative means of identifying modified nucleotides (Grosjean et al., 2004), generally requiring radioisotope labeling or large amounts of starting material. Each technique has its own distinct advantages and applications that are difficult to substitute by other methods. For instance, characterization of the three adjacent N^6-methyl adenosines, which are present in a minor proportion of the RNA sample (m/z 1722.3 signal in Fig. 1.3B), would be tricky, if not impossible, to achieve by any other technique than tandem-MS.

Some types of analyses present specific challenges for MS. For instance, uridine has the same mass as its structural isomer, pseudouridine, and the two nucleotides are thus indistinguishable by mass spectrometry. Pseudouridines are common rRNAs modifications (see later), and, therefore, a method has been devised, on the basis of specific cyanoethylation of pseudouridines, to facilitate their detection by mass spectrometry (Mengel-Jorgensen and Kirpekar, 2002). Cyanoethylation increases the mass of pseudouridine by 53.03 Da; this mass increase is maintained during RNase digestion and can be identified by MALDI tandem-MS under the standard experimental conditions described here.

In Eukarya, pseudouridinylation reactions and $2'$-O-methylations make up the bulk of rRNAs modifications and are guided by a variety of snoRNAs that function together with a limited set of enzymes (Decatur and Fournier, 2002; Kiss, 2002). Similar mechanisms are apparent in Archaea (Dennis and Omer, 2005). Such systems are absent in bacteria, where individual rRNAs modifications are added by a specific enzyme. There are a few examples of enzymes modifying more than one site (e.g., Gutgsell et al., 2001; Johansen et al., 2006; Madsen et al., 2005; Poldermans et al., 1979), but these are rare. Two main strategies are commonly used to characterize modification enzymes for bacterial rRNAs: the first requires the purification of a recombinant enzyme followed by modification in vitro and analysis of RNA transcripts (e.g., Agarwalla et al., 2002); whereas in the second approach, the gene for a putative modification enzyme is inactivated followed by a search for a missing rRNA modification (e.g., Madsen et al., 2003). Genome database searches that use the E. coli data reveal that some rRNA modification enzymes are present in all bacteria, whereas other enzymes are limited to related bacterial groups. At the time of writing, 24 of a total of 35 enzymes responsible for the housekeeping modifications in E. coli rRNA have been identified (Andersen and Douthwaite, 2006;

Lesnyak *et al.*, 2007; Okamoto *et al.*, 2007). The remainder will hopefully be characterized within the next few years, with MALDI-MS playing a role in the process.

ACKNOWLEDGMENTS

Support from the Danish Research Agency (FNU-grants #21-04-0505 and #21-04-0520), DABIC-Danish Post-Genome Initiative, and the Nucleic Acid Center of the Danish Grundforskningsfond is gratefully acknowledged.

REFERENCES

Agarwalla, S., Kealey, J. T., Santi, D. V., and Stroud, R. M. (2002). Characterization of the 23S ribosomal RNA m^{5}U1939 methyltransferase from *Escherichia coli*. *J. Biol. Chem.* **277,** 8835–8840.

Andersen, N. M., and Douthwaite, S. (2006). YebU is a m^{5}C methyltransferase specific for 16S rRNA nucleotide 1407. *J. Mol. Biol.* **359,** 777–786.

Andersen, T. E., Kirpekar, F., and Haselmann, K. F. (2006). RNA fragmentation in MALDI mass spectrometry studied by H/D-exchange: Mechanisms of general applicability to nucleic acids. *J. Am. Soc. Mass Spectrom.* **17,** 1353–1368.

Andersen, T. E., Porse, B. T., and Kirpekar, F. (2004). A novel partial modification at 2501 in *Escherichia coli* 23S ribosomal RNA. *RNA* **10,** 907–913.

Asara, J. M., and Allison, J. (1999). Enhanced detection of oligonucleotides in UV-MALDI-MS using the tetraamine spermine as a matrix additive. *Anal. Chem.* **71,** 2866–2870.

Bakin, A., and Ofengand, J. (1993). Four newly located pseudouridylate residues in *Escherichia coli* 23S ribosomal RNA are all at the peptidyltransferase center: Analysis by the application of a new sequencing technique. *Biochemistry* **32,** 9754–9762.

Ban, N., Nissen, P., Hansen, J., Moore, P. B., and Steitz, T. A. (2000). The complete atomic structure of the large ribosomal subunit at 2.4 Å resolution. *Science* **289,** 905–920.

Brimacombe, R., Mitchell, P., Osswald, M., Stade, K., and Bochkariov, D. (1993). Clustering of modified nucleotides at the functional center of bacterial ribosomal RNA. *FASEB J.* **7,** 161–167.

Cundliffe, E. (1989). How antibiotic-producing organisms avoid suicide. *Annu. Rev. Microbiol.* **43,** 207–233.

Decatur, W. A., and Fournier, M. J. (2002). rRNA modifications and ribosome function. *Trends Biochem. Sci.* **27,** 344–351.

Dennis, P. P., and Omer, A. (2005). Small non-coding RNAs in Archaea. *Curr. Opin. Microbiol.* **8,** 685–694.

Felden, B., Hanawa, K., Atkins, J. F., Himeno, H., Muto, A., Gesteland, R. F., McCloskey, J. A., and Crain, P. F. (1998). Presence and location of modified nucleotides in *Escherichia coli* tmRNA: Structural mimicry with tRNA acceptor branches. *EMBO J.* **17,** 3188–3196.

Gale, E. F., Cundliffe, E., Reynolds, P. E., Richmond, M. H., and Waring, M. J. (1981). "The Molecular Basis of Antibiotic Action." John Wiley and Sons, London.

Green, R., and Noller, H. F. (1999). Reconstitution of functional 50S ribosomes from *in vitro* transcripts of *Bacillus stearothermophilus* 23S rRNA. *Biochemistry* **38,** 1772–1779.

Grosjean, H., Keith, G., and Droogmans, L. (2004). Detection and quantification of modified nucleotides in RNA using thin-layer chromatography. *Methods Mol. Biol.* **265,** 357–391.

Gutgsell, N. S., Campo, M. D., Raychaudhuri, S., and Ofengand, J. (2001). A second function for pseudouridine synthetases: A point mutant of RluD unable to form pseudouridines 1911, 1915 and 1917 in *Escherichia coli* 23S ribosomal RNA restores normal growth to a RluD-minus strain. *RNA* **7,** 990–998.

Guymon, R., Pomerantz, S. C., Crain, P. F., and McCloskey, J. A. (2006). Influence of phylogeny on posttranscriptional modification of rRNA in thermophilic prokaryotes: The complete modification map of 16S rRNA of *Thermus thermophilus*. *Biochemistry* **45,** 4888–4899.

Hansen, L. H., Kirpekar, F., and Douthwaite, S. (2001). Recognition of nucleotide G745 in 23S ribosomal RNA by the *rrmA* methyltransferase. *J. Mol. Biol.* **310,** 1001–1010.

Hansen, L. H., Mauvais, P., and Douthwaite, S. (1999). The macrolide-ketolide antibiotic binding site is formed by structures in domains II and V of 23S ribosomal RNA. *Mol. Microbiol.* **31,** 623–631.

Hansen, M. A., Kirpekar, F., Ritterbusch, W., and Vester, B. (2002). Posttranscriptional modifications in the A-loop of 23S rRNAs from selected archaea and eubacteria. *RNA* **8,** 202–213.

Harms, J., Schluenzen, F., Zarivach, R., Bashan, A., Gat, S., Agmon, I., Bartels, H., Franceschi, F., and Yonath, A. (2001). High resolution structure of the large ribosomal subunit from a mesophilic eubacterium. *Cell* **107,** 679–688.

Johansen, S. K., Maus, C. E., Plikaytis, B. B., and Douthwaite, S. (2006). Capreomycin binds across the ribosomal subunit interface using *tlyA*-encoded 2′-O-methylations in 16S and 23S rRNAs. *Mol. Cell* **23,** 173–182.

Karas, M., and Hillenkamp, F. (1988). Laser desorption ionization of proteins with molecular masses exceeding 10,000 daltons. *Anal. Chem.* **60,** 2299–2301.

Kehrenberg, C., Schwarz, S., Jacobsen, L., Hansen, L. H., and Vester, B. (2005). A new mechanism for chloramphenicol, florfenicol and clindamycin resistance: Methylation of 23S ribosomal RNA at A2503. *Mol. Microbiol.* **57,** 1064–1073.

Khaitovich, P., Tenson, T., Kloss, P., and Mankin, A. S. (1999). Reconstitution of functionally active *Thermus aquaticus* large ribosomal subunits with *in vitro*-transcribed rRNA. *Biochemistry* **38,** 1780–1788.

Kirpekar, F., Douthwaite, S., and Roepstorff, P. (2000). Mapping posttranscriptional modifications in 5S ribosomal RNA by MALDI mass spectrometry. *RNA* **6,** 296–306.

Kirpekar, F., Hansen, L. H., Rasmussen, A., Poehlsgaard, J., and Vester, B. (2005). The archaeon *Haloarcula marismortui* has few modifications in the central parts of its 23S ribosomal RNA. *J. Mol. Biol.* **348,** 563–573.

Kirpekar, F., and Krogh, T. N. (2001). RNA fragmentation studied in a matrix-assisted laser desorption/ionisation tandem quadropole/orthogonal time-of-flight mass spectrometer. *Rapid Commun. Mass Spectrom.* **15,** 8–14.

Kirpekar, F., Nordhoff, E., Larsen, L. K., Kristiansen, K., Roepstorff, P., and Hillenkamp, F. (1998). DNA sequence analysis by MALDI mass spectrometry. *Nucleic Acids Res.* **26,** 2554–2559.

Kiss, T. (2002). Small nucleolar RNAs: An abundant group of noncoding RNAs with diverse cellular functions. *Cell* **109,** 145–148.

Korostelev, A., Trakhanov, S., Laurberg, M., and Noller, H. F. (2006). Crystal structure of a 70S ribosome-tRNA complex reveals functional interactions and rearrangements. *Cell* **126,** 1065–1077.

Kowalak, J. A., Bruenger, E., and McCloskey, J. A. (1995). Posttranscriptional modification of the central loop of domain V in *Escherichia coli* 23S ribosomal RNA. *J. Biol. Chem.* **270,** 17758–17764.

Kowalak, J. A., Pomerantz, S. C., Crain, P. F., and McCloskey, J. A. (1993). A novel method for the determination of post-transcriptional modification in RNA by mass spectrometry. *Nucleic Acids Res.* **21,** 4577–4585.

Krupp, G., and Soll, D. (1987). Simplified *in vitro* synthesis of mutated RNA molecules. An oligonucleotide promoter determines the initiation site of T7RNA polymerase on ss M13 phage DNA. *FEBS Lett.* **212,** 271–275.

Krzyzosiak, W., Denman, R., Nurse, K., Hellmann, W., Boubik, M., Gehrke, C. W., Agris, P. F., and Ofengand, J. (1987). *In vitro* synthesis of 16S ribosomal RNA containing single base changes and assembly into functional 30S ribosome. *Biochemistry* **26,** 2353–2364.

Lesnyak, D. V., Osipiuk, J., Skarina, T., Sergiev, P. V., Bogdanov, A. A., Edwards, A., Savchenko, A., Joachimiak, A., and Dontsova, O. A. (2007). Methyltransferase that modifies guanine 966 of the 16S rRNA: Functional identification and tertiary structure. *J. Biol. Chem.* **282,** 5880–5887.

Maden, B. E., Corbett, M. E., Heeney, P. A., Pugh, K., and Ajuh, P. M. (1995). Classical and novel approaches to the detection and localization of the numerous modified nucleotides in eukaryotic ribosomal RNA. *Biochimie* **77,** 22–29.

Madsen, C. T., Jakobsen, L., Buriankova, K., Doucet-Populaire, F., Pernodet, J. L., and Douthwaite, S. (2005). Methyltransferase Erm(37) slips on rRNA to confer atypical resistance in *Mycobacterium tuberculosis. J. Biol. Chem.* **280,** 38942–38947.

Madsen, C. T., Mengel-Jorgensen, J., Kirpekar, F., and Douthwaite, S. (2003). Identifying the methyltransferases for m^5U747 and m^5U1939 in 23S rRNA using MALDI mass spectrometry. *Nucleic Acids Res.* **31,** 4738–4746.

McCloskey, J. A., and Rozenski, J. (2005). The small subunit rRNA modification database. *Nucleic Acids Res.* **33,** D135–D138.

McLuckey, S. A., Van Berkel, G. J., and Glish, G. L. (1992). Tandem mass spectrometry of small multiply charged oligonucleotides. *J. Am. Soc. Mass Spectrom.* **3,** 60–70.

Mengel-Jorgensen, J., and Kirpekar, F. (2002). Detection of pseudouridine and other modifications in tRNA by cyanoethylation and MALDI mass spectrometry. *Nucleic Acids Res.* **30,** e135.

Mengel-Jorgensen, J., Jensen, S. S., Rasmussen, A., Poehlsgaard, J., Iversen, J. J., and Kirpekar, F. (2006). Modifications in *Thermus thermophilus* 23S ribosomal RNA are centered in regions of RNA-RNA contact. *J. Biol. Chem.* **281,** 22108–22117.

Nordhoff, E., Cramer, R., Karas, M., Hillenkamp, F., Kirpekar, F., Kristiansen, K., and Roepstorff, P. (1993). Ion stability of nucleic acids in infrared matrix-assisted laser desorption/ionization mass spectrometry. *Nucleic Acids Res.* **21,** 3347–3357.

Ofengand, J., and Del Campo, M. (2004). Modified nucleotides of *Escherichia coli* ribosomal RNA. *In* "EcoSal - *Escherichia coli* and *Salmonella*: Cellular and Molecular Biology" (R. Curtiss, ed.). ASM Press, Washington, DC. http://www.ecosal.org

Okamoto, S., Tamaru, A., Nakajima, C., Nishimura, K., Tanaka, Y., Tokuyama, S., Suzuki, Y., and Ochi, K. (2007). Loss of a conserved 7-methylguanosine modification in 16S rRNA confers low-level streptomycin resistance in bacteria. *Mol. Microbiol.* **63,** 1096–1106.

Poldermans, B., Roza, L., and Van Knippenberg, P. H. (1979). Studies on the function of two adjacent N^6,N^6-dimethyladenosines near the 3'-end of 16S ribosomal RNA of *Escherichia coli*. III. Purification and properties of the methylating enzyme and methylase-30S interactions. *J. Biol. Chem.* **254,** 9094–9100.

Rozenski, J., Crain, P. F., and McCloskey, J. A. (1999). The RNA Modification Database: 1999 update. *Nucl. Acids Res.* **27,** 196–197.

Schlünzen, F., Tocilj, A., Zarivach, R., Harms, J., Gluehmann, M., Janell, D., Bashan, A., Bartels, H., Agmon, I., Franceschi, F., and Yonath, A. (2000). Structure of functionally activated small ribosomal subunit at 3.3 angstroms resolution. *Cell* **102,** 615–623.

Schuwirth, B. S., Borovinskaya, M. A., Hau, C. W., Zhang, W., Vila-Sanjurjo, A., Holton, J. M., and Cate, J. H. (2005). Structures of the bacterial ribosome at 3.5 Å resolution. *Science* **310,** 827–834.

Selmer, M., Dunham, C. M., Murphy, F. V. T., Weixlbaumer, A., Petry, S., Kelley, A. C., Weir, J. R., and Ramakrishnan, V. (2006). Structure of the 70S ribosome complexed with mRNA and tRNA. *Science* **313,** 1935–1942.

Stern, S., Moazed, D., and Noller, H. F. (1988). Structural analysis of RNA using chemical and enzymatic probing monitored by primer extension. *Methods Enzymol.* **164,** 481–489.

Treede, I., Jakobsen, L., Kirpekar, F., Vester, B., Weitnauer, G., Bechthold, A., and Douthwaite, S. (2003). The avilamycin resistance determinants AviRa and AviRb methylate 23S rRNA at the guanosine 2535 base and the uridine 2479 ribose. *Mol. Microbiol.* **49,** 309–318.

Wu, K. J., Shaler, T. A., and Becker, C. H. (1994). Time-of-flight mass spectrometry of underivatized single-stranded DNA oligomers by matrix-assisted laser desorption. *Anal. Chem.* **66,** 1637–1645.

Zhang, Z., Jackson, G. W., Fox, G. E., and Willson, R. C. (2006). Microbial identification by mass cataloging. *BMC Bioinformatics* **7,** 117.

Zhu, Y. F., Chung, C. N., Taranenko, N. I., Allman, S. L., Martin, S. A., Haff, L., and Chen, C. H. (1996). The study of 2,3,4-trihydroxyacetophenone and 2,4,6-trihydroxyacetophenone as matrices for DNA detection in matrix-assisted laser desorption/ionization time-of-flight mass spectrometry. *Rapid Commun. Mass Spectrom.* **10,** 383–388.

CHAPTER TWO

Identification of Modified Residues in RNAs by Reverse Transcription-Based Methods

Yuri Motorin, Sébastien Muller, Isabelle Behm-Ansmant, *and* Christiane Branlant

Contents

Laboratoire de Maturation des ARN et Enzymologie Moléculaire, Nancy Université, Faculté des Sciences et Techniques, Vandoeuvre-les-Nancy, France

Methods in Enzymology, Volume 425
ISSN 0076-6879, DOI: 10.1016/S0076-6879(07)25002-5

Abstract

Naturally occurring modified residues derived from canonical RNA nucleotides are present in most cellular RNAs. Their detection in RNA represents a difficult task because of their great diversity and their irregular distribution within RNA molecules. Over the decades, multiple experimental techniques were developed for the identification and localization of RNA modifications. Most of them are quite laborious and require purification of individual RNA to a homogeneous state. An alternative to these techniques is the use of reverse transcription (RT)-based approaches. In these approaches, purification of RNA to homogeneity is not necessary, because the selection of the analyzed RNA species is done by specific annealing of oligonucleotide DNA primers. However, results from primer extension analysis are difficult to interpret because of the unpredictable nature of RT pauses. They depend not only on the properties of nucleotides but also on the RNA primary and secondary structure. In addition, the degradation of cellular RNA during extraction, even at a very low level, may complicate the analysis of the data. RT-based techniques for the identification of modified residues were considerably improved by the development of selected chemical reagents specifically reacting with a given modified nucleotide. The RT profile obtained after such chemical modifications generally allows unambiguous identification of the chemical nature of the modified residues and their exact location in the RNA sequence. Here, we provide experimental protocols for selective chemical modification and identification of several modified residues: pseudouridine, inosine, 5-methylcytosine, 2′-*O*-methylations, 7-methylguanosine, and dihydrouridine. Advice for an optimized use of these methods and for correct interpretation of the data is also given. We also provide some helpful information on the ability of other naturally occurring modified nucleotides to generate RT pauses.

1. INTRODUCTION

The presence of posttranscriptionally modified nucleotides derived from the canonical A, C, G, and U residues is a characteristic feature of most cellular RNAs (for review, Björk, 1995; Björk *et al.*, 1987; Massenet *et al.*, 1998;

Ofengand and Fournier, 1998; Internet resource: http://library.med.utah.edu/RNAmods/). These modified residues are formed during RNA maturation steps, and their formation is catalyzed by numerous specific enzymes (RNA: modification enzymes) or by RNP particles (for review, Ansmant and Motorin, 2001; Garcia and Goodenough-Lashua, 1998; Grosjean *et al.*, 1998; Internet resource: http://genesilico.pl/modomics/index2.pt).

Despite extensive studies of modified nucleotides in RNAs over more than 40 years, our knowledge on their presence, chemical identity, and localization remains limited to highly abundant cellular RNAs, like tRNAs (Sprinzl and Vassilenko, 2005; Internet resource: http://www.uni-bayreuth.de/departments/biochemie/trna/), rRNAs (McCloskey and Rozenski, 2005; Internet resource: http://medlib.med.utah.edu/SSUmods) and U snRNAs (Massenet *et al.*, 1998). In contrast, low abundance or recently discovered RNAs, such as snoRNAs or regulatory RNAs from eukaryotes and bacteria, as well as mRNAs, have not been studied yet from this point of view. In most cases, this lack of information is due to experimental difficulties in detecting and localizing modified residues in low-abundance RNAs.

Multiple experimental techniques allowing the detection and analysis of modified residues in RNAs have been developed in the past. The techniques of direct RNA sequencing and fingerprinting require the purification of homogeneous, pure RNA in sufficient amounts (at least 1–2 μg). However, these methods allow the detection of noncanonical residues and their precise localization (Branch *et al.*, 1989; Gupta and Randerath, 1979; Stanley and Vassilenko, 1978; Tanaka *et al.*, 1980). The application of these and related techniques in 1970–1980 allowed an extensive analysis of tRNAs (more than 500 species) from various sources and the constitution of a tRNA modification database (Sprinzl and Vassilenko, 2005). Posttranscriptional modifications were also detected and localized in several rRNAs and U snRNAs (Branlant *et al.*, 1981; Krol *et al.*, 1981; Veldman *et al.*, 1981; see Massenet *et al.*, 1998, for review). Later on, the methods based on high-performance liquid chromatography (HPLC) allowed fragmentation and analysis of longer RNAs, like rRNAs from various sources (Bruenger *et al.*, 1993; Gehrke and Kuo, 1989; Gehrke *et al.*, 1982; Smith *et al.*, 1992). More recently, the approaches that use HPLC and mass spectrometry (MS) have been successfully used for direct RNA sequencing and the identification of modified residues (for review, Meng and Limbach, 2006; Thomas and Akoulitchev, 2006; Douthwaite and Kirpekar, Chapter 1). Recently, the use of DNA chips was suggested for the detection of modified residues in RNAs (Hiley *et al.*, 2005). Despite a great interest in these powerful techniques for RNA analysis, the use of these methods is limited to highly purified RNAs and frequently requires significant amounts of analyzed material. These two impediments are quite important, because purification of low-abundance RNA species from total RNA always represents a difficult task. In addition, the application of MS approaches requires particular expertise in data interpretation and cannot be done routinely and at high throughput.

An alternative group of techniques capable to detect and to identify some of the modified nucleotides in RNAs is derived from the RNA reverse-transcription (RT) sequencing method (Brownlee and Cartwright, 1977; Lane *et al.*, 1988; Qu *et al.*, 1983). In contrast to direct RNA sequencing, HPLC, or MS, these methods do not require purification of the analyzed RNA species and may be directly applied to total RNA fractions extracted from the cells. The selection of the analyzed RNA is done by specific annealing of DNA oligonucleotides selected for primer extension (Fig. 2.1).

2. REVERSE TRANSCRIPTION (RT)–BASED METHODS FOR DETECTION OF MODIFIED RESIDUES

Despite these advantages, the use of reverse transcription for detection of modified nucleotides in RNAs is not very straightforward. In fact, the reverse transcription profile, even for an unmodified RNA, depends on several parameters, especially the presence of Py–A (C–A and U–A) bonds, which are very sensitive to nuclease cleavage, and also on the RNA secondary structure that can generate RT pauses. Therefore, the simple presence of a pause in a primer extension profile does not necessarily indicate the presence of a modified residue. In addition, reverse transcription profiles of modified RNAs are often prone to some kind of stuttering (doubling of the band on the gel) at the modified residue that blocks the reverse transcription (Bakin and Ofengand, 1993; Denman *et al.*, 1988; Gustafsson and Persson, 1998). The reasons for such a phenomenon are not yet clear, but this possibility complicates the detection of nearby modified residues and should be considered during analysis of reverse transcription patterns.

A significant improvement of the RT-based techniques was the development of specific chemical reagents for several modified nucleotides found in RNAs. Because these chemicals specifically react with a given modified nucleotide, the intensity of the pause observed in the RT profile should depend on the chemical's concentration and the treatment time. For this reason, for an unambiguous identification of modified nucleotides, it is recommended to vary these two parameters. The comparison of the data obtained under these different conditions allows verification that the observed pause can, indeed, be attributed to the action of the selected chemical (Fig. 2.1A). These techniques generally give quite reliable results.

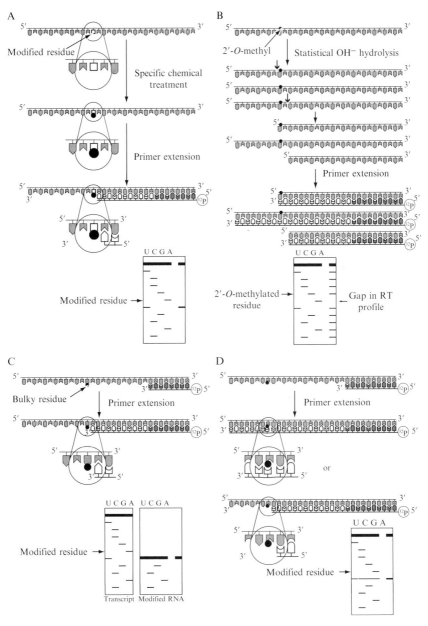

Figure 2.1 Schematic representation of the principle of reverse transcription–based approach used for the detection of modified nucleotides in RNA. (A) Detection of RNA modifications by use of specific chemical modification of modified residues present in natural RNAs. (B) Detection of 2′–O-methylated residues by statistical RNA cleavage by OH⁻. (C) Detection of bulky (or non-base pairing) modified residues in RNA that block primer extension. (D) Detection of modified residues that induce a pause in RT profile.

Hence, we will first describe the RT-based techniques based on preliminary chemical treatment of RNA that convert "silent" and otherwise invisible modified nucleotides into chemical adducts detectable by pauses or blockage of the primer extension. To get such extension arrest, the adduct may abolish or limit the base-pairing property of the modified residue or create some steric hindrance in the active site of the RT. Therefore, reverse transcription generally stops one nucleotide downstream of the modified residue (see Fig. 2.1A).

Information on the presence of modified residues in RNA can also be obtained by the natural ability of some modified nucleotides to induce a pause (or a complete block) of the primer extension by reverse transcriptase (Fig. 2.1C,D). Here again, this phenomenon may be related either to the inability to form a base pair with any canonical nucleotide or to steric hindrance limiting or excluding the recognition of the modified residue by the RT. However, as mentioned previously, because there can be multiple reasons for RT pausing, unambiguous localization of modified nucleotide requires additional information. One simple complementary experiment is the comparison of the reverse transcription profiles obtained for the *in vitro* RNA transcript and the authentic cellular RNA. Pauses (or blocks) observed in the cellular RNA but absent in the corresponding *in vitro* transcript may indicate the presence of modifications at these positions. However, extreme care should be taken to avoid nucleolytic degradation of fragile Py-A linkages during extraction of cellular RNA. In addition, modifications located too close to 5'- or 3'-extremity of RNA are difficult to detect because of natural RT pauses at the 5'-end of RNAs and to the necessity of annealing the oligonucleotide primer at the 3'-extremity of the analyzed molecule.

2.1. Reverse transcriptases used for RNA analysis

Viral RNA-dependent DNA polymerases (reverse transcriptases) are most frequently used for RT-based analysis of RNA structure and sequence (including modified nucleotides).

Avian Mieloblastosis virus (AMV) reverse transcriptase available from numerous suppliers is most frequently used for primer extension analysis. However, other reverse transcriptases may also be used, like Moloney murine leukemia virus (MMLV, M-MuLV) reverse transcriptase or its mutants deficient in RNAse H activity. The main advantage of AMV reverse transcriptase consists in its higher thermostability; indeed, in contrast to MMLV, it can be used up to $70°$ in the appropriate buffer (Fuchs *et al.*, 1999). In addition, AMV RT seems to be less sensitive to RNA secondary structures that may perturb primer extension.

2.2. General protocol for primer labeling and reverse transcription of RNA

2.2.1. Equipment

Water baths set to 37°, 42°, and 65°.

Heating block set to 96°.

Equipment for polyacrylamide gel electrophoresis (PAGE) and power supply: small vertical slab gels for primer purification and sequencing gels for primer extension analysis.

Vacuum drier for sequencing gels.

PhosphorImager (Typhoon 9410, GE Healthcare, formerly Amersham Biosciences, or equivalent) or screens, cassettes, and films for autoradiography.

2.2.2. Reagents

$[\gamma\text{-}^{32}P]ATP$ (3000 Ci/mmol; 10 mCi/ml) (purchased from GE Healthcare, formerly Amersham Biosciences).

T4 Polynucleotide kinase (PNK) (10 U/μl) (MBI Fermentas, Lituania or other source) and corresponding 10× reaction buffer (500 mM Tris-HCl, pH 7.6, at 25°, 100 mM MgCl$_2$, 50 mM DTT, 1 mM spermidine, and 1 mM EDTA).

3 M Sodium acetate, pH 5.4.

Elution buffer (0.5 M sodium acetate, pH 5.4, 1 mM EDTA).

RNAse free 96% ethanol.

Hybridization buffer 5× (250 mM Tris-HCl, pH 8.3, 300 mM NaCl, 50 mM DTT).

Mixture of dNTP (5 mM dATP, 5 mM dCTP, 5 mM dGTP, 5 mM dTTP).

ddNTP solutions (0.5 mM): ddATP, ddCTP, ddGTP, and ddTTP.

Primer extension buffer 10× (500 mM Tris-HCl, pH 8.3, 600 mM NaCl, 100 mM DTT, 60 mM MgCl$_2$).

AMV reverse transcriptase (MP Biomedicals, formerly QBiogene) (20 U/μl).

TBE buffer (89 mM Tris-borate, pH 8.3, 2 mM Na$_2$EDTA).

Formamide blue loading solution (20 mM EDTA, 0.05% bromophenol blue, 0.05% xylene cyanol blue in deionized formamide).

2.2.3. Procedure

1. The sequences of the oligonucleotides used as primers are chosen to hybridize to the target RNA at least 3-nt downstream from the region to be analyzed. Extension performed with one primer generally enables the analysis of an approximately 100–150-nt long RNA sequence. Thus, multiple primers have to be used to perform a complete analysis of a longer RNA. The region of the RNA targeted by primers must be

accessible and not hidden in a stable secondary structure, such as, for instance, a long hairpin. To optimize the hybridization, the lengths of primers are adjusted to obtain a Tm comprised between 55 and 70°. Furthermore, primer sequences have to be compared with genomic sequence databases, when available, to avoid nonspecific hybridization.

2. Approximately 100 ng of each primer are 5′-end labeled by use of 10 μCi of [γ-^{32}P]ATP (3000 Ci/mmol) and 10 U of T4 PNK in T4 PNK buffer. The purification of radiolabeled primers on 10% polyacrylamide denaturing gels usually improves the quality of the resulting cDNA profile. However, specific care must be taken to avoid primer contamination by polyacrylamide that can inhibit the RT.

3. The primer is eluted from the gel slice in the elution buffer at 4° overnight and ethanol precipitated. Approximately one fifth of the eluted primer (20 ng) is then used for one extension reaction. Hybridization of the primer is performed with 2–50 μg of total RNA (depending on the concentration of the target RNA within total RNA) or 2–50 pmol of purified transcript, in the hybridization buffer, without MgCl$_2$, to prevent stable folding of the target RNA and Mg^{2+}-dependent hydrolysis of RNA at 65°. The hybridization step is carried out in a final volume of 2.5 μl by incubation for 10 min at 65°, followed by a slow decrease to reach room temperature.

4. The extension step is performed by addition of a 2.5 μl mixture containing 40 μM of each dNTP (0.2 μl of stock solution), 0.5 U of AMV reverse transcriptase (0.025 μl), in 1× primer extension buffer (0.25 μl of 10× stock). The conditions are similar to the hybridization step except the presence of 3 mM MgCl$_2$ required for the enzymatic activity. The extension step is carried out in a 5 μl final volume by incubation for 1 h at 42°.

To prepare a sequencing ladder, four tubes are prepared as previously indicated, but 0.5 μl of either ddATP, ddCTP, ddGTP, or ddTTP (0.5 mM each) is added to the 5 μl final volume before incubation at 42°. The extension reaction is stopped by addition of 3 μl of formamide blue.

5. The extension mixture is heated for 2 min at 96° to denature secondary structure in the cDNA and placed on ice to prevent renaturation. Fractionation is achieved on a 7% denaturing (8 M urea) polyacrylamide (19:1 ratio acrylamide/bisacrylamide) gel by loading of 2 μl of the sample. The remaining 6 μl may be used later for another gel load. The electrophoresis is carried out in TBE buffer at 40–50° by use of a constant power of 100 W (approximatively 30–50 V/cm). The migration time is set according to the distance between the primer-annealing site and the sequence to be analyzed. Gels are then transferred onto Whatman 3 MM paper sheets and dried at 80° with a vacuum drier.

6. Resulting stops in cDNA synthesis are visualized by autoradiography or by exposure to PhosphorImager screens for a few hours.

2.2.4. Comments

Several preliminary tests must be performed to define the best oligonucleotide primers for reverse transcription. To test the primer quality, RNA sequencing is first done by primer extension. Only primers that give clearly readable sequencing profiles should be used for further experiments.

It is noteworthy that the analysis of highly structured G/C-rich RNA regions by primer extension may be rather difficult because the extension temperature does not exceed 42°. At this temperature, RT cannot denature stable RNA structure during elongation. Other RT enzymes that can be used at higher temperatures (50–55°, HIV RT and other RT mutants available from various producers) bring only a moderate improvement.

Naturally modified nucleotides in RNAs may also generate partial or total RT stops; thus the use of different primers that are located upstream or overlap those modified residues can be required.

When a large amount of total RNA is taken for primer extension, the use of a DNase-free RNase treatment before loading on the gel is recommended to reduce smearing of the cDNA pattern.

3. RNA Extraction from Various Cell Types

Numerous protocols exist for extraction of total RNA from different cell types. In this chapter, we will give two techniques: the first one uses guanidine thiocyanate and can be used for the RNA extraction from eukaryal and archaeal cells. The second one applies to the yeast *S. cerevisiae*. Extraction of bacterial (*Escherichia coli*) total RNA may be performed by hot phenol treatment (Aiba *et al.*, 1981). Other extraction procedures may also be used (see, for examples, Mangan *et al.*, 1997; Rivas *et al.*, 2001; Internet resource http://www.protocolonline.org./prot/Molecular_Biology/RNA/RNA_Extraction/Total_RNA_ Isolation/index.html). Whatever protocol is used, particular care has to be taken to avoid possible degradation of RNAs by nucleases. For this reason, the use of strong denaturing agents like guanidine thiocyanate and strong detergents generally provides total RNA of better quality.

3.1. Protocol for the extraction of total RNA from archaeal or eukaryal cells

This protocol is adapted from Chomczynski and Sacchi (1987).

3.1.1. Equipment

Vacuum drier (SpeedVac).
UV spectrophotometer for small volumes.

3.1.2. Reagents

2 *M* Sodium acetate, pH 4.0.
0.75 *M* Sodium citrate, pH 7.0, sterilized (Merck).
10% Sarcosine (dissolved at 65° in distilled water) sterilized (Sigma).
RNase free 96% ethanol.
Guanidine thiocyanate (Fluka).
2-mercaptoethanol (Merck).
Phenol-chloroform-isoamyl alcohol (125/24/1, v/v/v) saturated with water.

3.1.3. Procedure

1. The stock solution for extraction (solution S) is prepared by mixing of 100 g guanidine thiocyanate dissolved in 117.2 ml of sterilized water, with 7.04 ml of 0.75 *M* sodium citrate and 10.56 ml of a 10% sarcosine solution. This mixture can be stored at room temperature. The denaturing solution (solution D) is obtained by adding 72 μl of 2-mercaptoethanol to 10 ml of solution S. The solution D can be stored for 1 month.
2. A pellet of cells (from 1×10^8–1×10^{10} cells, depending on cell type), coming from cultures or tissues is resuspended in 4 ml of solution D. It is then separated into ~500-μl aliquots in eight Eppendorf tubes. Then 50 μl of 2 *M* sodium acetate, pH 4.0, and 500 μl of a phenol-chloroform-isoamyl alcohol mixture (125/24/1) are added to each tube. The tubes are vigorously vortexed and kept on ice for 15 min. The samples are then centrifuged at 15,000g for 20 min at 4°.
3. The aqueous phases are transferred in new tubes, and 2 volumes of 96% ethanol are added. The tubes are placed for more than 30 min at −20° and then centrifuged at 15,000g for 20 min at 4°.
4. The supernatant is removed and the pellet is dissolved in 50 μl of solution D. All the RNA solutions obtained are gathered into a single tube and 2 volumes of 96% ethanol are added. The tube is placed for more than 30 min at −20° and centrifuged at 4° (15,000g) for 10 min.
5. The pellet is washed by 500 μl of 70% ethanol and centrifuged at 15,000g for 10 min at 4°. The supernatant is discarded, and residual ethanol is completely evaporated with a SpeedVac.
6. The RNA pellet is dissolved in 80–160 μl of sterile water. The RNA concentration is determined by measuring the A_{260} and A_{280} with a UV spectrophotometer.

3.1.4. Comments

The presence of guanidine thiocyanate during extraction inhibits the activity of RNases. However, after the removal of guanidine thiocyanate one has to take special care to avoid contamination of samples by RNases coming from human skin. We recommend wearing gloves and using, when possible, sterilized material and reagents. It is also important to avoid unnecessary incubation at room temperature and also at 37°, the optimal temperature for human and bacterial RNases.

The quality of the RNA sample is estimated using the A_{260}/A_{280} ratio. RNA can be considered as pure for values between 1.8 and 2.

The use of guanidine thiocyanate requires particular precautions for safety reasons.

The use of an RNase-free DNase I can ensure the absence of DNA in the sample.

3.2. Protocol for the extraction of total RNA from yeast cells

3.2.1. Equipment

Vacuum drier (SpeedVac).
Vortex.
Water bath at 65°.
UV spectrophotometer.

3.2.2. Reagents

3 M Sodium acetate, pH 5.2.
RNase free 96% ethanol.
Phenol–chloroform–isoamyl alcohol (125/24/1, v/v/v) saturated with water.
5 M NaCl.
1 M Tris-HCl, pH 7.5.
0.5 M EDTA.
10% SDS.
Acid–washed glass beads (0.45–0.5 mm in diameter, SIGMA-ALDRICH ref# G8772).

3.2.3. Procedure

1. Prepare a stock solution of extraction buffer (50 mM Tris-HCl, pH 7.5, 100 mM NaCl, 10 mM EDTA).
2. A pellet of yeast cells, from a 25-ml culture in YPD medium grown to an A_{600} of 0.8 , is resuspended in 200 μl of extraction buffer. Add 300 μl of glass beads and vortex vigorously twice for 1 min with a break of 2 min on ice in between.
3. Add 200 μl of extraction buffer, 50 μl of 10% SDS and 400 μl of phenol-chloroform–isoamyl alcohol mixture. Carefully close the tubes (with a special cap that clamp the lid and the body of the tube or use Safe-Lock

Eppendorf Tubes), vortex 1 min and incubate 10 min at 65°. The tubes
are then centrifuged at 15,000g for 15 min at 4°.

4. The aqueous phase is transferred in a new tube, and a second phenol-
chloroform-isoamyl alcohol extraction is performed.

5. The RNAs present in the aqueous phase obtained after the second
extraction are then precipitated by addition of 1 ml of ethanol in the
presence of 100 mM sodium acetate (14 μl of a 3 M solution). The tube is
placed for 30 min at −80° and centrifuged at 15,000g for 15 min at 4°.

6. The pellet is washed with 500 μl of 70% ethanol and centrifuged at
15,000g for 10 min at 4°. The supernatant is discarded, and residual
ethanol is completely evaporated with a SpeedVac.

7. The pellet is resuspended in 50 μl of sterile water. The RNA concentration is determined by measuring the A_{260} and A_{280} with a UV
spectrophotometer.

4. RNA MODIFICATIONS DETECTABLE AFTER SPECIFIC CHEMICAL TREATMENT

This group of techniques uses preliminary chemical treatment of
RNA to achieve selective modification of specific RNA residues. Up to
now, reliable methods of this type have been developed only for a limited
number of modified residues (see Figs. 2.2 and 2.3 and Table 2.1). However, these modified residues are rather frequently encountered in many
cellular RNAs. It should also be mentioned that the development of
new specific chemical modification methods for the detection of other
types of modified nucleotides is an open field of research.

4.1. Detection of inosine using glyoxal treatment followed by RNAse T1 hydrolysis

The modified nucleotide inosine (I) is derived from adenosine (A) by
enzymatic deamination catalyzed by specific adenosine deaminases (Bass,
1995; Maas *et al.*, 2003; Schaub and Keller, 2002; Seeburg, 2002). RNAse
T1 does not distinguish inosine residues present in RNA from guanosine
residues and cleaves on their 3′ side. However, preliminary specific reaction
with glyoxal allows RNase T1 to distinguish G and I residues (Morse and
Bass, 1997). Indeed, glyoxal reacts with N1 and N2 of G residues and thus
abolishes their recognition by RNAse T1. The resulting covalent adduct is
stabilized by boric acid (Fig. 2.3A). In contrast, inosine cannot react in the
same way with glyoxal because of the absence of NH$_2$-group at position
2 and the glyoxal-inosine adduct is unstable. Thus, RNAse T1 cleavage of

glyoxal-treated RNA proceeds only at inosine residues and not at G residues. The cleavage positions may be detected by primer extension analysis or other methods.

When the genomic sequence is available, an alternative approach for inosine detection consists in direct RNA sequencing with RT and comparison of the cDNA sequence obtained with the corresponding genomic sequence. Because of the absence of amino-group at position 6, I base

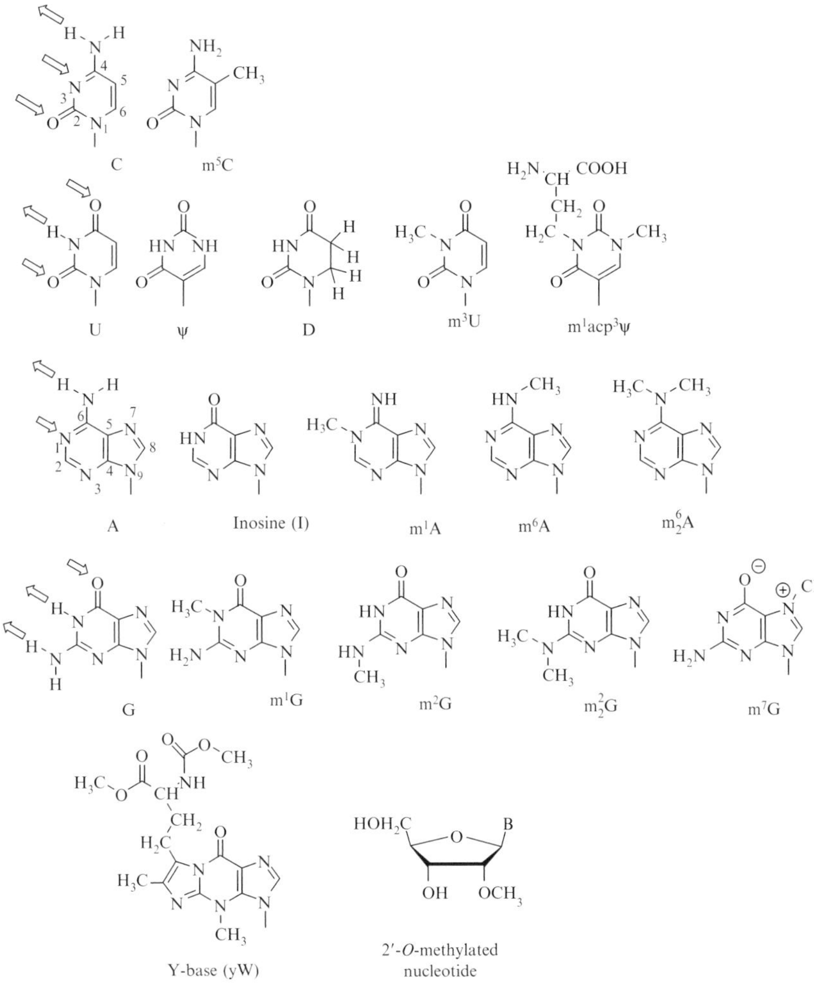

Figure 2.2 Chemical structures of selected modified nucleotides discussed in this review. Oxygen and nitrogen atoms used to form hydrogen bonds during reverse transcription are shown by arrows.

pairs with C residues instead of U residues and thus changes the sequence of the cDNA synthesized by RT. Thus one can detect unexpected A→G changes in the cDNA compared with the genomic sequence (Saccomanno and Bass, 1999). To confirm the data, it is also possible to compare the cDNA sequence obtained by direct sequencing of an unmodified transcript produced by *in vitro* transcription with the cDNA sequence obtained with the authentic RNA.

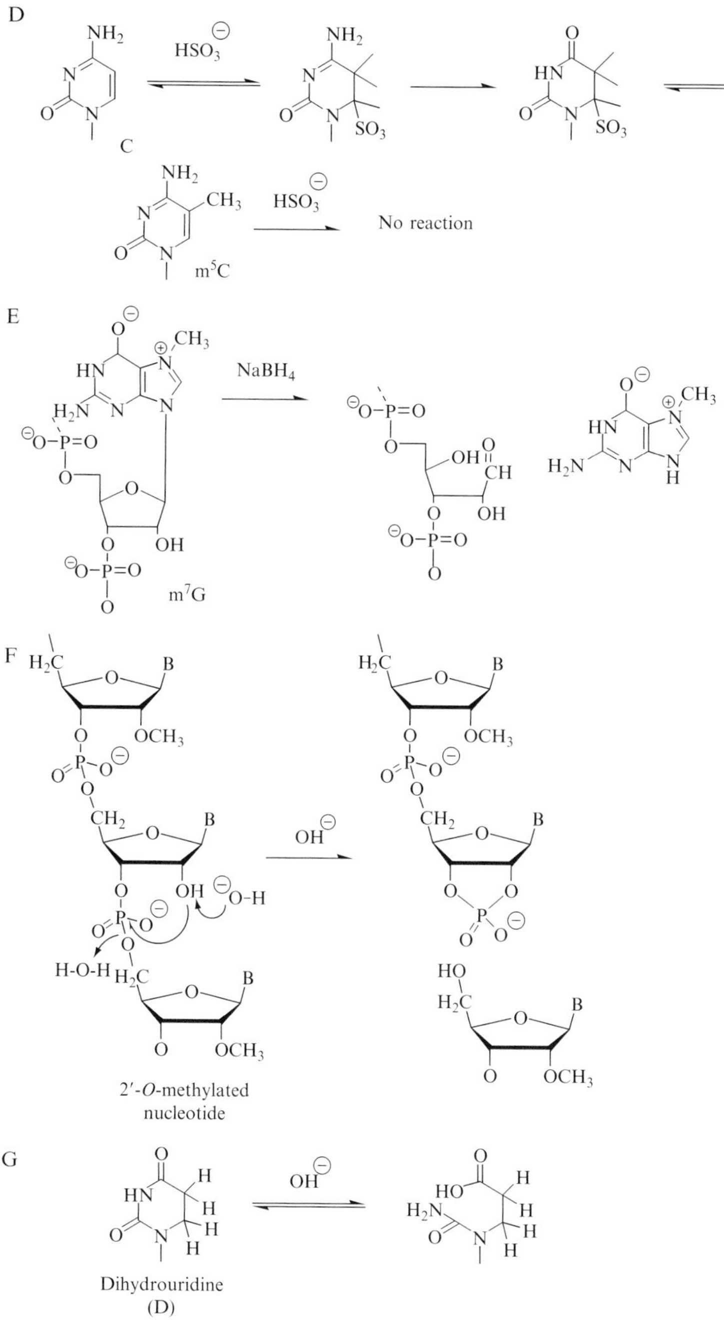

Figure 2.3 (*continued*)

4.2. Detection of pseudouridine (Ψ) residues

4.2.1. CMCT modification

Pseudouridine (5-[β-d-ribofuranosyl]uracil, abbreviation Ψ) residues are widespread in cellular RNAs. In contrast to all other known chemical modifications in RNAs, Ψ residues (and their derivatives) have an unusual C–C glycosidic bond linking the base to the ribose (Fig. 2.2, Cohn, 1960). Moreover, Ψ residues are formed by posttranscriptional isomerization of U residues, and thus both residues have the same molecular mass and base-pairing properties, which complicate experimental detection of Ψ residues. However, several methods of chemical treatment distinguishing U and Ψ have been developed.

The technique that is now generally used for detection of Ψ residues in RNA was initially developed for analysis of ribosomal RNA from various organisms (Bakin and Ofengand, 1993; 1998). RNA molecules are modified by a soluble carbodiimide (CMCT), which preferentially reacts with U and Ψ residues, and, to very limited extent, with G residues (Ho and Gilham, 1971; Metz and Brown, 1969a,b,c) (Fig. 2.3B). Extended incubation at alkaline pH (pH 10.3) of the CMCT-modified RNA allows the hydrolysis of U-CMT adducts, which are less stable than Ψ-CMC adducts. The remaining bulky CMC residues are then detected by primer extension using reverse transcriptase.

4.3. Protocol for the detection of pseudouridine residues in RNA using CMCT modification

4.3.1. Equipment

Water bath set to 37°.
Vacuum drier (SpeedVac).

4.3.2. Reagents

CMCT buffer (50 mM Bicine, pH 8.0, 7 M Urea, 4 mM EDTA).
CMCT solution (14 mg CMCT dissolved in 200 μl of CMCT buffer).

Figure 2.3 Chemical reactions specific to modified residues in RNAs that may be used for their detection. (A) Reaction of G or inosine (I) with glyoxal. (B) Modification of pseudouridine (Ψ) and uridine (U) residues by carbodiimide (CMCT). (C) Reactions of U and Ψ with hydrazine. (D) Deamination of cytosine (C) residues by reaction with bisulfite followed by alkaline treatment. (E) Depurination of 7-methylguanosine residue (m^{7}G) by NaBH$_4$. (F) Cleavage of phosphodiester bond under alkaline conditions, the linkage at the 3'- of 2'-O-methylated residue is resistant to cleavage. (G) Scission of the dihydrouridine (D) base ring under alkaline conditions.

Table 2.1 Common chemical modifications in tRNA, rRNA, snRNA and mRNA and their possible detection by RT-based techniques

N°	Chemical name	Abbreviation	RT block	RT pause	Chemical treatment
1	2′-O-Methyladenosine	Am	–	+ (low dNTP)	+
2	N^1-Methyladenosine	m^{1}A	+	+	–
3	N^2-Methyladenosine	m^{2}A	–	+	–
4	N^6-Methyladenosine	m^{6}A	–	+	–
5	N^6,N^6-Dimethyladenosine	m$^6_2$A	+	+	–
6	Inosine	I	–	–	+
7	2′-O-Methylguanosine	Gm	–	+ (low dNTP)	+
8	N^1-Methylguanosine	m^{1}G	+	+	+
9	N^2-Methylguanosine	m^{2}G	–	+	+
10	N^2,N^2-Dimethylguanosine	m$^2_2$G	+	+	–
11	N^7-Methylguanosine	m^{7}G	–	–	+
12	2′-O-Methyluridine	Um	–	+ (low dNTP)	+
13	N^3-Methyluridine	m^{3}U	+	+	–
14	5-Methyluridine (ribothymidine)	m^{5}U (T)	–	–	+ ?[a]
15	5-Carboxymethyluridine	cm^{5}U	–	–	+ ?[a]
16	Dihydrouridine	D	–	+	+
17	Pseudouridine	Ψ	–	–	+
18	N^1-Methylpseudouridine	m^{1}Ψ	–	–	+
19	N^3-Methylpseudouridine	m^{3}Ψ	+	–	+
20	N^1-Methyl-3-(3-amino-3-carboxypropyl)-pseudouridine	m^{1}acp^{3}Ψ	+	+	+
21	2′-O-Methylcytidine	Cm	–	+ (low dNTP)	+
22	N^3-Methylcytidine	m^{3}C	+	+	–
23	N^4-Methylcytidine	m^{4}C	–	+ ?[b]	–
24	N^4-Acetylcytidine	ac^{4}C	–	+ ?[b]	–
25	5-Methylcytidine	m^{5}C	–	–	+
26	5-Hydroxymethylcytidine	hm^{5}C	–	–	+ ?[c]

[a] m^{5}U and cm^{5}U may be modified by CMCT and the stability of respective adducts seems to be higher than for unmodified U.
[b] N^4-modified cytidines may provoke a RT-pause.
[c] hm^{5}C probably has the same properties as m^{5}C in the bisulfite modification reaction.
Comments: Complex modifications of both base and ribose, like Im, Ψm, m^{4}Cm, and ac^{4}Cm are not shown.

RNase-free 96% ethanol.

300 mM Sodium acetate, pH 5.2.

2 mg/ml Glycogen (Roche, USA).

Na$_2$CO$_3$ solution at pH 10.3 (500 mM; adjusted to pH 10.3 with 500 mM NaHCO$_3$). A precise adjustment of the pH to this value is very important for efficient hydrolysis of CMC-U adducts in RNA.

4.3.3. Procedure

1. Pseudouridine detection in RNA requires the comparison of the extension profiles obtained for four samples. One control experiment is done without CMCT treatment ("C") and another one without alkaline treatment ("C2") We recommend performing two assays on samples treated with CMCT for 10 min ("C10") and 20 min ("C20"), respectively. Each reaction is carried out on 2–200 μg total RNA (depending on the abundance of the analyzed RNA in total RNA) or on 2–200 pmol of purified transcripts (an amount of 20 pmol is recommended).

2. The CMCT solution should be freshly prepared a few minutes before use. The handling of CMCT requires particular care: the use of a hood and wearing gloves are recommended; 30 μl of CMC buffer is added to the control tube C, whereas 30 μl of the CMC solution is added to the C2, C10, and C20 tubes. Samples are incubated at 37°: C2 for 2 min, C10 for 10 min, C and C20 both for 20 min.

3. The CMCT treatment is stopped by incubation on ice and by addition of 100 μl of 300 mM sodium acetate at pH 5.2, 700 μl of ethanol and 1 μl of glycogen at 2 mg/ml. Glycogen is used as a carrier for cDNA precipitation. The tubes are kept for at least 1 h at −80° and centrifuged at 15,000g for 15 min at 4°. The supernatant is then removed, and the pellet is washed with 500 μl of 70% ethanol. The tubes are placed once again at −80° for at least 1 h and then centrifuged at 15,000g for 15 min at 4°. The supernatant has to be carefully and completely removed before the next step to avoid remaining CMCT in the further steps.

4. The C2 assay is kept at −80°, whereas the three remaining tubes are subjected to alkaline treatment.

5. The alkaline treatment is achieved by resuspending the RNA pellet in 40 μl of 50 mM Na$_2$CO$_3$ (10-fold diluted 500 mM stock solution) and incubating at 37° for 3 h (it is important to completely take up the RNA-glycogen precipitate in the Na$_2$CO$_3$ solution by pipetting up and down). After incubation, the CMC-modified RNAs are precipitated by addition of 1 μl of glycogen solution, 100 μl of 300 mM sodium acetate and 700 μl of 96% ethanol. After a 1 h precipitation at −80°, the RNA pellet is collected by centrifugation at 15,000g for 15 min at 4°. The

supernatant is completely removed and residual ethanol is evaporated with a SpeedVac.

6. The precipitated RNA is now ready for primer extension analysis (see previously). The extension products of the four reactions are fractionated in parallel on a denaturing 7% polyacrylamide-urea gel, together with a sequencing ladder obtained with the same primer.

4.3.4. Comments

Because the CMCT-modification protocol involves three successive ethanol precipitations, and because only small amounts of RNA are sometimes analyzed, particular care has to be taken to get an efficient RNA recovery at each precipitation step. To this end, the presence of glycogen, long precipitation times at $-80°$, centrifugation at $4°$, handling in ice to avoid RNA degradation, the use of gloves and sterilized material are recommended. Note that the use of total tRNA instead of glycogen as precipitation carrier is not recommended because primers might anneal to tRNAs.

4.3.5. Analysis pitfalls

First, the sequencing ladder has to be clearly readable to correlate the RT stops to a position in the sequence. It is important to remember that a CMCT-modified base prevents any Watson-Crick base pairing with dNTPs, and, therefore, the RT extension stops one base downstream from the CMCT modified base (Fig. 2.1). For that reason, there is a one-nucleotide shift between the sequencing ladder by use of ddNTPs and the stops corresponding to CMC-U and CMC-Ψ adducts in CMCT-treated RNA. Primer extension can display strong pauses, for a given position, in each of the sequencing ladder lanes and also within the C, C2, C10, and C20 lanes. These pauses are natural RT pauses and might result from RNA degradation or from the presence of stable RNA secondary structure. Therefore, strong natural RT pauses may prevent the identification of pseudouridine. The comparison of lane C2 with the control lane C reveals the presence of U and Ψ residues, whereas the comparison of lane C2 with lanes C10 and C20 enables the distinction between Ψ and U residues. Note that Ψ residues are modified to a lesser extent than U residues by CMCT. For this reason, the pauses generated by Ψ residues are quite faint in lane C2. However, because of the increased stability of the CMC-Ψ adduct, these residues give a stronger RT stop after alkaline treatment. As the CMCT modification of Ψ residue is not complete after 10 min, a stronger RT stop should be observed in lane C20 corresponding to a 20 min-step reaction. Finally, it is essential to verify the reproducibility of the results to confirm the identification of Ψ residues in RNA. We recommend performing three independent experiments.

Several examples of Ψ detection by CMCT-RT are presented in Fig. 2.4. Primer extension was performed on U2 snRNA and two tRNAs from *S. cerevisiae* by use of specific oligonucleotides. As one can see in these three panels, natural pauses of RT are rather frequent and may considerably

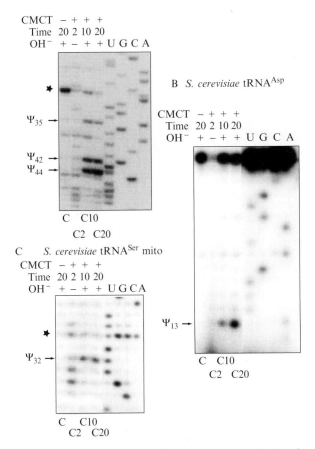

Figure 2.4 Examples of pseudouridine (Ψ) detection in RNA by the CMCT-RT approach. The primer extensions were performed on total *S. cerevisiae* RNA fractions that were treated with CMCT for 2, 10, and 20 min (lanes C2, C10, C20) followed or not by alkaline treatment (+ or −). Lane C is a control extension on RNA incubated for 20 min in CMCT buffer in the absence of CMCT and subjected to alkaline treatment. Oligonucleotide primers were specific for U2 snRNA (A) and two tRNAs (cytoplasmic tRNA^{Asp} and mitochondrial tRNA^{Ser}) (B and C). Sequencing ladders (lanes U, G, C, A) were prepared with the same primers. Natural RT pauses are indicated by stars. Arrows indicate the pauses corresponding to the identified Ψ residues.

complicate the analysis. However, the presence of Ψ residues is characterized by strong pauses in lanes C10 and C20 that are not detected in control lanes C and C2. Furthermore, in most cases, increased intensity of the RT pauses is seen in lane C20 compared with lane C10 (see Ψ_{44} in snRNA U2 or Ψ_{13} in tRNA^{Asp}). In some instances, the pauses in lanes C10 and C20 are accompanied by pauses in lanes C and C2. In these cases it may be difficult to conclude whether a Ψ residue is present at this position or not (see positions marked by an asterisk in primer extension profile for U2 snRNA and mitochondrial tRNA^{Ser}, Fig. 2.4A and C).

4.4. Complementary approaches for Ψ detection

Another approach has been used in the past for Ψ detection in RNA. After the development of CMCT-modification approach, its use is considerably reduced, but it still may be applied to confirm CMCT-RT results. The method is based on the different reactivities of U and Ψ residues to hydrazine (Peattie, 1979). Hydrazine reacts with the double C5=C6 bond in pyrimidine ring, leading to subsequent aniline cleavage of the RNA polynucleotide chain at every U residue (Fig. 2.3C). In contrast, Ψ residues do not react with hydrazine, and thus cleavages of the RNA are observed at U, but not at Ψ residues. Hydrazine cleavage was successfully used for the detection of two Ψ residues present at the 5′-end of U1 snRNA (Branlant *et al.*, 1980). Later, hydrazine treatment was combined with primer extension analysis to identify Ψ residues in U5 snRNA from *Physarum polycephalum* (Szkukalek *et al.*, 1996).

Alternative techniques for selective modification of pseudouridine residues in RNA have been developed recently (Emmerechts *et al.*, 2005; Mengel-Jorgensen and Kirpekar, 2002). In both cases, chemical modification is based on the particular reactivity of pseudouridine nitrogen N1 with the activated C=C bond of acrylonitrile (Mengel-Jorgensen and Kirpekar, 2002) or methylvinylsulfone (Emmerechts *et al.*, 2005). Both techniques were used for pseudouridine detection by mass spectrometry, but such a specific modification may probably also be used for primer extension analysis, although this has not been examined yet.

4.5. Detection of 5-methylcytosine (m5C)

The presence of base methylations at position 5 of pyrimidines in RNA is quite difficult to detect, because the CH_3 group at this position does not change the base-pairing properties of the nucleotide. The only methylation that is detectable using preliminary chemical treatment is that of m^{5}C. This modified nucleotide is quite common in various cellular RNAs and is also present in DNA. The techniques used for m^5C detection in DNA were developed a long time ago (for review, Rein *et al.*, 1998; Thomassin *et al.*, 1999). Besides m^5C-specific PCR and various HPLC- and MS-based techniques, DNA polymerase-based approaches that use primer extension were also proposed. These methods are based on the differential reactivity of m^5C compared with T (m^5U) and C. The detection of m^5C in DNA is performed using a specific oxidation of these residues by potassium permanganate ($KMnO_4$) (Hayatsu *et al.*, 1991) or by selective degradation of C and T, but not m^5C, residues by a hydrazine or bisulfite treatment. The reaction with bisulfite converts C (but not m^5C) into U residues. These techniques generally give reliable results for DNA, but their possible application to RNA has not been tested except for bisulfite treatment.

The bisulfite treatment technique is based on the low reactivity of m^5C compared with C residues with HSO_3^- ions in neutral or acidic conditions. Indeed, cytidine and some of its derivatives (like m^3C, Cm and ac^4C) form adducts, which are quite unstable and rapidly decompose at alkaline pH, concomitantly with chemical deamination of the cytosine ring. Such a chemical deamination converts these nucleotides to U or derivatives of U (Fig. 2.3D). The use of this technique for RNA was, until recently, limited by rather harsh reaction conditions (especially high pH), which were incompatible with the stability of phosphodiester bonds in RNA.

However, a technique based on bisulfite modification was recently proposed for m^5C mapping in RNA (Gu *et al.*, 2005). The bulk RNA is first incubated in the presence of sodium bisulfite under mildly acidic conditions (pH 5.1), followed by desalting and desulfonation at pH 9.0. These conditions lead to complete conversion of all C residues into U. The remaining m^5C residues in RNA are detected by primer extension by use of ddGTP. It should be noted that bisulfite treatment leads to complete conversion of all C residues in RNA into U residues. Thus, the sequence of the specific primer used for RNA analysis should be adapted to include the expected sequence changes because of this complete C→U conversion.

4.6. Detection of 7-methylguanine (m^7G)

To detect m^7G modifications in RNA, one can take advantage of their particular sensitivity to $NaBH_4$ reduction (Wintermeyer and Zachau, 1970, 1975). Such a treatment was used in the past for direct chemical RNA sequencing (Peattie, 1979). In this method, G residues in RNA are first methylated by dimethylsulfate (DMS) at position 7 and the resulting m^7G residues are cleaved by $NaBH_4$ reduction, followed by β-elimination with aniline (Fig. 2.3E). A similar technique without preliminary DMS methylation has been used for the detection of naturally present m^7G residues in yeast tRNAs (Alexandrov *et al.*, 2002).

4.7. Detection of 2'-O-methylated nucleotides using OH⁻ cleavage or 2'-OH reactivity

Alkaline hydrolysis of RNA polynucleotide chains proceeds by nucleophilic attack of the 2'-OH ribose group at the nearby 3'-phosphate. The resulting intermediate is unstable and rapidly decomposes into 2',3'-cyclic phosphate leading to phosphodiester bond cleavage (Fig. 2.3F). The methylation of ribose 2'-OH decreases the reactivity of the oxygen and thus prevents phosphodiester bond cleavage almost completely. This resistance to alkaline hydrolysis of the phosphodiester bonds on the 3'-side of 2'-O-methylated nucleotides can be detected by primer extension. The 2'-O-methylated residue appears as a "gap" in the regular ladder of OH⁻ cleavage (Fig. 2.1B). In practice, it is recommended that the primer extension profiles of both an unmodified RNA transcript and the natural modified RNA be compared. However, the presence of gaps in the RT profile is frequently hidden by natural pauses of RT at the corresponding positions (see Maden *et al.*, 1995; Maden, 2001 for discussion). Thus, an alternative approach for detection of 2'-O-methylations has been developed (see later).

4.8. Protocol for detection of 2'-O-methylated residues in RNA using OH⁻ cleavage

4.8.1. Equipment

Water bath set to 42°, heating block set to 90°.
Vacuum drier (SpeedVac).

4.8.2. Reagents

300 mM Sodium acetate, pH 5.2.
2 mg/ml Glycogen (Roche, USA).
Na_2CO_3 solution at pH 10.3 (500 mM; adjusted to pH 10.3 with 500 mM $NaHCO_3$).
RNase-free 96% ethanol.

4.8.3. Procedure

1. Each reaction is carried out on 5–200 μg of total RNA (depending on the concentration of the analyzed RNA in total RNA) or on 5–200 pmol of purified transcripts. The stock Na_2CO_3 solution is diluted five times to get a 100 mM final concentration. The volume of RNA solution is adjusted to 20 μl with RNase-free water, and 20 μl of the diluted Na_2CO_3 solution are added. The tube is then incubated at 96° for 5 min.

2. After incubation, the RNAs are precipitated by addition of 1 μl of glycogen solution, 100 μl of 300 mM sodium acetate and 700 μl of 96% ethanol. After at least 1 h incubation at $-80°$, the RNA pellet is collected by centrifugation at 15,000g for 15 min at 4°. The supernatant is completely removed and residual ethanol is evaporated using a SpeedVac.

3. The RNA is now ready for primer extension analysis (see previous). The products of the primer extensions are fractionated in parallel on a denaturing 7% polyacrylamide-urea gel together with a sequencing ladder obtained with the same primer.

4.8.4. Comments

Statistical RNA hydrolysis can also be performed by incubation at high temperature (95–100°) in neutral solutions (H_2O at pH 7.0–7.5). Incubation time ranges from 2–10 min and has to be adapted to each individual case.

4.8.5. Analysis pitfalls

Natural RT pauses may sometimes hide the absence of OH^- cleavage at a given position. The results obtained by OH^- hydrolysis should be verified by other methods, for example by use of RT with low concentrations of dNTPs (see later).

4.9. Detection of 2′-*O*-methylated nucleotides by use of low dNTP concentrations

The alternative approach for detection of 2′-*O*-methylated residues in RNA uses primer extension by reverse transcriptase at low dNTP Concentrations (Maden, 2001; Maden *et al.*, 1995).

This method is based on the observation that reverse transcriptase frequently slows down (or pauses) at such nucleotides, especially at low dNTP concentrations. These dNTP-concentration dependent stops seem to be specific to 2′-*O*-methylation and are rarely observed for other modified nucleotides in RNA. Primer extension with normal (unmodified) residues is much less sensitive to reduced dNTP concentration. However, the pauses observed in these conditions may correspond to regions where RT has a tendency to pause and not only to 2′-*O*-methylations of nucleotides. The comparison of RT profile for natural (modified) RNA and unmodified RNA transcript may help to distinguish between these two possibilities.

It is noteworthy that pauses observed at 2′-*O*-methylated residues are sequence dependent and, depending on the sequence context, no pause may occur at some of the 2′-*O*-methylated residues. Therefore, the method does not allow an exhaustive identification of all 2′-*O*-methylated residues in RNA molecules.

4.10. Protocol for detection of 2'-O-methylated residues in RNA by use of low dNTP concentrations

4.10.1. Equipment

Water bath set to 42°.

4.10.2. Procedure

1. Each primer extension reaction is carried out on 2–50 μg total RNA or 2–50 pmol of purified transcripts. Oligonucleotide labeling and purification are done as described previously.
2. Primer extension reactions are performed under the same conditions, except for the variable concentrations of dNTP in the mixture. Four concentrations are usually tested, ranging from 100–2 nM, with two intermediate concentrations of 20 nM and 10 nM.
3. A sequencing ladder, performed with the same oligonucleotide is prepared as described previously.
4. After denaturation at 94° for 2 min, 2 μl of each sample are loaded to 7% PAGE-urea gel.

4.10.3. Comments

The experiment can be performed by use of only the two most extreme concentrations of dNTPs (100 nM and 2 nM) instead of the four concentrations; nevertheless the two intermediate concentrations can be very helpful to interpret the results.

4.10.4. Analysis pitfalls

First, the sequencing ladder has to be clearly readable to correlate the RT stops to a position in the sequence. Here again the RT extension stops one base downstream from the 2'-O-methylated residue; therefore, there is a one nucleotide shift between the sequencing ladder using ddNTPs and the stops corresponding to the 2'-O-methylated residue. RNA degradation at sensitive Py–A bonds may easily induce pauses in the RT profile. One should always compare the RT profile obtained for unmodified RNA transcript, sequencing ladder, and RT profile at low dNTP concentration. Only pauses absent in the RNA sequencing ladder and in unmodified RNA may serve as an indication of a 2'-O-methylation.

An example of such an analysis is given in Fig. 2.5. Primer extension was performed on the naturally modified human U2 snRNA (panel B) and the corresponding unmodified RNA transcript (panel A). Multiple pauses are visible both in the T7 RNA transcript and the naturally modified RNA. However, additional strong RT stops that appear or are reinforced at low

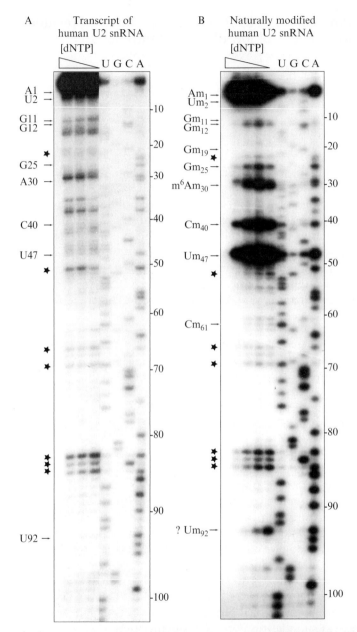

Figure 2.5 Examples of primer extension for the detection of $2'$-O-methylated residues in human U2 snRNA. (A) Primer extension was performed on an unmodified U2 snRNA transcript produced with the T7 RNA polymerase. (B) The extension was performed with the same primer on the authentic U2 snRNA contained in nuclear RNA fraction from HeLa cells. The sequencing ladder is shown on the right of each panel. Natural RT pauses are indicated by stars. Locations of the known $2'$-O-methylated residues

dNTP concentrations indicate the presence of 2′-O-methylated residues in human U2 snRNA. Note the great variability of the intensity of these stops; some of the 2′-O-methylations create very strong pauses whereas other ones give only a faint band in the RT profile.

4.11. Other chemical modification methods specific for 2′-O-methylations

Very recently, an alternative chemical treatment that should allow the distinction between unmethylated and 2′-O-methylated residues has been proposed (Merino et al., 2005). Free unconstrained ribose 2′-OH easily reacts with N-methylisatoic anhydride (NMIA), and the resulting adduct can be analyzed by primer extension. For the moment, the method was only applied to RNA structural analysis. However, comparisons of NMIA modification and primer extension profiles for unmodified transcript and 2′-O-methyl–containing RNA may probably be used as a tool for 2′-O-methylation mapping.

4.12. Detection of dihydrouridine (D) residues by alkaline hydrolysis

Chemical treatment allowing the detection of D residues in RNA has been developed to study the yeast *S. cerevisiae* RNA/dihydrouridine-synthases (DUS) (Xing et al., 2004). Dihydrouridine is particularly unstable at alkaline pH (pH ~13 for 5 min) and is converted into a β-ureidopropionic acid ribose derivative by pyrimidine cycle opening (Fig. 2.3G). The resulting product does not base pair with other nucleotides and may be easily detected by primer extension analysis. Another property of D residues that can be used for their detection is their stability to hydrazine treatment in contrast to uridine, but until now, this technique has not been used for primer extension analysis.

---

of human U2 snRNA are indicated on the left of (B). Altogether, 10 2′-O-methylated residues have been localized in human U2 snRNA (Reddy et al., 1981; Westin et al., 1984). Five of them can be unambiguously detected (Gm_{12}, Gm_{25}, m^6Am_{30}, Cm_{40}, and Um_{47}). Two residues are located too close to the RNA 5′-end (Am_1 and Um_2) to be detected by this approach, and three remaining residues (Gm_{11}, Gm_{19} and Cm_{61}) are not detected by RT at low (dNTP), because only very weak pauses are observed at the corresponding positions. Note the presence of (dNTP)-dependent stop at position 93, which may indicate the existence of an additional 2′-O-methylation at Um_{92}.

5. RNA MODIFICATIONS LEADING TO RT PAUSES WITHOUT PRELIMINARY CHEMICAL TREATMENT

Many modified nucleotides present in RNA molecules are not able to form base-pairing interactions with other nucleotides because of the presence of chemical groups on the Watson-Crick face of the base (positions 1, 2, and 6 in purines and 2, 3, and 4 in pyrimidines) (see Fig. 2.1 and Table 2.1). When the presence of such modified nucleotides is confirmed by the analysis of total RNA composition (HPLC or other techniques), the position of such nucleotides may be easily detected by primer extension. Examples include the localization of $acp^3\Psi$ in *D. melanogaster* 18S rRNA (Youvan and Hearst, 1981), m^6_2A in yeast 18S rRNA (Lafontaine *et al.*, 1995), m^1G_{745} in *E. coli* 23S rRNA (Gustafsson and Persson, 1998), and m^1A and m^3U in *H. marismortui* 23S RNA (Kirpekar *et al.*, 2005). Smaller RNAs, like tRNAs, were also analyzed by this method (for example, detection of m^2_2G in various yeast tRNAs [see Ansmant *et al.*, 2001; Behm-Ansmant *et al.*, 2004] and m^1G_9 or Y_{37}-base in yeast tRNA [Jackman *et al.*, 2003]).

Some other modified nucleotides may still base pair with complementary nucleotides, despite the presence of additional chemical groups at Watson-Crick positions. However, these additional groups may affect base-pairing stability, particularly when modification occurs at position 2 of G or 6 of A. Such modified residues are also detected by primer extension, but the intensity of the corresponding pause in the primer extension profile is much less pronounced than in the case of complete blockage of primer extension. Several examples of RT-based detection of this type have been described in literature, such as a kinetic pausing of RT by m^2G residues in rRNA (Brimacombe *et al.*, 1993; Youvan and Hearst, 1979) or various modifications of A at N6 (i^6A, m^6A, t^6A, etc.).

6. CONCLUSION

RT-based techniques applied to the analysis of RNA modification represent a powerful tool for the localization of modified residues in low-abundance cellular RNAs. At the current stage of development of these techniques, specific chemical treatment is only proposed for a few frequently encountered modified bases and 2'-O-ribose methylations. A specific treatment allowing the detection of other modified residues remains to be developed.

Despite a great interest of RT-based analysis, extreme care should be taken for unambiguous interpretation of the results. Indeed, the appearance of pauses in an RT profile may only serve as an indication of the possible occurrence of modified residues at these positions. The use of other complementary techniques of RNA analysis (like HPLC and MS) may be useful to confirm the RT results.

RT-based techniques for RNA analysis may also apply not only to the analysis of naturally occurring modified nucleotides but also to the detection and localization of chemically altered nucleotides in cellular RNAs. Like DNA, cellular RNA molecules may contain oxidized or alkylated residues resulting from the action of free radicals during oxidative stress. The presence of such spontaneous chemical modifications may alter RNA properties in mRNA translation. The development of RT-based techniques for these "nonnatural" modified residues may be extremely useful for better understanding these complex cellular phenomena.

REFERENCES

Aiba, H., Adhya, S., and de Crombrugghe, B. (1981). Evidence for two functional *gal* promoters in intact *Escherichia coli* cells. *J. Biol. Chem.* **256**, 11905–11910.

Alexandrov, A., Martzen, M. R., and Phizicky, E. M. (2002). Two proteins that form a complex are required for 7-methylguanosine modification of yeast tRNA. *RNA* **8**, 1253–1266.

Ansmant, I., and Motorin, Y. (2001). Identification of RNA-modification enzymes using sequence homology. *Mol. Biol. (Moscow)* **35**, 206–223.

Ansmant, I., Motorin, Y., Massenet, S., Grosjean, H., and Branlant, C. (2001). Identification and characterization of tRNA:Ψ_{31}-synthase (Pus6p) in yeast *S. cerevisiae*. *J. Biol. Chem.* **276**, 34934–34940.

Bakin, A., and Ofengand, J. (1993). Four newly located pseudouridylate residues in *Escherichia coli* 23S ribosomal RNA are all at the peptidyltransferase center: Analysis by the application of a new sequencing technique. *Biochemistry* **32**, 9754–9762.

Bakin, A. V., and Ofengand, J. (1998). Mapping of pseudouridine residues in RNA to nucleotide resolution. *Methods Mol. Biol.* **77**, 297–309.

Bass, B. L. (1995). RNA editing. An I for editing. *Curr. Biol.* **5**, 598–600.

Behm-Ansmant, I., Grosjean, H., Massenet, S., Motorin, Y., and Branlant, C. (2004). Pseudouridylation at position 32 of mitochondrial and cytoplasmic tRNAs requires two distinct enzymes in *Saccharomyces cerevisiae*. *J. Biol. Chem.* **279**, 52998–53006.

Björk, G. R. (1995). Biosynthesis and function of modified nucleosides. In "tRNA: Structure, Biosynthesis and Function" (D. Söll and U. RajBhandary, eds.), pp. 165–206. ASM Press, Washington.

Björk, G. R., Ericson, J. U., Gustafsson, C. E. D., Hagervall, T. G., Jonsson, Y. H., and Wikstrom, P. M. (1987). Transfer RNA modification. *Ann. Rev. Biochem.* **56**, 263–287.

Branch, A. D., Benenfeld, B. J., and Robertson, H. D. (1989). RNA fingerprinting. *Methods Enzymol.* **180**, 130–154.

Branlant, C., Krol, A., Ebel, J. P., Lazar, E., Gallinaro, H., Jacob, M., Sri-Widada, J., and Jeanteur, P. (1980). Nucleotide sequences of nuclear U1A RNAs from chicken, rat and man. *Nucleic Acids Res.* **8,** 4143–4154.

Branlant, C., Krol, A., Machatt, M. A., Pouyet, J., Ebel, J. P., Edwards, K., and Kossel, H. (1981). Primary and secondary structures of *Escherichia coli* MRE 600 23S ribosomal RNA. Comparison with models of secondary structure for maize chloroplast 23S rRNA and for large portions of mouse and human 16S mitochondrial rRNAs. *Nucleic Acids Res.* **9,** 4303–4324.

Brimacombe, R., Mitchell, P., Osswald, M., Stade, K., and Bochkariov, D. (1993). Clustering of modified nucleotides at the functional center of bacterial ribosomal RNA. *FASEB J.* **7,** 161–167.

Brownlee, G. G., and Cartwright, E. M. (1977). Rapid gel sequencing of RNA by primed synthesis with reverse transcriptase. *J. Mol. Biol.* **114,** 93–117.

Bruenger, E., Kowalak, J. A., Kuchino, Y., McCloskey, J. A., Mizushima, H., Stetter, K. O., and Crain, P. F. (1993). 5S rRNA modification in the hyperthermophilic archaea *Sulfolobus solfataricus* and *Pyrodictium occultum*. *FASEB J.* **7,** 196–200.

Chomczynski, P., and Sacchi, N. (1987). Single-step method of RNA isolation by acid guanidinium thiocyanate-phenol-chloroform extraction. *Anal. Biochem.* **162,** 156–159.

Cohn, W. E. (1960). Pseudouridine, a carbon-carbon linked ribonucleoside in ribonucleic acids: Isolation, structure, and chemical characteristics. *J. Biol. Chem.* **235,** 1488–1498.

Denman, R., Colgan, J., Nurse, K., and Ofengand, J. (1988). Crosslinking of the anticodon of P site bound tRNA to C-1400 of *E. coli* 16S RNA does not require the participation of the 50S subunit. *Nucleic Acids Res.* **16,** 165–178.

Emmerechts, G., Herdewijn, P., and Rozenski, J. (2005). Pseudouridine detection improvement by derivatization with methyl vinyl sulfone and capillary HPLC-mass spectrometry. *J. Chromat. B* **825,** 233–238.

Fuchs, B., Zhang, K., Rock, M. G., Bolander, M. E., and Sarkar, G. (1999). High temperature cDNA synthesis by AMV reverse transcriptase improves the specificity of PCR. *Mol. Biotechnol.* **12,** 237–240.

Garcia, G. A., and Goodenough-Lashua, D. M. (1998). Appendix 3: General properties of RNA-modifying and -editing enzymes. *In* "The Modification and Editing of RNA" (H. Grosjean and R. Benne, eds.), pp. 555–560. ASM Press, New York.

Gehrke, C. W., and Kuo, K. C. (1989). Ribonucleoside analysis by reversed-phase high-performance liquid chromatography. *J. Chromatogr.* **471,** 3–36.

Gehrke, C. W., Kuo, K. C., McCune, R. A., Gerhardt, K. O., and Agris, P. F. (1982). Quantitative enzymatic hydrolysis of tRNAs: Reversed-phase high-performance liquid chromatography of tRNA nucleosides. *J. Chromatogr.* **230,** 297–308.

Grosjean, H., Motorin, Y., and Morin, A. (1998). RNA-modifying and RNA-editing enzymes: Methods for their identification. *In* "The Modification and Editing of RNA" (H. Grosjean and R. Benne, eds.), pp. 21–46. ASM Press, New York.

Gu, W., Hurto, R. L., Hopper, A. K., Grayhack, E. J., and Phizicky, E. M. (2005). Depletion of *Saccharomyces cerevisiae* tRNA[His] guanylyltransferase Thg1p leads to uncharged tRNA[His] with additional m^5C. *Mol. Cell. Biol.* **25,** 8191–8201.

Gupta, R. C., and Randerath, K. (1979). Rapid print-readout technique for sequencing of RNA's containing modified nucleotides. *Nucleic Acids Res.* **6,** 3443–3458.

Gustafsson, C., and Persson, B. C. (1998). Identification of the *rrmA* gene encoding the 23S rRNA m^1G_{745} methyltransferase in *Escherichia coli* and characterization of an m^1G_{745}-deficient mutant. *J. Bacteriol.* **180,** 359–365.

Hayatsu, H., Atsumi, G., Nawamura, T., Kanamitsu, S., Negishi, K., and Maeda, M. (1991). Permanganate oxidation of nucleic acid components: A reinvestigation. *Nucleic Acids Symp. Ser.* **25,** 77–78.

Hiley, S. L., Jackman, J., Babak, T., Trochesset, M., Morris, Q. D., Phizicky, E., and Hughes, T. R. (2005). Detection and discovery of RNA modifications using microarrays. *Nucleic Acids Res.* **33,** e2.

Ho, N. W., and Gilham, P. T. (1971). Reaction of pseudouridine and inosine with N-cyclohexyl-N'-beta-(4-methylmorpholinium)ethylcarbodiimide. *Biochemistry* **10,** 3651–3657.

Jackman, J. E., Montange, R. K., Malik, H. S., and Phizicky, E. M. (2003). Identification of the yeast gene encoding the tRNA m^{1}G methyltransferase responsible for modification at position 9. *RNA* **9,** 574–585.

Kirpekar, F., Hansen, L. H., Rasmussen, A., Poehlsgaard, J., and Vester, B. (2005). The archaeon *Haloarcula marismortui* has few modifications in the central parts of its 23S ribosomal RNA. *J. Mol. Biol.* **348,** 563–573.

Krol, A., Branlant, C., Lazar, E., Gallinaro, H., and Jacob, M. (1981). Primary and secondary structures of chicken, rat and man nuclear U4 RNAs. Homologies with U1 and U5 RNAs. *Nucleic Acids Res.* **9,** 2699–2716.

Lafontaine, D., Vandenhaute, J., and Tollervey, D. (1995). The 18S rRNA dimethylase Dim1p is required for pre-ribosomal RNA processing in yeast. *Genes Dev.* **9,** 2470–2481.

Lane, D. J., Field, K. G., Olsen, G. J., and Pace, N. R. (1988). Reverse transcriptase sequencing of ribosomal RNA for phylogenetic analysis. *Methods Enzymol.* **167,** 138–144.

Maas, S., Rich, A., and Nishikura, K. (2003). A-to-I RNA editing: Recent news and residual mysteries. *J. Biol. Chem.* **278,** 1391–1394.

Maden, B. E. (2001). Mapping 2'-O-methyl groups in ribosomal RNA. *Methods* **25,** 374–382.

Maden, B. E., Corbett, M. E., Heeney, P. A., Pugh, K., and Ajuh, P. M. (1995). Classical and novel approaches to the detection and localization of the numerous modified nucleotides in eukaryotic ribosomal RNA. *Biochimie* **77,** 22–29.

Mangan, J. A., Sole, K. M., Mitchison, D. A., and Butcher, P. D. (1997). An effective method of RNA extraction from bacteria refractory to disruption, including mycobacteria. *Nucleic Acids Res.* **25,** 675–676.

Massenet, S., Mougin, A., and Branlant, C. (1998). Posttranscriptional modifications in the U small nuclear RNAs. *In* "The Modification and Editing of RNA" (H. Grosjean and R. Benne, eds.), pp. 201–227. ASM Press, New York.

McCloskey, J. A., and Rozenski, J. (2005). The Small Subunit rRNA Modification Database. *Nucleic Acids Res.* **33,** D135–D138.

Meng, Z., and Limbach, P. A. (2006). Mass spectrometry of RNA: Linking the genome to the proteome. *Brief Funct. Genomic Proteomic.* **5,** 87–95.

Mengel-Jorgensen, J., and Kirpekar, F. (2002). Detection of pseudouridine and other modifications in tRNA by cyanoethylation and MALDI mass spectrometry. *Nucleic Acids Res.* **30,** e135.

Merino, E. J., Wilkinson, K. A., Coughlan, J. L., and Weeks, K. M. (2005). RNA structure analysis at single nucleotide resolution by selective 2'-hydroxyl acylation and primer extension (SHAPE). *J. Am. Chem. Soc.* **127,** 4223–4231.

Metz, D. H., and Brown, G. L. (1969a). The investigation of nucleic acid secondary structure by means of chemical modification with a carbodiimide reagent. I. The reaction between N-cyclohexyl-N'-beta-(4-methylmorpholinium)ethylcarbodiimide and model nucleotides. *Biochemistry* **8,** 2312–2328.

Metz, D. H., and Brown, G. L. (1969b). The investigation of nucleic acid secondary structure by means of chemical modification with a carbodiimide reagent. II. The reaction between N-cyclohexyl-N'-beta-(4-methylmorpholinium)ethylcarbodiimide and transfer ribonucleic acid. *Biochemistry* **8,** 2329–2342.

Metz, D. H., and Brown, G. L. (1969c). The tertiary structure of transfer ribonucleic acid investigated by a chemical modification technique. *Biochem. J.* **114,** 35P.

Morse, D. P., and Bass, B. L. (1997). Detection of inosine in messenger RNA by inosine-specific cleavage. *Biochemistry* **36,** 8429–8434.

Ofengand, J., and Fournier, M. J. (1998). The pseudouridine residues of rRNA: Number, location, biosynthesis, and function. *In* "The Modification and Editing of RNA" (H. Grosjean and R. Benne, eds.), pp. 229–254. ASM Press, New York.

Peattie, D. A. (1979). Direct chemical method for sequencing RNA. *Proc. Natl. Acad. Sci. USA* **76,** 1760–1764.

Qu, H. L., Michot, B., and Bachellerie, J. P. (1983). Improved methods for structure probing in large RNAs: A rapid 'heterologous' sequencing approach is coupled to the direct mapping of nuclease accessible sites. Application to the 5' terminal domain of eukaryotic 28S rRNA. *Nucleic Acids Res.* **11,** 5903–5920.

Reddy, R., Henning, D., Epstein, P., and Busch, H. (1981). Primary and secondary structure of U2 snRNA. *Nucleic Acids Res.* **9,** 5645–5658.

Rein, T., DePamphilis, M. L., and Zorbas, H. (1998). Identifying 5-methylcytosine and related modifications in DNA genomes. *Nucleic Acids Res.* **26,** 2255–2264.

Rivas, R., Vizcaino, N., Buey, R. M., Mateos, P. F., Martinez-Molina, E., and Velazquez, E. (2001). An effective, rapid and simple method for total RNA extraction from bacteria and yeast. *J. Microbiol. Methods* **47,** 59–63.

Saccomanno, L., and Bass, B. L. (1999). A minor fraction of basic fibroblast growth factor mRNA is deaminated in *Xenopus* stage VI and matured oocytes. *RNA* **5,** 39–48.

Schaub, M., and Keller, W. (2002). RNA editing by adenosine deaminases generates RNA and protein diversity. *Biochimie* **84,** 791–803.

Seeburg, P. H. (2002). A-to-I editing: New and old sites, functions and speculations. *Neuron* **35,** 17–20.

Smith, J. E., Cooperman, B. S., and Mitchell, P. (1992). Methylation sites in *Escherichia coli* ribosomal RNA: Localization and identification of four new sites of methylation in 23S rRNA. *Biochemistry* **31,** 10825–10834.

Sprinzl, M., and Vassilenko, K. S. (2005). Compilation of tRNA sequences and sequences of tRNA genes. *Nucleic Acids Res.* **33,** D139–D140.

Stanley, J., and Vassilenko, S. (1978). A different approach to RNA sequencing. *Nature* **274,** 87–89.

Szkukalek, A., Mougin, A., Gregoire, A., Solymosy, F., and Branlant, C. (1996). A unique U5→A substitution in the *Physarum polycephalum* U1 snRNA: Evidence at the RNA and gene levels. *Biochimie* **78,** 425–435.

Tanaka, Y., Dyer, T. A., and Brownlee, G. G. (1980). An improved direct RNA sequence method; its application to *Vicia faba* 5.8S ribosomal RNA. *Nucleic Acids Res.* **8,** 1259–1272.

Thomas, B., and Akoulitchev, A. V. (2006). Mass spectrometry of RNA. *Trends Biochem. Sci.* **31,** 173–181.

Thomassin, H., Oakeley, E. J., and Grange, T. (1999). Identification of 5-methylcytosine in complex genomes. *Methods* **19,** 465–475.

Veldman, G. M., Klootwijk, J., de Regt, V. C., Planta, R. J., Branlant, C., Krol, A., and Ebel, J. P. (1981). The primary and secondary structure of yeast 26S rRNA. *Nucleic Acids Res.* **9,** 6935–6952.

Westin, G., Lund, E., Murphy, J. T., Pettersson, U., and Dahlberg, J. E. (1984). Human U2 and U1 RNA genes use similar transcription signals. *EMBO J.* **3,** 3295–3301.

Wintermeyer, W., and Zachau, H. G. (1970). A specific chemical chain scission of tRNA at 7-methylguanosine. *FEBS Lett.* **11,** 160–164.

Wintermeyer, W., and Zachau, H. G. (1975). Tertiary structure interactions of 7-methylguanosine in yeast tRNA^{Phe} as studied by borohydride reduction. *FEBS Lett.* **58,** 306–309.

Xing, F., Hiley, S. L., Hughes, T. R., and Phizicky, E. M. (2004). The specificities of four yeast dihydrouridine synthases for cytoplasmic tRNAs. *J. Biol. Chem.* **279,** 17850–17860.

Youvan, D. C., and Hearst, J. E. (1979). Reverse transcriptase pauses at N2-methylguanine during *in vitro* transcription of *Escherichia coli* 16S ribosomal RNA. *Proc. Natl. Acad. Sci. USA* **76,** 3751–3754.

Youvan, D. C., and Hearst, J. E. (1981). A sequence from *Drosophila melanogaster* 18S rRNA bearing the conserved hypermodified nucleoside amY: Analysis by reverse transcription and high-performance liquid chromatography. *Nucleic Acids Res.* **9,** 1723–1741.

тRNA MODIFICATIONS

DETECTION OF ENZYMATIC ACTIVITY OF TRANSFER RNA MODIFICATION ENZYMES USING RADIOLABELED tRNA SUBSTRATES

Henri Grosjean,[*] Louis Droogmans,[†] Martine Roovers,[†] and Gérard Keith[‡]

Contents

[*] Institut de Génétique et Microbiologie, Université Paris-Sud, Orsay, France
[†] Laboratoire de Microbiologie et Institut de Recherches, Université Libre de Bruxelles, Bruxelles, Belgique
[‡] Institut de Biologie Moléculaire et Cellulaire, Strasbourg, France

Methods in Enzymology, Volume 425
ISSN 0076-6879, DOI: 10.1016/S0076-6879(07)25003-7

Abstract

The presence of modified ribonucleotides derived from adenosine, guanosine, cytidine, and uridine is a hallmark of almost all cellular RNA, and especially tRNA. The objective of this chapter is to describe a few simple methods that can be used to identify the presence or absence of a modified nucleotide in tRNA and to reveal the enzymatic activity of particular tRNA-modifying enzymes *in vitro* and *in vivo*. The procedures are based on analysis of prelabeled or postlabeled nucleotides (mainly with [^{32}P] but also with [^{35}S], [^{14}C] or [^{3}H]) generated after complete digestion with selected nucleases of modified tRNA isolated from cells or incubated *in vitro* with modifying enzyme(s). Nucleotides of the tRNA digests are separated by two-dimensional (2D) thin-layer chromatography on cellulose plates (TLC), which allows establishment of base composition and identification of the nearest neighbor nucleotide of a given modified nucleotide in the tRNA sequence. This chapter provides useful maps for identification of migration of approximately 70 modified nucleotides on TLC plates by use of two different chromatographic systems. The methods require only a few micrograms of purified tRNA and can be run at low cost in any laboratory.

1. INTRODUCTION

Among cellular RNAs, transfer RNAs (tRNA) are by far the nucleic acids that are most frequently and diversely posttranscriptionally modified by specific RNA modification enzymes. These enzymes alter the chemical nature of a genetically encoded nucleotide by adding a chemical group (such as a methyl-, formyl-, acetyl-, or even a bigger compound such as isopentenyl- or threonyl carbamoyl group), by deamination, reduction, or thiolation of one atom of a base or $2'$-hydroxyl group of a ribose. Most of these reactions require the participation of metabolites (cofactors) originating from the central metabolism, such as ATP, GTP, S–adenosyl-L-methionine, methylene tetrahydrofolate, isopentenyl pyrophosphate, pyridoxal phosphate, NADH, and/or various amino acids such as L-cysteine, L-lysine, L-threonine, or L-glutamic acid. Of a total of 120 distinct modified nucleosides to which structures have been assigned, approximately 100 have been found in tRNAs (Dunin-Horkawicz *et al.*, 2006; Limbach *et al.*, 1994). However, the pattern of modifications (type and location) depends much on the tRNA isoacceptor considered and on the organism or the organelle they originate from. Transfer RNAs with lower numbers of modified nucleotides (2–6 per tRNA molecule) are found in mitochondria and *Mycoplasmas,* whereas cytoplasmic tRNAs from higher eukaryotic cells like mouse, bovine, or human (metazoan) can contain up to 22 modified nucleotides per molecule (for details, see Dunin-Horkawicz *et al.*, 2006; http://genesilico.pl/modomics/; Grosjean *et al.*, 1995; Rozenski *et al.*, 1999; http://medstat.med.utah.edu/RNAmods/; and Sprinzl Vassilenko, 2005; www.uni-bayreuth.de/departments/biochemie/trna/).

Enzymatic modification of precursor tRNA occurs concomitantly with many other processing events, such as $5'$ and $3'$ trimming, intron splicing, nucleotide addition at the $5'$- or $3'$-end, and sometimes tRNA editing. In eukaryotic cells, most of these posttranscriptional events occur in the nucleus, but a few additional steps of RNA modification take place in the cytoplasm or simultaneously with the transport of tRNA through the nuclear pores. A distinct RNA maturation apparatus is also present in mitochondria and chloroplasts; however, all of the corresponding enzymes in these organelles are the products of nuclear genes (reviewed in Hopper and Phizicky, 2003).

A large number of RNA modification enzymes (and their corresponding genes) have been identified and characterized, mostly from the bacteria *E. coli* (reviewed in Björk and Hägervall, 2005) and the yeast *Saccharomyces cerevisiae* (reviewed in Johansson and Byström, 2005). These enzymes are site-specific, multisite-specific, or dual-specific (acting on different types of RNA); some of them act at early steps of tRNA maturation, on pre-tRNA containing an intron, whereas others work only at later steps, on partially matured RNA of which intron has been removed (reviewed in Garcia and Goodenough-Lashua, 1998).

A major difficulty in working with tRNA modification enzymes is the identification of the target nucleotide and the minimal substrate on which these enzymes can work *in vitro* (specificity problem). In this chapter, we describe in detail a few simple procedures that we found particularly suited for detecting the activity and specificity of many different tRNA modification enzymes *in vitro*, as well as *in vivo,* and the identification of the corresponding tRNA modification products. This chapter also provides useful maps of the migration characteristics of approximately 70 naturally occurring modified nucleotides on thin-layer cellulose plates. These maps complete those that were already published in earlier articles or reviews (Grosjean *et al.*, 2004; Hashimoto *et al.,* 1975; Keith, 1995; Kuchino *et al.*, 1987; Nishimura, 1979). The method described in the following constitutes an alternative approach to other methods based on the use of reverse transcriptase (see Chapter 2) or of high-performance liquid chromatography coupled with mass spectrometry (HPLC/MS; see Chapters 1 and 9). Each of these techniques has advantages and drawbacks, and what cannot be solved with one method usually can be solved with an alternate technique. Nevertheless, at variance with the costly HPLC/MS method, the techniques described in the following, as well as in Chapter 2, can be run at low cost in any laboratory.

2. IDENTIFICATION OF MODIFIED NUCLEOTIDES IN tRNA

Figure 3.1 summarizes the various experiments that can be used to identity the presence or absence of a given modified nucleotide in tRNA extracted from a cell or after *in vitro* incubation of a tRNA substrate with tRNA modification enzymes. All tests described in the following exploit the possibility of identifying what is happening at a given atom of a tRNA substrate provided that the target nucleotide was preradiolabeled or post-radiolabeled with either [^{32}P], [^{14}C], or [^{3}H]. Identification of modified nucleotides is performed after complete digestion of a given tRNA molecule into mononucleotide or dinucleotide monophosphates by use of different types of nucleases and analysis of the resulting hydrolysates by two-dimensional thin-layer chromatography (2D-TLC). The use of pre- or post-[^{32}P]–labeling procedures greatly facilitates the identification of even tiny amounts of modified nucleotides. However, the method works well only if the RNA, like tRNA or rRNA fragments, are not longer than 100–150 nt. With longer RNA molecules, such as rRNA with several thousands of nucleotides, detection of the presence of a few modified

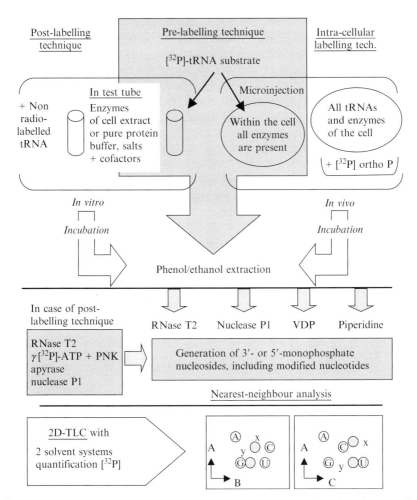

Figure 3.1 Overview of the different steps of the procedure to identify and eventually quantify the type of modified nucleotides produced by tRNA modification enzymes under *in vitro* or *in vivo* conditions. Depending on the main goal of the experiment, there are two main strategies: in the "prelabeling method," radiolabeled tRNA substrates are provided to the enzymes or microinjected into a living cell (e.g., *Xenopus laevis*), whereas in the "postlabeling method," the tRNA substrates are radiolabeled only after they have been modified by the modification enzymes. Alternately, cellular tRNAs can be radiolabeled *in vivo* by incubation of the microorganisms with radiolabeled metabolite precursors such as [^{32}P]-orthophosphate or [^{35}S]-sulfate. In all cases, the phenol/ethanol extract of RNA is digested with various nucleases to generate 3′- or 5′-monophosphate nucleosides, the composition of which is then analyzed by 2D-TLC.

nucleotides within a huge amount of unmodified ones is more difficult but, in principle, not impossible. For these long RNAs, it is preferable to cleave the molecule into pieces and analyze each fragment independently. The

most popular endonucleases allowing generation of fragments of RNA are
RNase T1 and RNase H, used in combination with various types of splints
(discussed in details in Cedergren and Grosjean, 1987; Grosjean *et al.*, 1998;
Moore and Query, 1998; Yu and Steitz, 1997; Zhao and Yu, 2004;
Zimmermann *et al.*, 1998). Because only tiny amounts of RNAs are needed
for postlabeling and prelabeling procedures, purification of enough fragments of RNAs can usually be performed by one-dimensional (1D) or 2D
electrophoresis or by any other purification technique (see Ikemura, 1989;
Suzuki *et al.*, 2002; Tanner, 1990; and Chapter 10).

3. TESTING THE ACTIVITY OF RNA MODIFICATION ENZYMES

Naturally occurring, fully modified RNAs isolated from cells are the
end products of the complete maturation process and thus cannot usually
serve as substrates for testing the activity of a putative RNA-modifying
enzyme *in vitro*, except in certain cases of heterologous reactions between
enzyme and tRNA from different organisms. However, the development of
recombinant DNA and RNA techniques, as well as of the chemical and
enzymatic synthesis of RNA *in vitro*, offers numerous alternatives for
obtaining synthetic or semisynthetic unmodified or partially modified pre-
RNA substrates (radiolabeled or not) suitable for testing tRNA modification enzymes *in vitro* or *in vivo*.

The first rather tricky method we developed was based on the earlier
work of Uhlenbeck and coworkers. They showed it was possible to replace
in vitro several nucleotides in the anticodon loop of naturally occurring
tRNA molecules by any of the four canonical nucleotides (Bruce and
Uhlenbeck, 1982). The method involved the use of various nucleases,
phosphatases, T4-polynucleotide kinase, and T4-RNA ligase. We originally used this "cut and paste" technology to introduce a [^{32}P]-labeled
phosphate 5'- or 3'-adjacent to the nucleoside of interest, simultaneously
to anticodon replacement in tRNA (reviewed in Grosjean *et al.*, 1998).
These site-specific [^{32}P]-radiolabeled tRNA substrates were then incubated
in vitro in the presence of an enzymatic preparation or microinjected into
Xenopus oocytes (*in vivo* experiments). After phenol extraction and purification by gel electrophoresis of the modified RNA products, the [^{32}P]-labeled
RNA was further subjected to complete digestion with nuclease P1 or
RNase T2 and analyzed by 2D TLC on cellulose plates (see later). After
autoradiography, the unique radiolabeled 5'- or 3'-[^{32}P]-nucleotide and its
modified derivative(s) appear as distinct and well-defined spots on the TLC
plates. The relative positions of the different radioactive spots on the
TLC were indicative of the chemical nature of the modified nucleotide
(compared with reference maps obtained with the same chromatographic

solvent systems). Evaluation of radioactivity in the TLC spot(s) allows one to calculate the molar amount of modified nucleotide within the RNA molecule and hence, if analysis is performed at different incubation times, to determine the rate of the enzymatic reaction *in vitro* or *in vivo*. Examples of data obtained with such strategies for identification of enzymatic activity related to formation of several modified nucleotides in the anticodon loop of tRNA (inosine, queuosine, wyosine, Gm, and t^6A) can be found in Carbon *et al.* (1982, 1983), Fournier *et al.* (1983), Haumont *et al.* (1984, 1987), Droogmans *et al.* (1986), Droogmans and Grosjean (1987, 1991), Grosjean *et al.* (1987), and Kretz *et al.* (1990).

Currently, site-specific introduction of a $[^{32}P]$ label (or any other radiolabeled nucleotide) inside an RNA is more easily performed by use of T4-DNA ligase than T4-RNA ligase, provided that the two RNA fragments, of which one is radiolabeled at its 5′-end, are brought together by use of an appropriate splint DNA (see Moore and Query, 1998; Zhao and Yu, 2004; Zimmerman *et al.*, 1998). This "cut-and-paste" method that uses the more efficient DNA ligase procedure has been used successfully, for example, to study the enzymatic formation of pseudouridines in U2-small nuclear RNA (snRNA, Ma *et al.*, 2003).

3.1. *In vitro* production of T7-runoff transcripts and nearest-neighbor analysis of modified nucleotides

In the pioneering work of Zeevi and Daniel (1976), *E. coli* RNA polymerase was successfully used to transcribe several *E. coli* tRNA genes *in vitro* in reaction mixtures containing one of the four NTP nucleotides (N stands for A, G, C, or U) radiolabeled with an $[\alpha\text{-}^{32}P]$. The resulting radiolabeled RNA transcripts, harboring $[^{32}P]$-label to only one of the four nucleosides according to the labeled NTP that is used, were incubated with *E. coli* S100 extract supplemented with several cofactors (ATP, *S*-adenosyl-L-methionine, isopentenyl-pyrophosphate, folinic acid, L-cysteine, L-threonine, pyridoxal phosphate). At the end of the incubation period, the resulting matured tRNAs were analyzed for the presence of modified nucleotides. This simple system allowed detection of the formation of several modified nucleotides (Psi, m^5U, D, s^4U, i^6A, t^6A, Gm, and m^7G) in tRNA transcripts. Similarly, *in vitro* transcription of $[5\text{-}^3H]$-uridine–labeled tRNA^{Tyr} precursor from a bacteriophage DNA carrying the *E. coli* tRNA^{Tyr} was successfully used to study the biosynthesis of pseudouridines at both positions 39 and 55 in tRNAs (Ciampi *et al.*, 1977; Schaefer *et al.*, 1973).

A variation of the preceding method consists in the use of commercially available purified recombinant polymerase from the bacteriophage T7 and appropriate linearized plasmids or synthetic DNA templates containing the T7-promotor (17 nucleotides long) upstream of the sequence of interest

(Melton *et al.*, 1984). This useful method now allows easy production of so-called *in vitro* "runoff RNA transcript" of any sequence (full length or a selected fragment corresponding to a wild-type or mutant tRNAs), radiolabeled with [^{32}P] at the only phosphates corresponding to the radiolabeled NTP used in the reaction mixture. After incubation with extracts containing tRNA modification enzymes or microinjection into the *Xenopus laevis* oocyte, we were able to reveal formation of modified nucleotides at several positions within the same tRNA substrate and also identify its 5′-adjacent "nearest neighbor." To this end, the [^{32}P]-runoff transcripts are digested first with nuclease P1 to generate 5′-monophosphate nucleosides, including [^{32}P]-NMP of the same type as the one used during transcription, and in another series of tests the same RNA transcripts were digested by RNase T2 to generate 3′-monophosphate nucleosides (see Fig. 3.2 and the following). Notice, instead of [^{32}P]-label, any other type of radiolabeled ([^{14}C] or [^{3}H]) nucleotide triphosphate can be used during *in vitro* transcription. It is also worthwhile to mention the existence of T7-RNA polymerase mutants, which are able to use both ribonucleotides and deoxyribonucleotides triphosphates, thus allowing synthesis of deoxyribose containing RNAs (Aphazizhev *et al.*, 1997; Kostyuk *et al.*, 1995).

Nowadays, *in vitro* T7-transcription can best be performed with various commercial transcription kits. The one we prefer is Riboprobe T7-transcription kit from PROMEGA (cat. no. P2075). It is provided with (1) highly active recombinant T7-polymerase at 15–20 U/μl in 20 mM K phosphate, pH 7.7, 1 mM EDTA, 10 mM DTT, 100 mM NaCl, 0.1% triton X-100, and 50% (v/v) glycerol; (2) TSC reaction buffer (5× concentrated) consisting of 200 mM Tris-HCl buffer, pH 7.9, 30 mM MgCl$_2$, 10 mM spermidine, 50 mM NaCl; (3) the set of 5′-ATP, 5′-GTP, 5′-CTP, and 5′-UTP solutions each at 10 mM (Na salts, pH 7.0); (4) the RNase inhibitor RNasin at 4U/μl; and (5) RNase-free water. To perform reactions, 50 μCi of the appropriate [α-^{32}P]-labeled radiolabeled XTP (where X stands for A, G, C, or U; 400 or 800 Ci/mmol from Amersham (Piscataway, NJ) are first lyophilized to dryness in siliconized Eppendorf tubes together with 1 μg of linearized plasmid. To each tube maintained at room temperature, to avoid any precipitation of the DNA with the spermidine contained in the Promega buffer, add 2 μl of the TSC buffer (5×), 1 μl of 100 mM of DTT, 3 μl of a mix 10 mM of each of the three non radiolabeled NTP, 1 μl of 1 mM of nonradiolabeled XTP (2 mM when GTP is the radiolabeled XTP, because of low affinity of T7-polymerase for that particular nucleotide), 1 μl of 100 mM 5′-GMP (to facilitate initiation of transcription), 1 μl (4 U) of RNAsin (optional). After spinning droplets down, transcription is initiated by addition of 1 μl of T7-RNA polymerase (usually at 15–20 U/μl), and each tube is incubated under mild agitation for 3 h at 37°. At the end of the incubation period, DNA can be digested by addition of a few units of

RNase-free DNase Q1 (provided in the Promega T7-transcription kit) followed by an extra incubation at 37° for 5 or 10 min (this step is optional, and in practice we never do it). The reaction is stopped by the addition of 0.5 μl of 0.5 M EDTA, pH 8.0, and 10 μl of transcription stop mix containing 8 M urea, 30% RNase-free sucrose (Roche Biochemical, cat. no 711454), 0.1% bromophenol blue, and 0.1% xylene cyanol. Reaction mixes are stored at −20° until use or better proceeded immediately by heating the tubes at 70° for 1 min, followed by application of each sample onto the well (1 cm width, 1-mm thick) of a 6% denaturing polyacrylamide gel (cross-link ratio 19:1, 15–20 cm high), cast in Tris (90 mM)-borate (90 mM) buffer containing 2 mM EDTA and 7 M urea, and previously run at room temperature for approximately 30 min with Tris-borate buffer. After sample application, electrophoresis is first run at low voltage (150 Volts) until all of the 20 μl sample has entered the gel, then a constant voltage of 400 V is applied until the bromophenol blue runs just off the gel together with the unincorporated "very hot" [α-^{32}P]-triphosphate nucleosides. During electrophoresis, the glass plates should become warm (usually 40–50°) to help the RNA to remain denatured (this might not be the case when transcript is obtained from DNA template harboring G + C–rich RNA sequence as in the case of tRNA of hyperthermophilic organisms). The full-length tRNA transcripts are located half way between the xylene cyanol and bromophenol blue, whereas DNA and a minor fraction (>1%) of the tRNA transcript still remain associated with the undigested polymerase/DNA at the top of the gel. The gel is removed from the apparatus and wrapped with Saran microfilm. Fluorescent stickers are put on each corner, and the gel is exposed to an X-ray film (Kodak X-omatAR5) for 5–10 min or 1–2 min to a PhosphorImager screen. The film or the screen is developed, and the radioactive spots corresponding to the full-length [^{32}P]-tRNA transcripts are cut out of the gel (into one piece using a razor blade, do not crush the gel to avoid later insolubilization of acrylamide particles together with the RNA after ethanol treatment, see later), put in 2-ml Eppendorf tubes for passive elution (2 times, 3–6 hours each time) at room temperature with 350 μl of 500 mM NH$_4$ acetate, 10 mM MgCl$_2$, 0.1 mM EDTA, 0.1% SDS, or more simply in the cold, overnight in 400 μl of water. The eluted tRNA (usually up to 75% of the tRNA initially entrapped in the gel) is precipitated with 2 volumes of cold ethanol (100%) in the presence of 0.3 M Na acetate (pH 5.0). At this step (optional), 2 μg of bulk yeast tRNA (Na salt, Sigma, cat.no.109495) can be added in each tube as carrier for better quantitative precipitation of the RNA. The Eppendorf tubes are kept at −20° for at least 1 h (or overnight), and the RNA transcripts are then collected by centrifugation (12,000 rpm for 20–30 min at 4°). When removing the ethanolic solution, check with a Geiger counter whether all radiolabeled [^{32}P]-material is found at the bottom of the tube.

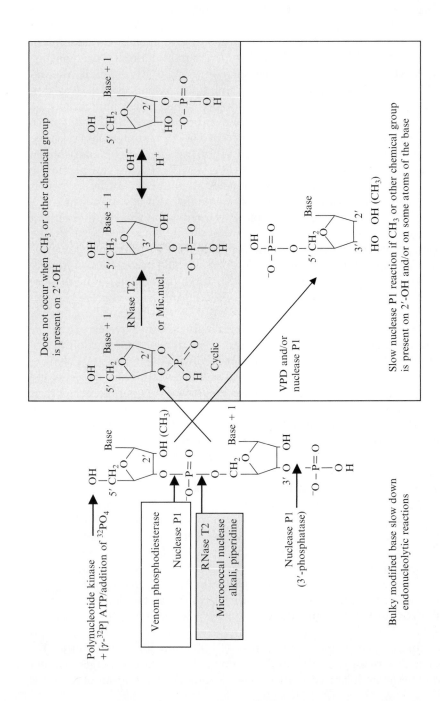

The radioactive pellets are washed 1 or 2 times with cold 70% ethanol/water (v/v) to remove completely SDS and free salts. They are finally dissolved in 10–15 μl of RNase-free (DEPC-treated) water. The amount of RNA transcripts will be accurately evaluated by counting the radioactivity of a 0.2-μl sample with a scintillation counter (eventually further diluted 10 times, use siliconized tips). Knowing the specific activity of the radiolabeled [^{32}P]-nucleotide used in the transcription mix and the efficiency of counting radioactivity of the scintillation counter, the amount of tRNA transcript in the aliquot can be evaluated. The total amount of tRNA recovered from the gel (and the minor fraction remaining in the piece of eluted gel) should correspond to an amplification ratio during transcription of at least 50–100 RNA molecules per molecule of plasmid. The presence of radioactive bands on the X-ray film or the PhosphorImager screen at position of the gel corresponding to shorter than expected RNA transcript points out possible problems during transcription, such as: (1) the template is too G + C rich (causing premature termination during transcription and/or unfolding of the RNA transcript as often occurs with tDNA sequences of hyperthermophilic tRNAs), (2) one of the nucleoside triphosphate concentrations is too low (inadequacy of the relative concentration of NTP, especially of the radiolabeled XTP with the sequence to be transcribed), (3) inactivation of the T7-polymerase (too old or too frequently thawed), or worse (4) because of RNase contamination at one or the other steps of the procedure (usually when talking and spitting while cutting out the polyacrylamide gels and transfer to Eppendorf tubes). Do not forget that the most frequent RNase contamination comes from human skin and spit from the mouth.

When substrates shorter than a normal size of tRNA (75–96 nt in length) have to be produced or analyzed, use a 10% or even 15% denaturing-polyacrylamide gel instead of 6%. Remember also that T7-transcripts obtained after purification from denaturing gel may not be folded in the correct conformation (Uhlenbeck, 1995). Renaturation can be made by incubation of the RNA sample contained in siliconized Eppendorf tubes (to avoid irreversible sticking of RNA on the plastic wall) for 2 min at 80°, usually in the salt buffer used for the incubation mix (see later), then chilled on ice before use.

Figure 3.2 The different cleavage procedures that are used to generate the mononucleotides from purified RNA. As a rule, the RNA molecule should not be too long (maximum length, 150–200 nt) because of difficulties in accurately determining the presence of only one or a few modified nucleotides among the entire RNA sequence. For long RNAs, the molecule is cleaved into appropriate smaller fragments and purified by one-dimensional (1D) or 2D electrophoresis or by any other purification techniques.

3.2. Experimental conditions for testing activity of tRNA modification enzymes *in vitro*

With such [^{32}P]-labeled transcripts, activity of many tRNA modification enzymes can be tested *in vitro* by incubation with cell-free extract or better with purified (recombinant) enzymes. The easiest enzymes to test are tRNA pseudouridine synthases that do not require any cofactor beside salts and buffer and tRNA methyltransferases that require only a methyl donor, generally *S*-adenosyl-L-methionine. Our standard reaction mixture contains 100 mM Tris-HCl buffer, pH 8.0; 100 mM NH$_4$ acetate, 5 mM MgCl$_2$, 0.1 mM EDTA, usually prepared as 5× concentrate stock mix, filtered through 0.2-μm Millipore or Sartorius membrane and kept at $-20°$ until use. For each data point, the reaction mix (10 μl final volume) contains in siliconized Eppendorf tubes, 2 μl of 5× reaction mix, 1 μl of DTT 10 mM, and [^{32}P]-tRNA substrate (approximately 100–300 cps Cerenkov, usually diluted with the same nonradiolabeled tRNA to reach concentration approaching nM or even μM, depending on the enzyme to be tested; the binding constant of most tRNA modification enzymes is usually very high, in the range of 1–100 nM. The tRNA solution is first heated for 2 min at 80°, then chilled at 0° (tRNA renaturation). One (or 2) μl of 0.2 mM *S*-adenosyl-L-methionine (Sigma, cat. no. A7007-chloride form, freshly prepared from a stock solution at 20 mM in 5 mM H$_2$SO$_4$, kept at $-70°$ in 5- or 10-μl aliquots) in RNase-free water is eventually added, immediately followed by the addition of the enzyme preparation (approximately 1 μg of crude cell extract S10 or S100 or, of course, much less of the purified recombinant enzyme, previously diluted just before its use with enzyme dilution buffer D10, containing 50 mM Tris-HCl, pH 7.5, 1 mM MgCl$_2$, 10% glycerol, and 1 mg/ml of RNase-free bovine serum albumin molecular biology grade from Roche, cat. no. 711454). In kinetic experiments, all volumes to be added in one tube are multiplied by a factor corresponding to the number of points on the final kinetic curve and 10-μl aliquots of the mix are withdrawn at various time intervals within 30 min or 1 h maximum. A blank containing everything but the enzyme is always included to evaluate the stability of RNA substrate, absence of nucleases, and to identify any abnormality in the radiolabeled tRNA while inspecting the radiolabeled spots on the TLC (see later). The mix of tRNA with enzyme is incubated usually for 30 min at the desired temperature (30° for yeast enzymes, 37° for *E. coli* enzymes, and 50° or even 70° for enzymes from archaeal hyperthermophilic organisms; see also Chapter 10). At the end of the incubation time, the reaction is stopped by adding (or diluting) the 10-μl sample with 200 μl of 0.3 M Na acetate buffer (pH 6.0), immediately followed by the addition of 220 μl of a mix of phenol/chloroform/isoamyl alcohol (25/24/1) buffered at pH 6.0 with 25 mM Tris-HCl. Tubes are vortexed and left at room temperature for no longer than 20 min, until a set of tubes can be centrifuged together for 10 min at 12,000 rpm. From each tube, 200 μl of the clear upper (aqueous) phase is transferred to new (not siliconized) Eppendorf tubes. Verify that the interphase and lower (phenol) phase retain less than 10–15% of total radioactivity (some modification enzymes like RNA:m⁵U/m⁵C methyltransferases form cross-link adducts as an intermediate step of the reaction that will remain trapped in the denatured protein interphase; see Chapter 5). If problems occur, try to extract the phenol phase again with another 100 μl of 0.3 M Na acetate buffer and combine the new aqueous supernatant with the first one, but this might not be necessary if enough [^{32}P]-counts (at least half of the initial amount of counts initially engaged in the reaction) are obtained in the first supernatant. To each supernatant (200 or 300 μl), add 2 μl of carrier polyU or better bulk yeast tRNA (stock solution at 2 mg/ml in water; Roche cat. no.109495) and finally 500 or 750 μl of cold ethanol 100%. Mix well and keep the tubes in dry ice for 30 min or a few hours at $-20°$. Precipitated nucleic acids are collected by centrifugation for 30 min at 4°, washed once with cold ethanol/water 70% (v/v), and quickly dried under vacuum, after verification that all labeled [^{32}P]-RNA is present at the bottom of the tubes. The samples are kept at $-20°$ until further digestion with nucleases and analysis of base composition by 2D-TLC (see later).

3.3. Optimization of enzymatic reaction conditions

Having demonstrated that the *in vitro* reaction works with one type of enzyme, optimize experimental conditions before starting a more systematic study of the mechanism and specificity of the enzyme. Indeed, beside the necessary cofactor(s) of the chemical reaction, some enzymes may require special assay conditions, such as the presence or strict absence of Mg^{2+} (in that case the presence of EDTA is required), the presence of particular ions such as NH_4 instead of Na^+ or K^+, sometimes salt concentrations as high as 300 mM, polyamines (see later) or more acidic pH (such as 6.5, instead of the usual 8.0–8.5). Strict absence of nuclease in the enzyme preparation is also important to verify. As a rule, the first time an enzyme is incubated with its radiolabeled substrate, the integrity of the tRNA after incubation compared in a control experiment performed in the absence of enzyme (blank) should be verified by gel electrophoresis of the reaction mix on a 15% PAGE. In case of problems, the addition of RNase inhibitor, such as RNasin (1 or 2 U/10 μl of assay) and/or addition of poly U (few μg/10 μl) or eventually carrier tRNA may help reduce degradation of the radiolabeled RNA substrate (mass effect). Stability of the enzyme and the cofactor(s)—especially S-adenosyl-L-methionine, FAD, and/or tetrahydrofolate, generally in excess in the mix—also has to be checked. As a rule, the reaction mixture always contains 1 mg/ml bovine serum albumin and 5 or 10% glycerol to prevent inactivation of enzyme by excessive dilution. For enzymatic reactions performed at elevated temperature (up to 50°), these controls are particularly important. In this latter case, the tRNA transcripts should be G + C rich in stems (as naturally occurring tRNA from hyperthermophiles [reviewed in Grosjean and Oshima, 2007]) and addition of linear polyamines like cadaverine or branched polyamines like tetrakis (3-aminopropyl-amonium) should be explored (Oshima *et al.*, 1987; Terui *et al.*, 2005; Wildenauer *et al.*, 1974). This is particularly important for the tRNA modification enzymes requiring the integrity of the 3D architecture of the tRNA substrate.

Radiolabeled T7-runoff transcripts for detecting the activity of several tRNA modification enzymes and for studying various aspects of their mechanism and specificity have been extensively and successfully used in our laboratories in France and Belgium. Typical examples of successful use of radiolabeled T7-runoff transcripts can be found in Grosjean *et al.* (1990, 1996), Szweykowska-Kulinska *et al.* (1994), Auxilien *et al.* (1996), Becker *et al.* (1997, 1998), Jiang *et al.* (1997), Motorin *et al.* (1997, 1998), Motorin and Grosjean (1999), Morin *et al.* (1998), Brulé *et al.* (1998), Constantinesco *et al.* (1999a,b), Droogmans *et al.* (2003), Roovers *et al.* (2004, 2006), and Urbonavicius *et al.* (2006).

4. COMPLETE DIGESTION OF RNA WITH VARIOUS NUCLEASES

4.1. Use of RNase T2 to generate 3'-monophosphate nucleosides

To the dried pellets of modified [^{32}P]-tRNA recovered from reaction mixes described previously (or half of the sample, in case the second half has to be analyzed by other methods, see later), add 10 μl of 50 mM NH$_4$ acetate (pH 4.5) containing 0.2 U of RNase T2 from *Aspergillus oryzae* (Gibco-BRL, Gaithersburg, MD/cat. no. 18031–113; stock solution in water at 5 U/μl kept at −20°). Mix well, and spin down before incubation for 4–6 h at 37° (or better overnight in an incubator hood, not in a water bath, to avoid evaporation and condensation of water in the Eppendorf cap). This will produce 3'-[^{32}P]-XMP (abbreviated in Xp, where X is any nucleotide, modified or not), except for 2'-O-methylated derivatives and the 2-O-ribosylpurine phosphate derivatives, for which the corresponding diphosphate dinucleosides XmpYp and XrpYp will be generated. Notice also that RNase T2 activity is slowed down and inefficient in the case of certain methylations of the base moiety, such as m^{1}G, m$^2_2$G, m^{7}G, m^{1}A, or wybutosine, leading to 2'-3'-cyclic derivatives, the normal intermediates of the enzymatic hydrolytic mechanism (see Fig. 3.2). These should not be confused with another modified nucleotide after chromatography on the cellulose plate. To circumvent this problem, one could either use 10 times more RNase T2 and digest for a longer time (overnight) or treat the digest at pH 1.0 (0.1 M HCl) for several hours at room temperature to allow the cyclic structure to open before the hydrolysates will be subjected to TLC analysis. In fact, duplicated experiments with and without acidic treatment are recommended, but remember that during such acidic treatment, some modified bases are unstable and may create other problems, such as m^1A, which will isomerize into m^6A; m^3C, which will slowly deaminate into m^3U; s^4U, which will oxidize into U (especially if a trace amount of H_2O_2 is added; Watanabe, 1980); ac^4C, which will deacylate into C; and chain breakage at m^7G (Wintermeyer and Zachau, 1970) and at the hypermodified wybutosine nucleotide (Philippsen et al., 1968). If an elevated temperature is maintained ($>80°$), m^5C will start to deaminate (Ehrlich et al., 1986; Lindahl, 1967; Lindahl and Nyberg, 1974), and purines start to be cut off from the nucleic acid (depurination, Lindahl and Nyberg, 1972). These phenomena are even worse with DNA than RNA. These properties might serve as additional criteria for identification of these modified nucleotides, by comparing 2D-chromatograms of RNA hydrolysates that were exposed or not for several hours to low pH (<2.0) or a few minutes at high temperature ($80°$ or higher). In case $2'$-O-methylated derivatives or the $2'$-O-ribosylpurine phosphate derivatives are expected to occur after incubation of the RNA substrate with the RNA modification enzyme, one trick is to further treat the radiolabeled $[^{32}P]$-T2 digests containing the diphosphate dinucleosides XmpYp or XrpYp with 1 U of alkaline phosphatase (Roche, Indianapolis) for 30 min at $37°$. In this way, all terminal $3'$-phosphate will be hydrolyzed, and only internal $[^{32}P]$-phosphate will remain such as in $Xm[^{32}P]pY$ or $Xr[^{32}P]pY$, thus allowing an easier and unambiguous detection of such compounds after TLC chromatography (see for example, Wilkinson et al., 2007).

4.2. Homemade RNase mix

Recently it has been difficult to obtain commercial RNAse T2. Instead, a brownish enzyme prepared from Takadiastase powder of A. oryzae (several years ago it was bought from Sankyo and was called Sanzyme R) can be easily prepared after the procedure described by Hiramaru et al. (1966). This "homemade" RNase preparation, free of $3'$-phosphatase activity, cleaves RNA into $3'$-monophosphate nucleosides much more efficiently than any commercial RNase T2. This is probably due to contamination of such enzyme preparation by other RNase activities (e.g., RNase T1, RNase A), which all produce $3'$-monophosphate nucleosides. Some work better at certain modified bases than others and therefore complement nicely the activity of RNase T2. Thus, when inefficient hydrolysis is observed with commercial RNase T2, addition of small amounts of RNase T1 and RNase

A in the mix might help, unless the problem comes from the presence in the RNA sample of interfering molecules like the presence of residual SDS, phenol, or excess salts, especially phosphate buffer because of insufficient washing of the RNA pellets (see previously).

4.3. Use nuclease P1 to generate 5′-monophosphate nucleosides

To the dried pellets of uniformly labeled tRNA or remaining sample after testing RNase T2, add 10 μl of 50 mM NH$_4$ acetate (pH 5.3) containing 0.2 μg of nuclease P1 from *Penicillium citrinum* (Roche; cat. no. 236225 or US Biomedicals, cat. no. 195352). Mix well, and spin down before incubation for at least 2 h at 37° (usually overnight in an incubator hood to avoid evaporation and condensation of water in the Eppendorf cap). This will produce 5′-[^{32}P]-XMP (abbreviated as pX, where X is any nucleoside, including 2′-O-methylated nucleoside monophosphates, pXm). However, at a lower concentration of nuclease P1 (0.02 μg/10 μl) and shorter incubation time (1 h at 37°), 2′-O-methylated dinucleotides (pXmpY) will accumulate because of the slower rate of hydrolysis of phosphodiester bonds when a 2′-O-methyl group is present. Likewise, when a pseudouridine is present or a bulky group on a base (like in wyosine, queuosine derivatives, N^6-threonylcarbamoyl- and N^6-isopentenyl-derivatives), the activity of nuclease P1 is lower and verification of completeness of RNA digestion should be checked at low versus high ratios of RNA to nuclease. Significant differences between the two types of hydrolysates should alert you to the presence of a group on 2′-hydroxyl ribose or of a bulky group on one or the other of a base in the RNA sequence. This can also be identified by alternative reverse transcriptase–based methods (see the end of this chapter and Chapter 2).

4.4. Use of venom phosphodiesterase VPD, eventually in combination with nuclease P1 to generate 5′-monophosphate nucleosides

To the dried pellets of radiolabeled RNA, add 10 μl of volatile 0.1 M triethylammonium carbonate buffer (pH 8.0; Sigma; cat. no. 7408) or 0.1 M Tris-HCl (pH 8.0) containing 50 mU of VPD from *Crotalus* snake (Worthington, cat. no. 31B217J). Mix well, and spin down before incubation for at least 2 h at 37°. VPD is an exonuclease that efficiently cleaves RNA (and DNA, as does nuclease P1) into 5′-monophosphate nucleosides provided that there is a phosphate-free 3′-end. It cleaves 2′-O-methyl-containing nucleotide bonds and any other bonds adjacent to nucleotides that are resistant to nuclease P1. In practice, combined digestion of RNA (or DNA) with nuclease P1 and VPD is the best solution to avoid most drawbacks.

This can be done by addition of 40 μl 0.1 M triethylammonium carbonate buffer (pH 8.0) and 100 mU of VPD to the 10 μl of nuclease P1 hydrolysate described in the preceding section. Mix well, and spin down before incubation for at least 2 h at 37° (usually overnight in an incubator hood). In this way, exclusively 5'-monophosphate nucleosides will be produced.

4.5. Use of piperidine to generate mixes of 2'- and 3'-monophosphate nucleosides

Complete digestion of RNA into a mixture of the isomeric forms 2'-,3'-monophosphate nucleosides is performed under alkaline conditions by use of volatile piperidine. This type of hydrolysis allows testing for the thermo-alkali instability of certain modified nucleotides, such as dihydrouridine (D, House and Miller, 1996), m^1I, m^7G, m^1A, m^3C, and s^4U, which then may help to confirm their identities. The lyophilized RNA sample (20 μg or less) is incubated in 10 μl of 15% piperidine (v/v; Merck; cat. no. 9724) for 1 h at 95°. Piperidine is preferred over NaOH or KOH, because it is volatile, will not destroy the cellulose layer, and evaporates completely when spotting on the TLC plates. It also allows avoiding Na^+ or K^+ salts, which favor smearing of spots during migration on thin-layer plates. Because the 2'- and 3'-monophosphate nucleosides do not migrate at the same rate on the TLC, one always obtains two "tandem" spots for each nucleotide after chromatography, one migrating as expected for 3'-phosphate derivatives (as obtained after RNase T2 hydrolysis) and the other migrating nearby (2'-derivative; see later).

5. ANALYSIS OF RNA DIGEST PRODUCTS BY THIN-LAYER CHROMATOGRAPHY (TLC)

5.1. Preparation of the chromatography tanks and plates

1. Prepare in advance, solution of homemade "P1-nucleotide marker mix" by incubating 1 mg of total baker's yeast tRNA (Sigma, cat. no. R-8759; stock solution at 20 mg/ml in water) in 250 μl of 50 mM NH$_4$ acetate buffer at pH 5.3 with approximately 5 μg of nuclease P1 and incubate overnight in an incubator hood at 37°. Likewise, preparation of home-made solution of "T2-nucleotide marker mix" is prepared by incubating 1 mg of bulk baker's yeast tRNA in 250 μl of 50 mM NH$_4$ acetate buffer (pH 4.5) with 5 U of RNase T2 and incubate overnight in an incubator hood at 37°. Alternately, a mix of equimolar amounts of each of the four commercially available 5'-, or 3'-monophosphate nucleosides (AMP, GMP, UMP, CMP from Sigma approximately 1 mg in total) in 250 μl of water, adjusted at pH 7.0, can also be used instead of a P1- or T2-digest

of tRNA. However, when performing TLC of tRNA hydrolysates, it is preferable to use markers derived from baker's yeast tRNA hydrolysates, because it already contains some modified nucleotides that help migration of [^{32}P]-homologs during TLC chromatography. All solutions are kept at $-20°$ until used.

2. Prepare in advance the following three chromatographic solvents: Solvent A (also designated N1 in our earlier articles; should be freshly made): isobutyric acid/concentrated ammonia/water (66/1/33 [v/v/v]). Mix 165 ml of isobutyric acid (Merck; cat. no. 800472) with 2.5 ml of ammonia solution (25% concentrated ammonia solution Merck, cat. no. 105432; the purest quality as possible, need quasi absence of contaminating divalent cations) and 82.5 ml of water (in case of divalent metal contamination from commercial products, add 0.5 ml of 0.5 M EDTA, pH 8.0). Solvent B (also designated R2 in our earlier articles): phosphate buffer/NH$_4$ sulfate/n-propanol (100/60/2 [v/w/v]). Prepare by adding 150 g of (ultrapure) NH$_4$ sulfate (Merck; cat. no. 101217, or better ultrapure NH$_4$ sulfate Biomol quality of Gibco-BRL, cat. no. 5501UA) to 250 ml of 100 mM Na phosphate buffer (pH 6.8). Mix well and stir at $50-60°$ until all NH$_4$ sulfate is dissolved. Allow solution to cool to room temperature, and then add 5 ml of 1-propanol (Merck; cat. no. 100997, this is not isopropanol!). Continue stirring until the cloudy solution becomes clear. Solvent C (also designated N2 in our earlier articles): isopropanol/concentrated HCl/water (68/18/14 [v/v/v]). Mix 170 ml of isopropanol (2-propanol, Merck; cat. no. 109634), 45 ml of concentrated fuming HCl (37% concentrated, Merck, cat. no. 100317; the purest quality as possible, need quasi absence of contaminating divalent cations), and 35 ml of water. All chromatographic solvents are kept at room temperature in a ventilated area. This is particularly important for solvent A, which has a strong pungent and unpleasant odor, and solvent C (N2), which can corrode metallic zones and damage scientific equipment. Small volumes of solvent A (N1), B (R2), and C (N2) are each poured into a different tank, so that the height of the solution in the bottom of the tank is 4–5 mm maximum. The tanks are kept hermetically closed under a well-ventilated hood to ensure good saturation with solvent vapors.

3. Use commercial TLC plates (20 × 20 cm) with cellulose (0.1-mm thick) coated on glass or plastic plates. Glass-coated plates are easier to handle, but plastic-coated plates are cheaper, do not break, and they can be cut into smaller sizes (10 × 10 cm or even 6 × 6 cm), when the chromatographic pattern of the modified nucleotides to be analyzed is not too complex. This allows one to perform 2D chromatography in only a few hours for each dimension, instead of 2 days for the large 20 × 20 cm plates. Of the commercially available TLC plates, the ones sold by Macherey-Nagel (Polygram CEL-300–10; cat. no. 808–013) and Merck (cat. no. 105730–001) are satisfactory, the latter showing slower

solvent migration rates than the former ones, especially with acidic solvent C (N2). Do not use plates with UV indicator, because the latter acquires a strong dark blue color when acidic solvent C is used, which hinders detection of marker controls by UV light.

4. The origin of the deposit is marked on one corner (at 1.5–2 cm from each edge) of the plates with a soft pencil, and the information needed for plate identification is written at the opposite corner. Before spotting the radioactive material, approximately 10–20 μg of the unlabeled monophosphate nucleosides marker mix (see earlier, never confuse "T2 mix or T2 sample" with "P1 mix or P1 sample"!) is spotted to allow further assignment of the nucleotides under a short-wavelength UV lamp with 254-nm filter (use UV-blocking eyewear). These must be applied on the TLC plates approximately 0.3–0.5 μl at a time to give the smallest spots as possible, not more than 3–4 mm in diameter. More than 40 μg deposit will produce smears during migration. Between successive applications, use a moderately warm air stream from a hair dryer (some modified bases are unstable at high temperature, see previously). In certain cases, 1D chromatography allows the separation and clear-cut visualization of the [^{32}P]-compounds after autoradiography. In that case, different samples can be spotted at 1 cm from each other onto the same plate along a horizontal line.

5. If two types of solvent systems have to be tested, each radiolabeled RNA sample will have to be spotted on two identical plates. Aliquots of the [^{32}P]-labeled "RNase T2 or nuclease P1" hydrolysates (ideally approximately 200–300 cpm/Cerenkov counts) are spotted on top of the dried spots corresponding to either "T2 or P1" marker mix. The tiny amount of sample is allowed to dry between each successive application. This operation will take 5–10 min per plate, but usually several plates are prepared in parallel (up to 16–20). We suggest the use of a Geiger counter to evaluate the amount of radioactivity spotted on each plate; try to spot approximately the same amount of cpm for better comparison of the distribution of the different radioactive spots.

5.2. First dimensional chromatography with solvent A (N1)

This is always developed overnight (15–18 h/20 cm; or 3–4 h/10 cm) with the isobutyric solvent designated A (N1). Once the solvent reaches the top of the plate, some extra time for migration is allowed with this solvent to maximize resolution of the different nucleotides along the migration lane. Several TLC plates (five) can be run in parallel in the same 20 × 20 cm commercial chromatography tank (with *ad hoc* nonmetallic adaptor) or up to 45 plates when 10 × 10 cm plastic-coated plates and special homemade chromatographic tanks are used. After migration, the plates are withdrawn from the tank and dried thoroughly under a well-aerated fume hood for several hours (or overnight). Do not dry wet plates with solvent A and HCl-containing solvents within the same area, because of formation of NH_4Cl fume. To avoid the influence of isobutyric acid solvent on migration of nucleotides in the second dimension, be sure that all of the acid has evaporated before starting the migration of the second dimension (eventually use a hair dryer with moderately warm air under a ventilated hood; remain at temperature $<40°$ to avoid chemical destabilization of certain modified nucleotides). After the plates have been dried (as tested from its smell), information about migration of the nucleotides in the first dimension can be obtained by quick inspection of the plates under UV light with a filter at 254 nm (protect eyes with UV-blocking eyewear). Normally three dark blue spots should be seen corresponding to [GMP + UMP], CMP, and AMP (from bottom to top of the migration line, with AMP at 3/4 of the height of the plate).

5.3. Second dimensional chromatography with solvent B (R2) or solvent C (N2)

After extensive drying of the plates developed with solvent A (N1), one of the duplicated plates is now run in the perpendicular direction to the first dimension in a second tank with neutral (pH 6.8) solvent B (R2). This will take approximately 8–10 h for 20×20 cm plates (3–4 h for 10×10 cm plates), depending on which commercial plates are used. The second of the duplicated dried plates is run in the direction perpendicular to the first dimension in a third chromatographic tank with the acidic solvent C (N2). This latter solvent migrates rather slowly and requires 18–22 h to reach the top of the 20×20 cm plate (6–8 h for 10×10 cm plates). Once solvent B (R2) in one of the two chromatographic tanks reaches the top, the plate is withdrawn and dried at $37–50°$ in a well-ventilated oven to avoid slow crystallization and swelling of the NH_4 sulfate as a powder at the surface of the coated side of the plates. In this case, the resulting distribution of the modified nucleotides after 2D chromatography results from a combination of solvent A + B (System I, also referred as system N1/R2 in some of our earlier articles). Later, when solvent C (N2) in the second chromatographic tank finally reaches the top, the plate is withdrawn and dried thoroughly under a ventilated hood to eliminate acid vapor, which otherwise will darken X-ray films, corrode any metallic and other items in the room, and destroy crystals of the costly PhosphorImager screens. Complete dryness of the plates can easily be obtained if they are maintained at $35–45°$ for several hours.

The dried cellulose plates are visualized in a dark room under UV light with a filter at 254 nm to detect the nucleotide "T2 or P1" markers. In both types of chromatographic solvents, AMP, CMP, and UMP appear as dark blue spots, whereas GMP yields a characteristic light blue fluorescent spot (only in chromatographic system II, because of solvent C (N2)). Normally, 3–5 μg of a nucleotide can be adequately visualized under UV light. Use a marker pen to indicate positions of these spots on the back of the plates or a soft pencil on the cellulose coat. The positions of the radiolabeled nucleotides are next obtained after exposure of the plates to autoradiographic X-ray films (20 × 40 cm from Kodak), using X-ray intensifying screens or PhosphorImager screens. The spots corresponding to the four major canonical 5'- or 3'-[^{32}P]-ribonucleotides always give large dark spots on the film, thus limiting resolution of any faint spot corresponding to minor modified nucleotide(s) in their closest vicinity. This problem is much less serious when [^{14}C]- or [^{3}H]-labeled nucleotides (or nucleosides) are analyzed by 2D TLC (less lateral radiation). Also, because solvent C (N2) is strongly acidic (pH < 1.0), it causes depurination of the nucleotides during migration in that particular solvent. As a result, characteristic [^{32}P] smears ahead of each G- and A-derivative, with the corresponding degradation products (ribosylphosphate and inorganic [^{32}P]-phosphate) migrating faster than the parent nucleotides, will always be observed with solvent C (N2), especially when 20 × 20 cm plates are run for such a long time.

5.4. Final identification of modified nucleotides on thin-layer plates

Detection and identification of the radiolabeled [^{32}P]-modified nucleotides in RNA digests is performed by comparing the autoradiographic and UV pattern against reference maps established in the two TLC systems (system I: solvents A + B [N1 + R2]; system II: solvents A + C [N1 + N2]) described previously. These reference maps were established by compiling (1) the published localization of modified nucleotides that were found in hundreds of RNAs (mostly tRNAs) sequenced in many laboratories over the last 40 years, and (2) many published and unpublished results obtained by us in sequencing dozens of tRNAs or by studying tRNA modification *in vitro* and *in vivo*. The resulting maps for these two most commonly used chromatographic solvent systems are shown in Figs. 3.3–3.6. They indicate schematically the relative location of the various 5'- mononucleotides in the 2D TLC systems. They are grouped into four different sets according to the major nucleotide from which they derive. Altogether, positions for 70 modified nucleotides are shown. Those for which the positions are not known are indicated in the figure legends. Figure 3.7 shows the positions of dinucleoside diphosphate containing a methyl group on the 2'-hydroxyl group of the internal ribose. It also illustrates positions of some dinucleoside diphosphates resulting from difficult cleavages by nuclease P1 or RNase T2 as pointed out previously. They also arise from incomplete digestion by RNase T2 and nuclease P1 because of peculiar chemical structure of the purine or pyrimidine ring and because of the presence of interfering

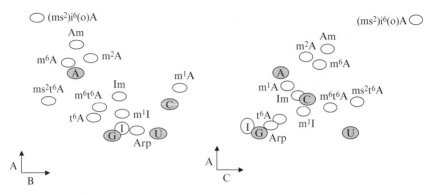

Figure 3.3 Diagrams of chromatographic mobility of 5′-monophosphate nucleosides derived from adenosine (A) after 2D-TLC. Positions of the standard nucleotides are indicated by gray ellipses, with A, G, C, and U corresponding to positions for 5′-phosphate nucleoside markers. Abbreviations for modified nucleotides are those defined in Rozenski *et al.* (1999) and Sprinzl and Vassilenko (2005). Details about their occurrence and formula can be found in Limbach *et al.* (1994). Arrows A/B and A/C correspond to solvent systems I and II, respectively (for details see text). Chromatographic mobilities of g⁶A, hn⁶A, ms²hn⁶A, and ms²m⁶A present at position 37 in the anticodon loop of certain tRNAs, and m¹Im in the T-loop of archaeal tRNAs and of m₂⁶A, m⁶Am, and m₂⁶Am present in certain rRNAs, mRNAs, or snRNAs, are lacking. Note that i⁶A, ms²i⁶A, i(o)⁶A, and ms²i(o)⁶A present at position 37 of the anticodon loop of certain tRNAs migrate at the same place in both types of solvent, we use one symbol: (ms²)i⁶(o)A.

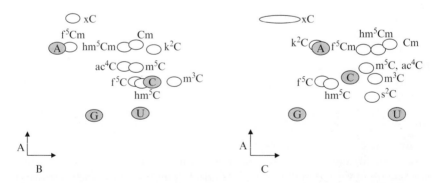

Figure 3.4 Diagrams of chromatographic mobility of modified 5′-monophosphate nucleosides derived from cytosine (C) after 2D-TLC. See also text and legend of Fig. 3.3. xC corresponds to the modified cytidine found in the *H. volcanii* tRNA-Ile (anticodon CAU; see Gupta, 1984). In solvent system A/C, xC gives a smear, probably because of its instability under acidic conditions. We lack information for chromatographic mobilities of ac⁴Cm, m⁵Cm present in certain tRNAs and of m⁴C, m⁴Cm present in certain rRNAs. The position of s²C (from tRNA) is shown only in solvent system A/C.

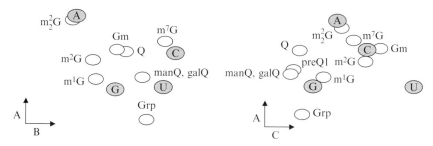

Figure 3.5 Diagrams of chromatographic mobility of modified 5′-monophosphate nucleosides derived from guanosine (G) after 2D-TLC. See also text and legend of Fig. 3.3. The chromatographic mobilities of mimG, imG (wyosine or Yt), yW (wybutosine, also called Y), oHyW (hydroxy-Y), oHyW* (Ye), o2yW (peroxy-Y or Yr), and oQ (epoxyqueuosine) from position 37 in the anticodon loop of certain tRNAs, and m²Gm, m₂²Gm from position 10 or 26 in hyperthermophilic archaeal tRNAs and gQ (archaeosine) normally present at position 15 in most archaeal tRNAs are missing. Also those of m²,²,⁷G and m²,⁷G present in certain mRNAs and snRNAs are missing. Note that preQo and preQ1 are precursors of Q base (queuine).

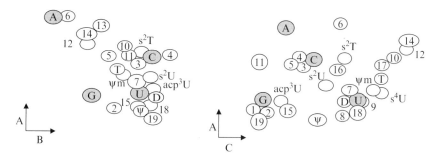

Figure 3.6 Diagrams of chromatographic mobility of 5′-monophosphate nucleosides derived from uridine (U) after 2D-TLC. See also text and legend of Fig. 3.3. For clarity of the figure, some U derivatives are designated by numbers: 1 for cmnm⁵U (alias cmam⁵U); 2 for cmnm⁵s²U (alias cmam⁵s²U); 3 for cmnm⁵Um; 4 for mnm⁵U (alias mam5U); 5 for mnm⁵s²U (alias mam⁵s²U); 6 for mnm⁵s²Um; 7 for m¹Psi; 8 for cmo⁵U (also named V); 9 for mo⁵U; 10 for Um; 11 for mcm⁵U; 12 for mcm⁵s²U; 13 for mcm⁵Um; 14 for Tm (m⁵Um); 15 for ncm⁵U; 16 for ncm⁵Um; 17 for m¹Psdm; 18 for cm⁵U; and 19 for Tau-m⁵U and Tau-m⁵s²U (taurine derivatives that migrate at the same place in the two solvent systems; see Suzuki et al., 2002). Chromatographic mobilities of a given ribonucleotide in both types of solvent systems are not always known. In addition, the possibility exists for misannotation or mispositioning of a few U derivatives of this diagram because of much confusion in the scientific literature about these U derivatives. Chromatographic mobilities for ho⁵U, mcmo⁵U (also called mV), chm⁵U, mchm⁵U, nm⁵s²U, and mnm⁵se²U present at position 34 of certain tRNAs; s²Um from tRNAs; and m³U, m³Um, and m¹acp³Psi present in certain rRNAs are not known yet. Note that T = m⁵U and s²T = s²m⁵U.

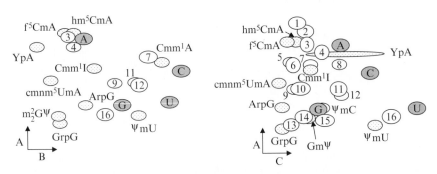

Figure 3.7 Diagrams of chromatographic mobility of diphosphate dinucleotides that are resistant to RNase T2 and alkali digestions owing to the presence of a methyl group on the 2′-hydroxyl of the internal ribose (see also Fig. 3.2). Numbers correspond as follows: 1, AmAp; 2, AmCp; 3, CmAp; 4, AmUp; 5, AmGp; 6, GmAp; 7, CmCp; 8, UmAp; 9, CmGp; 10, GmCp; 11, CmUp; 12, UmCp; 13, GmGp; 14, GmUp; 15, UmGp; and 16, UmUp (see also Hashimoto *et al.*, 1975). A few diphosphate dinucleotides that contain, in addition to the 2′-methylribose, a modification on one of the two bases are also indicated. In solvent system A(N1)/B(R2), not all diphosphate dinucleotides have been identified. The dinucleotide YpA, which is derived from yeast tRNA^Phe recovered from *Xenopus laevis* oocytes (Droogmans and Grosjean, 1987), gives a smear in solvent A(N1)/C(N2) because of the instability of the Y derivative under acidic conditions (depurination).

group (methyl, ribosyl-phosphate) on the 2′-hydroxyl group of the ribose. Notice, the 5′-ribonucleotides generated by nuclease P1 and/or VPD do not migrate at the same rate as their 3′-ribonucleotide counterparts generated by RNase T2 digestion in any of the three solvent systems. However the "relative" migration positions of the different 5′- and 3′-derivatives display closely similar global characteristic patterns on the TLC plates.

When these maps are used, we have to remember that most of the modified nucleotide monophosphates identified in Figs. 5.3–5.7 have never migrated together, because they do not coexist in the same RNA molecule or in RNAs of the same organism. Therefore, the true relative position of each spot on the TLC plates in a given experiment could be slightly different compared with the reference maps that have to be used only to "guide" for identification of putative modified nucleotides. The only unambiguous way to identify a modified nucleotide by TLC should be based on chromatographic comigration with the corresponding true UV markers; unfortunately, most modified nucleotide markers are not commercially available. Nevertheless, to be correctly assigned, the putative modified nucleotide should migrate to the expected position in both chromatographic solvent systems I and II. If not, this generally means that the spot may correspond to a modified nucleotide of unknown chemical structure, and other types of experiments, such as testing different combinations and concentrations of nucleases or behavior to the acido-, alkali-, thermotreatments, have to be performed. Slight differences in migration of given modified nucleotide compared with reference maps can also arise from differences in "room" temperature, incomplete saturation of the tank with vapor, impurities present in the chemicals used to prepare solvents, the origin of the commercial cellulose layer, and the age of the solvents. As a rule, do not use the same solvents more than three to four times, and be sure your chromatographic tanks are tightly closed.

5.5. Quantification of modified nucleotides within an RNA fragment

Evaluation of the radioactivity in each individual spot of a TLC chromatogram, as visualized from the PhosphorImager, is performed using ImageQuant computer program coupled with Excel algorithms. Alternately, the amount of radioactive compounds in each individual spot detected after exposure of the TLC plates to X-ray films can be measured after scrapping the corresponding area of cellulose from the glass plate or by cutting off the area of the plastic sheet and transfer of materials into vials containing scintillation mixture. The radioactivity is measured with a scintillation counter. If the nucleotide sequence of the uniformly labeled RNA is known, the relative molar amount of each radiolabeled compound can easily be calculated.

Notice, a spot corresponding to $[^{32}P]$-orthophosphate is usually always observed on the TLC. In solvent A (N1), it migrates at the same level as UMP and GMP and thus interferes with identification and quantification of radioactivity present in these monophosphate nucleosides, whereas in both solvent B (R2) and C (N2), it migrates with the front of the solvent, thus far away from the monophosphate nucleosides. Theoretically, this spot should be absent or at least present only in trace amount (few percent of total radioactivity from the $[^{32}P]$-tRNA transcript analyzed). An unusually high proportion of $[^{32}P]$-orthophosphate relative to $[^{32}P]$-phosphate present in nucleotides may reflect instability of some modified nucleotides because of misuse of experimental conditions (acidic-, alkali- or thermodegradation) but most probably because of presence of undesirable contaminating nuclease(s) and/or phosphatase(s) in the enzyme extract in one of the solutions used and sometimes also in certain batches of commercial nuclease P1 or T2-RNase. Such contaminating nuclease(s)/phosphatase(s) may differentially affect the radiolabeled nucleotides; as a result, important errors in evaluation of the relative abundance of modified nucleotides in a given RNA digest can be made. One way to limit this drawback is to add carrier tRNA to the radioactive RNA to be digested (see earlier). This is important because very low amounts of labeled RNAs are analyzed (high nuclease-to-RNA ratio). Carrier will protect (because of dilution) the radiolabeled material against contaminating phosphatases or other undesirable RNase activities. It also helps migration of nucleotides and serves as visual control spots on TLC when performing UV shadowing.

6. DETECTION OF MODIFIED NUCLEOTIDES IN UNIFORMLY LABELED [^{32}P]-RNA OR [^{35}S]-CONTAINING RNA EXTRACTED FROM MICROORGANISMS

One way to identify the activity of a tRNA modification enzyme of which the gene is known (or being tested) within a cell is to delete the gene and check whether the corresponding putative modified nucleotide(s) is (are) lacking in the bulk tRNA isolated from the mutant strain compared with bulk tRNA isolated from the wild-type strain. Conversely, once the gene for this putative enzyme has been cloned in a plasmid suitable for further transformation of a bacteria or a yeast strain (the mutant species as produced previously or an heterologous species in which the identified enzyme is lacking), one can test whether the tRNAs are now modified again and contain the expected modified nucleotide. With microorganisms like *E. coli* or yeast that can be grown in the presence of highly radioactive material, such as [^{32}P]-orthophosphate or [^{35}S]-sulfate, this is an easy task. Indeed, enough highly radioactive bulk tRNA can be extracted from just a few milliliters of cell culture by direct phenol extraction of intact cells.

6.1. Preparation of low-phosphate medium for culture of microorganisms in the presence of [^{32}P]-orthophosphate

The method for yeast culture is from Warner (1991): to 100 ml of medium containing 1 g of Bacto yeast extract and 2 g of Bacto peptone, add 1 ml of 1 M MgSO$_4$ followed by 1 ml of concentrated aqueous ammonia (25%, until the final pH is approximately 8.5). Gently stir and allow the Mg(NH$_4$)PO$_4$ to precipitate at room temperature (or in a cold room at 4°) for a few hours. Remove the precipitate by centrifugation or filtration through a 0.45-μm Millipore filter. Adjust the pH of the filtrate to 5.8 with HCl and sterilize by autoclaving. To 100 ml of the sterile low-phosphate medium, add 10 ml of a sterile glucose solution at 20% (v/w).

The method for *Escherichia coli* culture medium is from Landy *et al.* (1967): prepare 100 ml of low-phosphate Bacto peptone (2 g/100 ml) as described previously. Prepare another medium containing 1.5 g/liter of KCl, 5.0 g/liter of NaCl, 1.0 g/liter of NH$_4$Cl, 2.0 g/liter of vitamin-free casamino acids (Difco, Detroit, MI), 100 ml/liter of the low-phosphate bacto-peptone, and 12.1 g/liter of Tris base in water and autoclave. To every 100 ml of this sterile medium, add 2 ml of sterile 20% (v/w) glucose and 1 ml of sterile 0.1 M MgSO$_4$.

For medium allowing efficient incorporation of [^{35}S]-sulfate see Sullivan and Bock (1985) and Laten *et al.* (1983).

6.2. Preparation and purification of uniformly [³²P]-labeled RNA

Because highly radioactive material is handled, experiments (at least at their initial stages) must be performed in a special radioactive safety room, behind Plexiglas screens, using gloves and disposable materials for protection against β-irradiation. Usually, microorganisms are first grown in normal phosphate-containing medium. A small volume (0.1–0.5 ml) of the culture is then transferred into a 50-ml Falcon tube containing 3–5 ml of the low-phosphate medium described previously. When cells start to grow exponentially, 0.5–1.0 mCi of [³²P]-orthophosphate (carrier-free grade) is added. The solution is incubated for several hours at 30° (yeast) or 37° (E. coli) until most of the [³²P]-orthophosphate (60–80%) has been incorporated into the macromolecules. The uptake of [³²P]-orthophosphate is measured as follows: an aliquot of the culture (10 μl) is passed through a nitrocellulose Millipore filter (0.2 μm). The filter is washed with an isotonic solution, and the radioactivity that remains on the filter is measured by scintillation counting. When most of the [³²P]-orthophosphate has been incorporated into the cells, 0.5 ml of Na phosphate buffer (pH 7.0) is added to the culture and it is incubated for another 1 or 2 h. This reduces the amount of [³²P]-labeled immature RNA in the RNA preparation. At the end of the incubation period, cells are harvested by centrifugation, washed twice with isotonic buffer to eliminate most of the unincorporated [³²P]-orthophosphate, and transferred into 1.5-ml sure-lock Eppendorf tubes for further purification. After suspension of highly radioactive cells in 0.5 ml of Tris-HCl buffer, pH 7.5, containing 10 mM MgCl$_2$ and 0.5 M Na acetate, an equal volume of pure liquid phenol (without chloroform and isoamyl alcohol to avoid lipidic cell membrane destruction) is added for RNA extraction (10 min under mild agitation at room temperature) of intact cells. This will generally allow only the small cellular RNA (e.g., tRNAs, snRNAs, 5S-RNA, snoRNAs) to diffuse out of the permeabilized cells, whereas rRNA within the ribosomes and DNA within the nucleosomes will remain trapped inside the cells together with the denatured proteins (Brubacker and McCorquodale, 1963; Monier et al., 1960). After centrifugation for 15 min at 12,000 rpm, the bulk tRNA in the aqueous supernatant is precipitated by ethanol, centrifuged, and the pellet is extensively washed two or three times with ethanol/water 70% (v/v). This highly radioactive material is further purified by step elution on DEAE-cellulose or monoQ, either by batch elution in the Eppendorf tube or after placing the chromatographic resin inside a small plastic disposable column. The resin is first washed with Tris-HCl buffer, pH 7.5, containing 0.25 M NaCl to eliminate unincorporated "very hot" mononucleotides and small polynucleotides, then the bulk highly radiolabeled tRNA (each phosphate group is radiolabeled) is eluted with the same buffer containing 1 M NaCl. If any rRNA was present in the phenol-extracted material, it will remain in the DEAE/ monoQ resin. Bulk [^{32}P]-tRNA is collected by precipitation with ethanol, washed at least two times with 70% ethanol/water to get rid of as much of coprecipitated NaCl as possible. Commercial kits from Clontech and Qiagen for RNA purification can be satisfactorily used for the same purpose as DEAE/MonoQ purification of tRNA. The purified uniformly labeled [^{32}P]-nucleic acids can be further purified by 1D or 2D polyacrylamide gel electrophoresis (for examples, see Ikemura, 1989; Keith, 1990), or by affinity chromatography on a small avidine column containing a selected biotinylated oligonucleotide complementary to the RNA of interest (Mörl *et al.*, 1994; Tsurui *et al.*, 1994; see also Chapter 10 by Suzuki *et al.*). It is finally dissolved in RNase-free water and aliquoted into tRNA samples of approximately 500–5000 cpm each (Cerenkov counts) for further base composition analysis. Each sample can be mixed with 1 μl of carrier "cold" unlabeled tRNA solution (approximately 20 μg from stock solution of Baker yeast tRNA at 20 mg/ml in water) in a siliconized Eppendorf tube and lyophilized. The number of tubes depends on the different types of RNA digestions to be tested. All the rest of the procedure is identical as described previously, including TLC analysis of the various RNA digests and interpretation of data from the reference maps.

Notice, because position 54 of naturally modified tRNA of most bacteria and lower eukaryotes is ribothymidine, generally present in 1 mol/mol of tRNA, ribothymidine monophosphate on the TLC plates (pT or pm^{5}U in Fig. 3.6) is a useful reference for the determination of the relative amount of a given modified nucleotide. However, for some tRNAs from hyperthermophilic bacteria, a mixture of pm^{5}U and thiolated derivative pm^{5}s^{2}U can be found, whereas in tRNAs from higher eukaryotic cells, a mixture of pT (pm^{5}U) and pTm (pm^{5}Um) will be detected. In archaeal tRNAs of Thermococcales phyla, uridine at position 54 is m^{5}U and m^{5}s^{2}U, whereas in all other Archaea it is m^{1}Psi or Um (Dunin-Horkawicz *et al.*, 2006).

Instructive examples of detection of various modified nucleotides from uniformly labeled RNA ([^{32}P], [^{14}C], or [^{35}S]) extracted from either wild-type or mutant *E. coli*, *Saccharomyces cerevisiae*, *Halobacterium volcanii*, or *Mycoplasma* can be found in Sakano *et al.* (1974), Nishikura and De Robertis (1981), Sullivan and Bock (1985), Adanchi *et al.* (1989), Laten *et al.* (1983), Cavaillé *et al.* (1999), Gupta (1984), Suzuki *et al.* (2002), Pintard *et al.* (2002a), and Purushothaman *et al.* (2005).

7. BASE COMPOSITION ANALYSIS OF RNA BY POSTLABELING PROCEDURES

This procedure allows analysis of base composition from unlabeled starting material. It was initially described by Szekely and Sanger (1969), Silberklang *et al.* (1979), and RajBhandary (1980). We have only slightly improved the protocol by adapting it to the use of new available commercial products. Each tRNA sample to be analyzed (bulk tRNA from wild or mutant strains), or better an affinity-purified tRNA species as obtained with procedures described by Suzuki *et al.* (Chapter 10), is dissolved in water. The concentration of residual salts should be as minimal as possible, and it should be free of any trace of detergent or phenol that will interfere with the subsequent "all in one tube" enzymatic reactions described later. This is usually obtained by precipitating the RNA by ethanol, washing two to three times with 70% (v/v) ethanol/water and dissolving in DEPC-water.

7.1. Step 1: complete digestion of RNA (or RNA fragment) with RNase T2

1. Lyophilize 1–3 μg of RNA sample (<0.1 A260) in a siliconized Eppendorf tube.
2. Add 10 μl of 50 mM NH$_4$ acetate buffer, pH 4.5, containing 0.2–0.5 U of RNase T2 from *Aspergillus oryzae* (cat. no. 18031–113; Gibco-BRL/ stock solution in water at 5 U/μl). This will cleave the RNA into 3′-monophosphate nucleosides (and dinucleosides when a 2′-hydroxyl of a ribose is occupied by a methyl group or ribose phosphate group). Mix well and spin down.
3. Incubate the 10 μl reaction mix overnight in an incubator hood at 37° (which prevents the small volume of sample from drying at the bottom of the tube owing to evaporation and condensation of water in the Eppendorf cap). Keep at $-20°$ until used. For each experiment described next, we use only 1 μl at a time. Various different samples are run in parallel.

7.2. Step 2: [^{32}P]-labeling of RNase T2 digest into 3′,5′-diphosphate nucleosides with PNK

1. Lyophilize 1–2 μCi of 5′-[γ-^{32}P]-ATP (5000 Ci/mmol) together with 250–2500 pmol of nonradiolabeled ATP (from a stock solution of 5′-ATP-Na$_2$ [SigmaUltra; cat. no. A-7699] at 0.5 mg/ml in water neutralized to pH 7.0). This amount of nonradiolabeled ATP should correspond to a slight excess of the molar amount of nucleotides of the starting RNA to be labeled. Specific radioactivity of ATP will be reduced accordingly.
2. To the lyophilized ATP, add 1.5 μl of RNase-free water followed by 1 μl of 5× labeling buffer containing 100 mM bicine-HCl, pH 8.2 (Sigma; cat. no. B-3876), 50 mM MgCl$_2$, 50 mM DTT, 5 mM RNase-free spermidine, pH 8.2 (cat. no. S-0266; Sigma), 1 μl of the RNA hydrolysate obtained in step 1 previously, and finally 1.5 μl (1.5 U) of a solution of polynucleotide kinase from bacteriophage T4 (T4-PNK from New England Biolabs, cat. no. 201S) delivered at 10 U/μl in 50 mM KCl, 10 mM Tris-HCl, pH 7.4, 0.1 mM EDTA, 1 mM dithiothreitol (DTT), 0.1 μM

ATP, and 50% glycerol. Mix well the resulting 5 μl reaction mix and spin down any droplets on the wall. The enzyme of New England Biolabs is free of any 3$'$-phosphatase activity. However, because the optimal activity of this usually contaminating 3$'$-phosphatase is pH 5.0–7.0, which is below the optimal activity of the kinase reaction (7.0–9.0), such contamination, if any, might not be so important after all. Be sure that the working solution contains 50% glycerol, because PNK is inactivated by freezing/tawing. Notice that PNK does not show equal efficiency of [^{32}P]-incorporation into each of the different mononucleotides and especially some of the hypermodified nucleotides derived from t^{6}A, i^{6}A, queuosine, and wyosine, thus determination of the relative amount of nucleotides (modified or not) is not possible with this postlabeling method.

3. Incubate the 5 μl reaction mix for 30 min to 1 h at 37°, and then proceed to the step 3 of this protocol.

7.3. Step 3: hydrolysis of excess ATP with a mix of ATPase and ADPase (Apyrase)

1. Just before use, dilute the stock solution of apyrase at 500 U/ml (approximately 2 mg/ml in water) 200-fold in RNase-free water (2.5 mU/μl final). This enzyme, partially purified from potato—*Solanum tuberosum*—is obtained from Sigma (grade VI cat. no. A-6410). It always has to be tested for possible contamination by various 3$'$-,5$'$-phosphatases that are harmful to the labeled nucleotides; unfortunately, some commercial batches are more contaminated than others. This is why phosphate buffer and carrier 5$'$-monophosphate nucleosides are routinely added to the reaction mix (see later) to limit dephosphorylation of the [^{32}P]-nucleotides (but not completely avoiding it). Apyrase catalyzes the two-step degradation of ATP into ADP + Pi and ADP into AMP + Pi. Thus unused [γ-^{32}P] from radiolabeled ATP in the reaction mix will become free [^{32}P]-orthophosphate and will not interfere with the remaining radioactivity to be detected in nucleotides.

2. Premix 2 μl (5 mU) of the apyrase dilution with 1 μl of 10 mM Na phosphate buffer, pH 7.0, and 2 μl of 20 mg/ml of 5$'$-monophosphate nucleosides mix (5 mg of each AMP, GMP, UMP, CMP, Na salt from Sigma; Cat. no. A-1752, C-1006, G-8377, U-6375) in 1 ml of water and the pH of the resulting solution adjusted to 7.0.

3. Add this apyrase mix (5 μl) to the mix of 5$'$-[^{32}P] 3$'$,5$'$-diphosphate nucleosides obtained in the preceding step of the protocol. Mix well and spin down.

4. Incubate the 10 μl reaction mix for 15–30 min at 37°.

5. If you are sure that your RNA sample does not contain any thermolabile modified nucleotide like D, m^{7}G, and m^{5}C, quickly raise the temperature to 85–90° for one or two min to inactivate both PNK and apyrase.

6. Chill tubes in ice and proceed immediately to the next step.

7.4. Step 4: removal of 3′-phosphate of diphosphate nucleosides with nuclease P1

1. Prepare a fresh dilution of nuclease P1 (from a 10× stock solution of nuclease P1 of *Penicillium citrinum;* Roche; cat. no. 236225) at 0.1 mg/ml in 200 mM NH$_4$ acetate, pH 4.5.
2. Add 5 μl of the diluted nuclease P1 (0.5 μg) to the 10 μl hydrolysate mix obtained above. Mix well and spin down.
3. Incubate the final 15 μl mix for 2 or 3 h at 37°. Only a few microliters of this final hydrolysate will be used for TLC analysis. Keep the remaining frozen at −20° for further identification tests.

This nuclease P1 has two activities: as an endonuclease it cleaves RNA (and DNA as well) into the corresponding 5′-monophosphate nucleosides, and as a 3′-nucleotidase it removes the phosphate groups at the 3′ end of nucleotides (Fig. 3.2). However, when 2′-O-methylnucleotides or bulky nucleotides (e.g., wyosine, queuosine derivatives, bulby N^6-purine derivatives) are present in the RNA molecule, the nuclease activity becomes slow. As a result, dinucleoside diphosphates of the type pXmpY may be present in the P1-hydrolysate. If this might be a problem, increase the amount of enzyme in the reaction mix (10 times) or increase the incubation time (overnight), or, alternately couple the P1 reaction with *Crotalus* snake venom phosphodiesterase VPD (Worthington; cat. no.31B217J; dissolved in water at 1 mg/ml see earlier section "Complete digestion of RNA with various nucleases").

7.5. TLC analysis

After digestion, TLC analyses are performed exactly as described previously with either glass or plastic plates coated with cellulose. Examples of identification of modified nucleotides in RNA by the postlabeling technique followed by TLC analysis can be found in Kuchino *et al.* (1987); Xue *et al.* (1993); and Paul and Bass (1998).

8. RNA SEQUENCING OF PURIFIED NONRADIOLABELED tRNA SPECIES

A logical extension of the postlabeling procedure as described previously is the "single-hit random labeling RNA sequencing method" that allows one to obtain the full sequence, including the position of each modified nucleotides from only few μg (or even less) of purified tRNAs (Fig. 3.9, see also RajBhandary, 1980). Here, the purified, nonradiolabeled RNA (at least 95% pure, maximum 50–150 nt in length), usually purified by electrophoresis on denaturing polyacrylamide gel, is first cleaved randomly under mild conditions into oligonucleotide fragments by incubation for only a few minutes at high temperature (95°). Under such experimental conditions, only one phosphodiester bond is hydrolyzed on average per RNA molecule (step 1 in Fig. 3.9). However, some thermolabile modified nucleotides might be affected during this procedure (see later for an alternate approach that allows avoidance of this problem). Each resulting RNA fragment will have a 5'-hydroxyl group and a phosphate group at its a 3'-end. RNA fragments are next radiolabeled with [^{32}P] at their 5'-end with T4-polynucleotide kinase (T4-PNK) and (γ-^{32}P)-ATP (step 2 in Fig. 3.9, for experimental procedure, see preceding). The oligonucleotides of the hydrolyzed RNA are separated according to their sizes at "single-nucleotide resolution" by 1D electrophoresis on a denaturing polyacrylamide gel to generate the so-called ladder (usually 15 or even 20% gel, 50–80 cm in length, sequencing gels; step 3 in Fig. 3.9). After electrophoresis, the oligonucleotides in each band are visualized by autoradiography. Notice, the presence of 2'-O-methylated nucleosides or any nucleoside that hinders the initial thermo-induced hydrolytic cleavage of RNA, is signaled by "gaps" in the ladder on the gel. Comparison of data with RNA transcript lacking modified nucleotides can be used as control. Regions of the gel containing RNA fragments that you want to analyze for presence of 5'-[^{32}P]–modified nucleotides are cut out of the gel, eluted, and treated subsequently with nuclease P1 and/or VDP as described previously (step 4a in Fig. 3.9). In general, the sequence of the tRNA used as substrate for detecting the formation of a given modified base is known. The resulting RNA digests, containing postlabeled 5'-[^{32}P]-nucleotide monophosphates, are analyzed by 2D-TLC using 6 × 6 cm or 10 × 10 cm plates and one or both types of solvent systems as described previously (step 5a in Fig. 3.9). The identity of the detected 5'-[^{32}P]-modified nucleotide on the TLC plate is obtained after comparison with reference maps (Figs. 3.3–3.8), whereas the position of the detected modified nucleotide corresponds to the position of the cut band in the ladder of the gel and, therefore, to the position in the RNA sequence.

Alternately, each fragment of the gel as obtained in step 3 can be hydrolyzed by *in situ* action of RNase T2 to generate nonradiolabeled nucleoside 3'-phosphate, except for the nucleotide at the 5'-end of the RNA that will become a 3'-,5'-diphosphate nucleoside with the radiolabeled [^{32}P]-phosphate at the 5' end (step 4b in Fig. 3.9). Contact transfer of the diphosphate nucleotides from the gel on a special PEI (polyethylenimine) cellulose plate is then performed, followed by 1D chromatography with a special NH$_4$ formate buffer (step 5b in Fig. 3.9). The plate is autoradiographed and identification of each radiolabeled terminal 3'-, 5'-dinucleotide phosphate is simply obtained from its position on the plate compared with those of published reference maps (see in Gupta and Randerath, 1979). Again, because the fragments are separated on the polyacrylamide gel according to their lengths, inspection of the PEI data

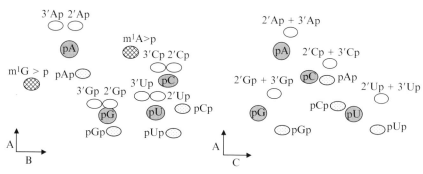

Figure 3.8 Effect of phosphate position on chromatographic mobility of monophosphate nucleosides. pN (N being A, G, C, or A) corresponds to nucleotides with the phosphate group on the 5'-hydroxyl (gray ellipses); 3'- or 2'-Np to the monophosphate nucleosides harboring a phosphate group at position 3'- or 2'-, respectively (open ellipses); and pNp to diphosphate nucleosides harboring one phosphate group on the 5'-hydroxyl and the second one on the 3'-hydroxyl (light gray ellipses). Also shown are the positions for some of the cyclic monophosphate nucleosides (2'-, 3'-, designated N›, hatched ellipses) that often appear on the chromatograms after RNA digestion by RNase T2, especially at a low ratio of RNase to substrate (this happens with other modified nucleotides too). In case of doubt, hydrolyze an aliquot of the same sample with 10 times more RNase T2 (or raise the pH to 1.0 for few minutes before spotting on the TLC; see text) and compare the chromatographic profiles.

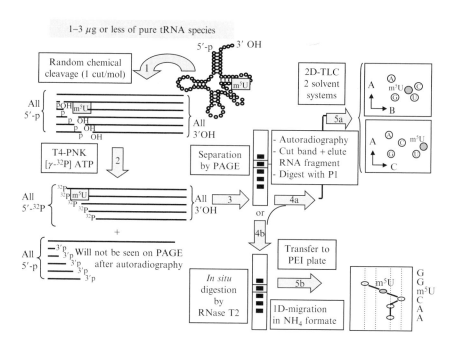

reveal not only the nature but also the position of each individual nucleotide in the studied sequence.

It is worthwhile to mention that the preceding "single-hit random labeling RNA sequencing method," on the basis of thermally induced cleavages of RNA, can be adapted by use of commercially available very pure specific endonucleases (RNase T2, RNase T1, RNaseA), provided that the enzymatic experimental conditions allow only one cut per RNA molecule on average and that the resulting oligonucleotide fragments harbor a 5'-hydroxyl group and a phosphate group at the 3'-end. The advantage of using RNase T1 (cut after G) or RNase A (cut after U and C) is that they allow us to explore in the RNA sequence the presence or absence of modified nucleotides derived solely from G (in the case of RNase T1) or from U and C (in the case of RNase A). However, as mentioned previously, the rate of enzymatic RNA cleavage at positions where a bulky modified base is present (for example) might be too slow to allow its detection. Nevertheless for most simple modified nucleotides, the use of site-specific endonucleases, especially when thermal degradation of a particular modified nucleotide is suspected, is of great advantage.

For details concerning the RNA sequencing strategies described previously, see Gupta *et al.* (1976), Stanley and Vassilenko (1978), Kuchino *et al.* (1987), Gupta and Randerath (1979), RajBhandary (1980), Nishimura and Kuchino (1983), and Kuchino *et al.* (1987).

9. Identification of Modified Residues in RNAs by Reverse Transcriptase–Based Methods

Besides the various methods described previously, there is at least one other powerful method to localize certain modified nucleotides in RNA. It is based on the use of reverse transcriptase (usually avian myeloblastosis virus reverse transcriptase [RT]) and specific 5'-end-[^{32}P]–labeled oligonucleotide primers (15–20 nt in length) that are complementary to a RNA present in a mixture of unfractionated RNAs (primer extension analysis of RNA;

Figure 3.9 Schematic of a procedure for direct sequencing of tRNA or any fragment of RNA shorter than 150 nt containing modified nucleotides (e.g., m^{5}U). RNA is first cleaved randomly under mild conditions (single cut per molecule). Each fragment is then radiolabeled with [^{32}P] at the 5' end, and the resulting radiolabeled RNA fragments are separated by electrophoresis on polyacrylamide gel. The fragments containing modified nucleotide at its 5' end are analyzed by either 2D-TLC or 1D-PEI chromatography. For details and references corresponding to steps 1–5, see text. Identification of each radiolabeled terminal 5'-nucleotide phosphate is obtained from its position on the TLC (step 5a) or PEI plates (step 5b) compared with those of published reference maps, whereas its position in RNA is deduced from its position of the band in the "ladder" of the gel.

see Lane *et al.* [1988] and Boorstein and Craig [1989]). Here the method takes advantage of the fact that certain modified nucleotides (those for which the Watson-Crick pairings is prevented or hindered), block or slow down the progress of reverse transcription at the nucleotide preceding the modified residue, an effect that can easily be amplified by reducing the concentrations of dNTPs. This is the case for most modifications at atom $N1$ or $N6$ of adenine (such as m^1A, m^1I, m^6_2A, and some "bulky" hyper-modified A, such as ms^2t^6A and ms^2i^6A), N^1 or N^2 of guanine (m^1G, m^2G, m^2_2G and "bulky" hypermodified G, such as wyosine and wybutosine), N^3 of cytidine (m^3C), N^3 of uridine and pseudouridine (m^3U, acp^3U, m^3Psi, m^1acp^3Psi), and dihydrouridine (D), which cannot base pair with A (see for examples: Wittig and Wittig [1978], Youvan and Hearst [1979, 1981], Maden *et al.* [1995], Liu *et al.* [1995], Lafontaine *et al.* [1995], Gustafsson and Persson [1998], Auxilien *et al.* [1999]), Pintard *et al.* [2002b], Jackman *et al.* [2003], and Kirkepar *et al.* [2005]). The existence of "strong stops" or "pauses" of chain elongation generated by such an abnormal situation during transcription, as identified after autoradiography of 1D denaturing gel electrophoresis of the *in vitro* transcription products, may only indicate the presence of a putative modified nucleotide, not its identity. Also, certainty about the existence of a modified base in RNA from simple inspection of an X-ray film is not always evident. Indeed, a "pause" or a "strong stop" can result from stable secondary structure of the RNA ("false positive") or any other physical parameter of the RNA molecule that interferes with the normal progression of RT along the RNA molecules, a situation that makes the detection of certain modified nucleotides in that particular regions of RNA a real challenge, especially with G + C−rich RNA, as those originating from hyperthermophilic organisms. Moreover, depending on the "degree" of modification of a given nucleotide (partially or fully modified) within the RNA population, the arrest of RT is not easily detectable <20% of nucleotide modifications, and appropriate controls always have to be done. Moreover, evaluation of the mole amount of a given modified base in RNA by the RT-based methods will never be possible.

In certain cases, localization of a given modified nucleotide within an RNA population by reverse transcription can be facilitated either after introducing a chemical (eventually bulky) group on a given modified nucleotide, thus impairing its base pairing property, such as N-cyclohexyl-N'-beta-(4-methylmorpholinium) carbodiimide on N^3 of pseudouridine (Bakin and Ofengand, 1993, 1998; Ho and Gilham, 1971; Ofengand *et al.*, 2001), or after scission of the phosphodiester backbone of RNA resulting from selective chemical alteration of m^7G, dihydrouridine (D), inosine, or wybutosine of tRNA (Alexandrov *et al.*, 2002; Beltchev and Grunberg-Manago, 1970; Cerutti *et al.*, 1968; Gu *et al.*, 2005; Morse and Bass, 1997; Peattie, 1979; Stern *et al.*, 1988; Thiebe and Zachau, 1971; Wilkinson *et al.*, 2006; Wintermeyer and Zachau, 1970, 1974, 1975). In the first case, the progress of the polymerase is impaired by the strict inability of the template nucleotide to base pair. In the second case, cDNA synthesis stops simply when the polymerase run off the template RNA. Comparison of transcription products pattern before and after the chemical treatment, also, whenever possible, comparison with T7 transcript of the same RNA sequence thus lacking all modified nucleotides, can better indicate the presence of a given modified nucleotide in the sample RNA. However, detection of the modified bases within an RNA population will depend on whether the chemical treatment changing the structure of the base and/or inducing selective cleavage of the phosphodiester bond can be performed quantitatively without affecting the stability of certain modified nucleotides or the phosphodiester bonds at other sites of the tRNA molecule.

Despite the many difficulties in properly handling this RT-based technique, it can complement nicely those we described previously for base composition and nearest neighbor analysis, because it allows mainly the identification of the "position" of certain modified bases in RNA. The main advantage of the primer reverse extension assay is that it can be used with very small amounts (ng level) of nonradiolabeled and completely unfractionated RNA of any length, as present for example in a small volume of a crude cell extract. For more details about this technique and description of the procedures to be used, see Chapter 2.

10. DISCUSSION AND OUTLOOK

Transfer RNA modification enzymes are a fascinating class of enzymes to study. In the past their studies have always been hampered by the difficulties encountered in purifying these enzymes, which are present in low amounts in the cells, and the lack of adequate substrates to test their activities *in vitro*. Not only did they hinder progress of research on RNA modification, they also discouraged many generations of scientists from working with these RNA modification enzymes. Fortunately, recent developments of recombinant DNA and RNA technologies, together with techniques allowing *in vitro* chemical and enzymatic synthesis of RNA substrates of any sequence, considerably facilitate the development of research on RNA modification enzyme. This area of research now becomes again "à la mode," as judged by the increasing numbers of laboratories around the world working at explaining the structure and functions of modified nucleotides in RNA, as well as of the corresponding RNA modification enzymes (Grosjean, 2005). However, identification and quantification of a given modified nucleotide in naturally occurring RNA, or after incubation of a synthetic or semisynthetic tRNA substrate with cell extract or purified recombinant enzyme, often remains difficult. In this chapter, we have presented a few techniques that allow facile measurement of the activity of a given modification enzyme *in vitro* or *in vivo*. The methods are based on use of preradiolabeled or postradiolabeled RNA substrates and 2D TLC to identify the modified nucleotides in an RNA hydrolysate. These simple tests allowed recent important progress in the identification and characterization of many modification enzymes (references are listed previously). In the case of enzymes involved in the formation of hypermodified nucleotides or those that are dependent on the presence of other modified nucleotides in tRNA, microinjection of radiolabeled substrates into the *Xenopus laevis* oocytes, or any other types of cells, can be used to advantage. However, synthetic (or semisynthetic) RNA substrates completely devoid of modified nucleotides may not necessarily be suitable for studying the latest events in a complex sequential and possibly interdependent tRNA maturation process. A few "early" modification events may facilitate the attainment of a proper pre-RNA conformation by preventing misfolding, by stabilizing a particular domain, or by allowing correct splicing events and other subsequent maturation steps (see, for example, Helm and Attardi [2004]; Steinberg and Cedergren [1995]; Urbonavicius *et al.* [2006]; and as reviewed in Helm [2006]). The absence of these "early" modified nucleotides (as other maturation processes such as intron removal) may, therefore, compromise the action of some "late" modification enzymes that depend on such structural or conformational parameters. Also, most of the enzymes that now remain to be identified are those for which (1) the corresponding genes are difficult to identify; (2) the recombinant enzymes are not easily produced *in vitro*, because the putative gene is not easily expressed in transformed cells, or recombinant gene products are insoluble or instable; (3) the required cofactors needed to test their activities *in vitro* are not known; (4) association of the modification enzyme with other proteins, or more complex modification machinery, or subcellular structure(s) may be required.

It is hoped that access to complete genome sequences of many different organisms (see Chapter 7 and Bujnicki *et al.* [2004a]), coupled with routine cloning techniques, identification, and production of purified recombinant enzymes from almost any organism (see Chapters 6 and 8), open an extraordinary avenue to study this huge family of enzymes. It is now possible to compare them at the level of their primary and tertiary structures (when obtained in the crystallized form) and explore their putative evolutionary origins (see e.g., Anantharaman *et al.* [2002]; Bujnicki *et al.* [2004b] in relation to that of tRNA (Crécy-Lagard *et al.*, 2007; Marck and Grosjean, 2002).

ACKNOWLEDGMENTS

We thank Jonatha M. Gott for advice and improvements on the manuscript. The early parts of the works of H. G. reported in this chapter were supported by research grants from the Centre National de la Recherche Scientifique (CNRS); the Ministère de l'Education Nationale, la Recherche Scientifique et de la Technologie (Programme Interdépartemental de Géomicrobiologie des Environnements Extrêmes). Only recently has it been supported from funds of University of Orsay to Prof. J. P. Rousset (IGM. Bat 400, Orsay-France). L. D. was supported by the Belgian Fonds de la Recherche Fondamentale Collective (FRFC), M. R. by the "Commission Communautaire Française," and G. K. by the CNRS in Strasbourg.

NOTE

Any original paper(s) difficult to obtain from the web or from your favorite libraries could be asked for from Henri Grosjean (*henri.Grosjean@igmors.u-psud.fr*).

REFERENCES

Adanchi, Y., Yamao, F., Muto, A., and Osawa, S. (1989). Codon recognition patterns as deduced from sequences of the complete set of tRNA species in *M. capricolum. J. Mol. Biol.* **209**, 37–54.

Alexandrov, A., Martzen, M. R., and Phizicki, E. M. (2002). Two proteins that form a complex are required for 7-methylguanosine modification of yeast tRNA. *RNA* **8**, 1253–1266.

Anantharaman, E. M., Koonin, E. V., and Aravind, L. (2002). Comparative genomics and evolution of proteins involved in RNA metabolism. *Nucleic Acids Res.* **30**, 1427–1464.

Aphazizhev, R., Theobald-Dietrich, A., Kostyuk, K. D., Kochetkov, S. N., Kisselev, L., Giegé, R., and Fasiolo, F. (1997). Structure and aminoacylation capacities of tRNA transcripts containing deoxyribonucleotides. *RNA* **3**, 893–904.

Auxilien, S., Crain, P., Trewyn, R. W., and Grosjean, H. (1996). Mechanism, specificity and general properties of the yeast enzyme catalyzing the formation of inosine-34 in the anticodon of transfer RNA. *J. Mol. Biol.* **262**, 437–458.

Auxilien, S., Keith, G., Le Grice, F. J., and Darlix, J.-L. (1999). Role of posttranscriptional modifications of primer tRNA^{Lys3} in the fidelity and efficacy of plus strand DNA transfer during HIV-1 reverse transcription. *J. Biol. Chem.* **274**, 4412–4420.

Bakin, A. V., and Ofengand, J. (1993). Four newly located pseudouridylate residues in *E. coli* 23S ribosomal RNA are all at the peptidyl center: Analysis by the application of a new sequencing technique. *Biochemistry* **32**, 9754–9762.

Bakin, A. V., and Ofengand, J. (1998). Mapping of pseudouridine residues in RNA to nucleotide resolution. *In* "Methods in Molecular Biology. Protein Synthesis: Methods and Protocols" (R. Martin, ed.), Vol. 77, pp. 297–309. Humana, Totowa, NJ.

Becker, H. F., Motorin, Y., Sissler, M., Florentz, C., and Grosjean, H (1997). Major identity determinants for enzymatic formation of ribothymidine-54 and pseudouridine-55 in Tpsi loop of yeast tRNAs. *J. Mol. Biol.* **274**, 505–518.

Becker, H. F., Motorin, Y., Florentz, C., Giegé, R., and Grosjean, H. (1998). Pseudouridine and ribothymidine formation in the tRNA-like domain of Turnip Yellow Mosaiuc Virus RNA. *Nucleic Acids Res.* **26**, 3991–3998.

Beltchev, B., and Grunberg-Manago, M. (1970). Preparation of pG-fragment from yeast tRNA^{Phe} by chemical scission at the dihydrouracil and inhibition of yeast tRNA^{Phe} charging by this fragment when combined with the -CCA half of this tRNA. *FEBS Lett.* **12**, 24–27.

Björk, G. R., and Hägervall, T. G. (2005). Transfer RNA modification. *In* "*E. coli* and *Salmonella*. Cellular and Molecular Biomol" (R. Curtis, A. Böck, J. L. Ingrahan,

J. B. Kaper, S. Maloy, F. C. Neidhardt, M. M. Riley, C. L. Squires, and B. L. Wanner, eds.), p. 4.6.2. ASM Press, Washington DC.

Boorstein, W. R., and Craig, E. A. (1989). Primer extension analysis of RNA. *Methods Enzymol.* **180,** 347–369.

Brubacker, L. H., and McCorquodale, D. J. (1963). The preparation of amino acid tRNA from *E. coli* by direct phenol extraction of intact cells. *Biochem. Biophys. Acta* **76,** 48–53.

Bruce, A. G., and Uhlenbeck, O. C. (1982). Enzymatic replacement of the anticodon of yeast phenylalanine tRNA. *Biochemistry* **21,** 855–861.

Brulé, H., Grosjean, H., Giegé, R., and Florentz, C. (1998). A pseudoknotted tRNA is a substrate for m⁵C methyltransferase from *X. laevis*. *Biochimie* **80,** 977–985.

Bujnicki, J. M., Droogmans, L., Grosjean, H., Purushothaman, S. K., and Lapeyre, B. (2004a). Bioinformatics-guided identification and experimental characterization of novel RNA methyltransferases. *In* "Nucleic Acids and Molecular Biology" (J. M. Bujnicki, ed.), Practical Bioinformatics Series, Vol. 15, pp. 139–168. Springer-Verlag, Berlin.

Bujnicki, J. M., Feder, M., Ayres, C. L., and Redman, K. L. (2004b). Sequence-structure function studies of tRNA:m⁵C methyltransferase Trm4p and its relationship to DNA: m⁵C and RNA:m⁵U methyltransferases. *Nucleic Acids Res.* **32,** 2453–2463.

Carbon, P., Haumont, E., De Henau, S., Keith, G., and Grosjean, H. (1982). Enzymatic replacement *in vitro* of the first anticodon base of yeast tRNA^Asp: Application to the study of tRNA maturation *in vivo*, after microinjection into frog oocytes. *Nucleic Acids Res.* **10,** 3715–3730.

Carbon, P., Haumont, E., Fournier, M., de Henau, S., and Grosjean, H. (1983). Site-directed *in vitro* replacement of nucleosides in the anticodon loop of tRNA: Application to the study of structural requirements for queuine insertase activity. *EMBO J.* **2,** 1093–1097.

Cavaillé, J., Chetouani, F., and Bachellerie, J. P. (1999). The yeast *S. cerevisiae* YDL112w ORF encodes the putative 2′-O-ribose methyltransferase catalyzing the formation of Gm18 in tRNAs. *RNA* **5,** 66–81.

Cedergren, R., and Grosjean, H. (1987). RNA design by *in vitro* RNA recombination and synthesis. *Biochem. Cell. Biol.* **65,** 677–692.

Cerutti, P., Holt, J. W., and Miller, N. (1968). Detection and determination of 5,6--dihydrouridine and 4-thiouridine in transfer RNA from different sources. *J. Mol. Biol.* **34,** 505–518.

Ciampi, M. S., Arena, F., and Cortese, R. (1977). Biosynthesis of pseudouridine in the *in vitro* transcribed tRNA^Tyr precursor. *FEBS Lett.* **77,** 75–82.

Constantinesco, F., Motorin, Y., and Grosjean, H. (1999a). Transfer RNA modification enzymes from *P. furiosus*: Detection of the enzymatic activities *in vitro*. *Nucleic Acids Res.* **27,** 1308–1315.

Constantinesco, F., Motorin, Y., and Grosjean, H. (1999b). Characterization and enzymatic properties of the trNA (guanosine-26, *N2,N2*) dimethyltransferase (Trm1p) from *P. furiosus*. *J. Mol. Biol.* **291,** 375–392.

Crécy-Lagard, V., Marck, C., and Grosjean, H. (2007). Comparative RNomics, and modomics in mollicutes/mycoplasmas: Prediction of gene function and evolutionary implications. *IUMB Life,* submitted.

Droogmans, L., Haumont, E., De Henau, S., and Grosjean, H. (1986). Enzymatic 2′-O-methylation of the wobble nucleoside of eukaryotic tRNA^Phe: Specificity depends on the structural elements outside the anticodon loop. *EMBO J.* **5,** 1105–1109.

Droogmans, L., and Grosjean, H. (1987). Enzymatic conversion of guanosine 3′-adjacent to the anticodon of yeast tRNA^Phe to *N1*-methylguanosine and the Wye nucleoside: Dependence on the anticodon sequence. *EMBO J.* **6,** 477–483.

Droogmans, L., and Grosjean, L. (1991). 2′-O-methylation and inosine formation in the wobble position of anticodon-substituted tRNA^Phe in a homologous yeast *in vitro* system. *Biochimie* **73,** 1021–1025.

Droogmans, L., Roovers, M., Bujnicki, J., Tricot, C., Hartsch, T., Stalon, V., and Grosjean, H. (2003). Cloning and characterization of tRNA (m^1A58 -TrmI) from *Thermus thermophilus* HB27, a protein required for cell growth at extreme temperatures. *Nucleic Acids Res.* **31,** 2148–2156.

Dunin-Horkawicz, S., Czerwoniec, A., Gajda, M. J., Feder, M., Grosjean, H., and Bujnicki, J. M. (2006). Modomics: A database of RNA modification pathways. *Nucleic Acids Res.* **34,** D145–D149.

Ehrlich, M., Norris, K. F., Wang, R. Y., Kuo, K. C., and Gehrke, C. W. (1986). DNA cytosine and heat-induced deamination. *Biosci. Rep.* **6,** 387–393.

Fournier, M., Haumont, E., De Henau, S., Gangloff, J., and Grosjean, H. (1983). Transcriptional modification of the wobble nucleotide in anticodon-substituted yeast $tRNA^{ArgII}$, after microinjection into *X. laevis* oocytes. *Nucleic Acids Res.* **11,** 707–718.

Garcia, G. A., and Goodenough-Lashua, D. E. M. (1998). Mechanism of RNA-modifing and editing enzymes. *In* "Modification and Editing of RNA" (H. Grosjean and R. Benne, eds.), pp. 135–168. ASM Press, Washington DC.

Grosjean, H. (2005). Modification and editing of RNA: Historical overview and important facts to remember. *Topics in Current Genetics* **12,** 1–22.

Grosjean, H., Doi, T., Yamane, A., Ohtsuka, E., Ikehara, M., Beauchemin, N., Nicoghosian, K., and Cedergren, R. (1987). The *in vivo* stability, maturation and aminoacylation of anticodon-substituted *E. coli* initiator methionine tRNA. *Eur. J. Biochem.* **166,** 325–332.

Grosjean, H., Droogmans, L, Giegé, R., and Uhlenbeck, O. C. (1990). Guanosine modifications in runoff transcripts of synthetic $tRNA^{Phe}$ genes microinjected into *X. laevis* oocytes. *Biochem. Biophys. Acta* **1050,** 267–273.

Grosjean, H., Sprinzl, M., and Steinberg, S. (1995). Posttranscriptionally modified nucleotides in tRNA: Their locations and frequencies. *Biochimie* **77,** 139–141.

Grosjean, H., Edqvist, J., Straby, K. B., and Giegé, R. (1996). Enzymatic formation of modified nucleosides in tRNA: Dependence on tRNA architecture. *J. Mol. Biol.* **255,** 67–85.

Grosjean, H., Motorin, Y., and Morin, A. (1998). RNA-modifying and RNA-editing enzymes: Methods for their detection. *In* "Modification and Editing of RNA" (H. Grosjean and R. Benne, eds.), pp. 21–46. ASM Press, Washington, DC.

Grosjean, H., Keith, G., and Droogmans, L. (2004). Detection and quantification of modified nucleotides in RNA using thin-layer chromatography. *Methods Mol. Biol.* **265,** 357–391.

Grosjean, H., and Oshima, T. (2007). How nucleic acids cope with high temperature? *In* "Physiology and Biochemistry of Extremophiles" (C. Gerday and N. Glansdorff, eds.). ASM Press. In press.

Gu, W., Hurto, R. L., Hopper, A. K., Grayhack, E. J., and Phizicky, E. M. (2005). Depletion of *Saccharomyces cerevisiae* tRNA-His guanylyltransferase Thg1P leads to uncharged tRNA-His with additional m^5C. *Mol. Cell. Biol.* **25,** 8191–8201.

Gupta, R. (1984). *H. volcanii:* Identification of 41 tRNAs covering all amino acids and the sequences of 33 class I tRNAs. *J. Biol. Chem.* **259,** 9461–9471.

Gupta, R. C., and Randerath, K. (1979). Rapid readthrough technique for sequencing of RNAs containing modified nucleotides. *Nucleic Acids Res.* **6,** 3443–3458.

Gupta, R. C., Randerath, E., and Randerath, K. (1976). An improved separation procedure for nucleoside monophosphates on polyethylenimine-(PEI)-cellulose thin layers. *Nucleic Acids Res.* **3,** 2915–2922.

Gustafsson, C., and Persson, B. C. (1998). Identification of the *rrnA* gene encoding the 23S rRNA m^1G745 methyltransferase in *Escherichia coli* and characterization of an m^1G745-deficient mutant. *J. Bacteriol.* **180,** 359–365.

Hashimoto, S., Sakai, M., and Muramatsu, M. (1975). $2'$-O-methylated oligonucleotides in ribosomal 18S and 28S RNA of a mouse hepatoma, MH134. *Biochemistry* **14,** 1956–1964.

Haumont, E., Fournier, M., De Henau, S., and Grosjean, H. (1984). Enzymatic conversion of adenosine to inosine in the wobble position of yeast tRNA^{Asp}: The dependence on the anticodon sequence. *Nucleic Acids Res.* **12**, 2705–2715.

Haumont, E., Droogmans, L., and Grosjean, H. (1987). Enzymatic formation of glycosyl queuosine in yeast tRNAs microinjected into *X. laevis* oocytes: The effect of the anticodon loop sequence. *Eur. J. Biochem.* **168**, 219–225.

Helm, M. (2006). Post-transcriptional nucleotide modification and alternative folding of RNA. *Nucleic Acids Res.* **34**, 721–733.

Helm, M., and Attardi, G. (2004). Nuclear control of cloverleaf structure of human mitochondrial tRNA^{Lys}. *J. Mol. Biol.* **337**, 545–560.

Hiramaru, M., Ushida, T., and Egami, F. (1966). Ribonuclease preparation for the base analysis of polyribonucleotides. *Anal. Biochem.* **17**, 135–142.

Ho, N. W. Y., and Gilham, P. T. (1971). Reaction of pseudouridine and inosine with *N*-cyclohexyl-*N'*-β-(4-methylmorpholinium)ethylcarbodiimide. *Biochemistry* **10**, 3651–3657.

Hopper, A. K., and Phizicky, E. M. (2003). tRNA to the limelight. *Genes Dev.* **17**, 1672–1680.

House, C. H., and Miller, S. L. (1996). Hydrolysis of dihydrouridine and related compounds. *Biochemistry* **35**, 315–320.

Ikemura, T. (1989). Purification of RNA molecules by gel techniques. *Methods Enzymol.* **180**, 14–25.

Jackman, J. E., Montange, R. K., Malik, H. S., and Phizicky, E. M. (2003). Identification of the yeast gene encoding the tRNA m^{1}G methyltransferase responsible for modification at position 9. *RNA* **9**, 574–585.

Jiang, H.-Q., Motorin, Y., Jin, Y.-X., and Grosjean, H. (1997). Pleiotropic effects of intron removal on base modification pattern of yeast tRNA^{Phe}: An *in vitro* study. *Nucleic Acids Res.* **25**, 2694–2701.

Johansson, M. J. O., and Byström, A. S. (2005). Transfer RNA modifications and modifying enzymes in *Saccharomyces cerevisiae*. In "Fine-Tuning of RNA Functions by Modification and Editing" (H. Grosjean, ed.), pp. 87–119. Springer-Verlag, Heidelberg.

Keith, G. (1990). Nucleic acid chromatographic isolation and sequence methods. In "Chromatography and Modifications of Nucleosides" (C. W. Gehrke and K. C. Kuo, eds.), Vol. 45A, pp. A103–A141. Elsevier, The Netherlands.

Keith, G. (1995). Mobilities of modified ribonucleotides on two-dimensional cellulose thin-layer chromatography. *Biochimie* **77**, 142–144.

Kostyuk, D. A., Dragan, S. M., Lyakhov, D. L., Rechinsky, V. L., Tunitskaya, V. L., Chernov, B. K., and Kotcetkov, S. N. (1995). Mutants of T7 RNA polymerase that are able to synthesize both RNA and DNA. *FEBS Lett.* **369**, 165–168.

Kretz, K. A., Trewyn, R. W., Keith, G., and Grosjean, H. (1990). Site-directed replacement of nucleotides in the anticodon loop of tRNA: Application to the study of inosine biosynthesis in yeast tRNA^{Ala}. In "Chromatography and Modifications of Nucleosides" (C. W. Gehrke and K. C. Kuo, eds.), Vol. 45B, pp. B143–B171. Elsevier, The Netherlands.

Kuchino, Y., Hanyu, N., and Nishimura, S. (1987). Analysis of modified nucleosides and nucleotide sequence of tRNA. *Methods Enzymol.* **155**, 379–396.

Lafontaine, D., Vandenhaute, J., and Tollervey, D. (1995). The 18S rRNA dimethylase Dim1p is required for pre-ribosomal RNA processing in yeast. *Genes Dev.* **9**, 2470–2481.

Landy, A., Abelson, J., Goodman, H. M., and Smith, J. D. (1967). Specific hybridization of tyrosine tRNA with DNA from transducing bacteriophage Phi80 carrying the amber suppressor gene suIII. *J. Mol. Biol.* **29**, 457–471.

Lane, D. J., Field, K. G., Olsen, G. J., and Pace, N. R. (1988). Reverse transcriptase sequencing of ribosomal RNA for phylogenetic analysis. *Methods Enzymol.* **167**, 138–144.

Laten, H. M., Cramer, J. H., and Rownd, R. H. (1983). Thiolated nucleotides in yeast transfer RNA. *Biochem. Biophys. Acta* **741,** 1–6.

Limbach, P. A., Crain, F. C., and McCloskey, J. A. (1994). Summary: The modified nucleosides of RNA. *Nucleic Acids Res.* **12,** 2183–2196.

Lindahl, T. (1967). Irreversible heat inactivation of tRNA. *J. Biol. Chem.* **242,** 1970–1973.

Lindahl, T., and Nyberg, B. (1972). Rate of depurination of native deoxyribonucleic acid. *Biochemistry* **11,** 3610–3618.

Lindahl, T., and Nyberg, B. (1974). Heat-induced deamination of cytosine residues in deoxyribonucleic acid. *Biochemistry* **13,** 3405–3410.

Liu, J., Zhou, W., and Doetsch, P. W. (1995). RNA polymerase bypass at sites of dihydrouracil: Implications for transcriptional mutagenesis. *Mol. Cell. Biol.* **15,** 6729–6735.

Ma, X., Zhao, X., and Yu, Y. T. (2003). Pseudouridylation (Psi) of U2 snRNA in *S. cerevisiae* is catalyzed by an RNA-independent mechanism. *EMBO J.* **22,** 1889–1897.

Maden, B. E. H., Corbett, M. E., Heeney, P. A., Pugh, K., and Ajuh, P. M. (1995). Classical and novel approaches to the detection and localization of the numerous modified nucleotides in eukaryotic ribosomal RNA. *Biochimie* **77,** 22–29.

Marck, C., and Grosjean, H. (2002). TRNomics: Analysis of tRNA genes from 50 genomes of Eukarya, Archaea and Bacteria reveals anticodon-sparing strategies and domain-specific features. *RNA* **8,** 1189–1232.

Melton, D. A., Krieg, P. A., Rebagliati, M. R., Maniatis, T., Zinn, K., and Green, M. R. (1984). Efficient in vitro synthesis of biologically active RNA and DNA hybridization probes from plasmids containing a bacteriophage SP6 promotor. *Nucleic Acids Res.* **12,** 7035–7056.

Monier, R., Stephenson, M. L., and Zamecnick, P. C. (1960). The preparation and some properties of a low molecular weight RNA from Baker's yeast. *Biochem. Biophys. Acta* **43,** 1–8.

Moore, M. J., and Query, C. C. (1998). Uses of site-specifically modified RNAs constructed by RNA ligation. *In* "RNA-Protein Interactions: A Practical Approach" (C. Smith, ed.), pp. 75–108. IRL, Oxford, UK.

Morin, A., Auxilien, S., Senger, B., Tewari, R., and Grosjean, H. (1998). Structural requirements for enzymatic formation of threonylcarbamoyladenosine (t^6A) in tRNA: An *in vivo* study with *Xenopus laevis* oocytes. *RNA* **4,** 24–37.

Mörl, M., Dörner, M., and Pääbo, S. (1994). Direct purification of tRNAs using oligonucleotides coupled to magnetic beads. *In* "Advances in Biomagnetic Separation" (M. Uhlén, E. Hornes, and O. Olsik, eds.), pp. 107–111. Eaton, Natik, MA.

Morse, D. P., and Bass, B. L. (1997). Detection of inosine in messenger RNA by inosine-specific cleavage. *Biochemistry* **36,** 8429–8434.

Motorin, Y., Bec, G., Tewari, R., and Grosjean, H. (1997). Transfer RNA recognition by the *E. coli* isopentenyl-pyrophosphate-tRNA isopentenyl transferase: Dependence on the anticodon arm structure. *RNA* **3,** 721–733.

Motorin, Y., Keith, G., Simon, C., Foiret, D., Simos, G., Hurt, E., and Grosjean, H. (1998). The yeast tRNA: Pseudouridine synthase (Pus1p) displays a multiple substrate specificity. *RNA* **4,** 856–869.

Motorin, Y., and Grosjean, H. (1999). Multisite-specific tRNA:m^5C-methyltransferase (Trm4) in yeast *S. cerevisiae*: Identification of the gene and substrate specificity of the enzyme. *RNA* **5,** 1105–1118.

Nishikura, L., and De Robertis, E. M. (1981). RNA processing in microinjected *Xenopus laevis* oocytes: Sequential addition of base modification in a spliced tRNA. *J. Mol. Biol.* **154,** 405–420.

Nishimura, S. (1979). Chromatographic mobilities of modified nucleotides. *In* "Transfer RNA, Structure, Properties, and Recognition" (P. R. Schimmel, D. Söll, and J. N. Abelson, eds.) pp. 547–552. Cold Spring Harbor Laboratory Press, Cold Spring Harbor, NY.

Nishimura, S., and Kuchino, Y. (1983). Characterization of modified nucleosides in tRNA. *In* "Methods of DNA and RNA Sequencing" (M. S. Weissman, ed.), pp. 235–255. Praeger, New York.

Ofengand, J., Del Campo, M., and Kaya, Y. (2001). Mapping pseudouridines in RNA molecules. *Methods* **25**, 365–373.

Oshima, T., Hamasaki, N., Senshu, M., Kakinuma, K., and Kuwajima, I. (1987). A new naturally occurring polyamine containing a quaternary ammonium nitrogen. *J. Biol. Chem.* **262**, 11979–11981.

Paul, M. S., and Bass, B. L. (1998). Inosine exists in mRNA at tissue-specific levels and is most abundant in brain mRNA. *EMBO J.* **17**, 1120–1127.

Peattie, D. A. (1979). Direct chemical method for sequencing RNA. *Proc. Natl. Acad. Sci. USA* **76**, 1760–1764.

Philippsen, P., Thiebe, R., Wintermeyer, W., and Zachau, H. G. (1968). Splitting tRNA^{Phe} into half molecules by chemical means. *Biochem. Biophys. Res. Comm.* **33**, 922–928.

Pintard, L., Lecointe, F., Bujnicki, J. M., Bonnerot, C., Grosjean, H., and Lapeyre, B. (2002a). Trm7p catalyses the formation of two 2'-O-methylriboses in yeast tRNA anticodon loop. *EMBO J.* **21**, 1811–1820.

Pintard, L. F., Bujnicki, C., Lapeyre, B., and Bonnerot, J. M. (2002b). MRM2 encodes a novel yeast mitochondrial 21S rRNA methyltransferase. *EMBO J.* **21**, 1139–1147.

Purushothaman, S. K., Bujnicki, J. M., Grosjean, H., and Lapeyre, B. (2005). Trm11p and Trm112p are both required for the formation of N^2-methylguanosine at position 10 in yeast tRNA. *Mol. Cell. Biol.* **25**, 43–59.

RajBhandary, U. L. (1980). Recent developments in methods for RNA sequencing using *in vitro* [^{32}P]-labeling. *Fed. Proc.* **39**, 2815–2821.

Roovers, M., Hale, C., Tricot, C., Terrns, M. P., Terns, R. M., Grosjean, H., and Droogmans, L. (2006). Formation of the conserved pseudouridine at position 55 in archaeal tRNA. *Nucleic Acid Res.* **34**, 4293–4301.

Roovers, M., Wouters, J., Bujnicki, J. M., Tricot, C., Stalon, V., Grosjean, H., and Droogmans, L. (2004). A primordial RNA modification enzyme: The case of tRNA (m^{1}A58) methyltransferase. *Nucleic Acids Res.* **32**, 465–476.

Rozenski, J., Crain, P. F., and McCloskey, J. A. (1999). The RNA modification database-1999 update. *Nucleic Acids Res.* **27**, 196–197.

Rozenski, J., *et al.* (1999). Determination of nearest neighbors in nucleic acids by mass spectrometry. *Anal. Chem.* **71**, 454–459.

Sakano, H., Shimura, Y., and Ozeki, H. (1974). Selective modification of nucleosides of tRNA precursors accumulated in a temperature sensitive mutant of *E. coli*. *FEBS Lett.* **48**, 117–121.

Schaefer, K. P., Altman, S., and Söll, D. (1973). Nucleotide modification *in vitro* of the precursor of tRNA of. *E. coli. Proc. Natl. Acad. Sci. USA* **70**, 3626–3630.

Silberklang, M. O., Gillum, A. M., and RajBhandary, U. L. (1979). Use of *in vitro* ^{32}P labeling in the sequence analysis of non-radioactive tRNAs. *Methods Enzymol.* **LIX**, 58–109.

Sprinzl, M., and Vassilenko, M. (2005). Compilation of tRNA sequences and sequences of tRNA genes. *Nucleic Acids Res.* **31**, D130–D140.

Stanley, J., and Vassilenko, S. (1978). A different approach to RNA sequencing. *Nature* **274**, 87–89.

Steinberg, S., and Cedergren, R. (1995). A correlation between *N2*-dimethylguanosine presence and alternative tRNA conformers. *RNA* **1**, 886–891.

Stern, S., Moazed, D., and Noller, H. F. (1998). Structural analysis of RNA using chemical and enzymatic probing monitored by prime extension. *Methods Enzymol.* **164**, 481–489.

Sullivan, M. A., and Bock, R. M. (1985). Isolation and characterization of antisuppressor mutations in *Escherichia coli. J. Bacteriol.* **161**, 377–384.

Suzuki, T., Suzuki, T., Wada, T., Saigo, K., and Watanabe, K. (2002). Taurine as a constituent of mitochondrial tRNAs: New insights into the functions of taurine and human mitochondrial diseases. *EMBO J.* **21,** 6581–6589.

Szekely, M., and Sanger, F. (1969). Use of polynucleotide kinase in fingerprinting non radioactive nucleic acids. *J. Mol. Biol.* **43,** 607–617.

Szweykowska-Kulinska, Z., Senger, B., Keith, G., Fasiolo, F., and Grosjean, H. (1994). Intron-dependent formation of pseudouridines in the anticodon of *S. cerevisiae* tRNA^{Ile}. *EMBO J.* **13,** 4636–4644.

Tanner, K. (1990). Purifying RNA by column chromatography. *Methods Enzymol.* **180,** 25–29.

Terui, Y., Ohnuma, M., Hiraga, K., Kawashima, E., and Oshima, T. (2005). Stabilization of nucleic acids by unusual polyamines produced by an extreme thermophile. *T. thermophilus. Biochem. J.* **388,** 427–433.

Thiebe, R., and Zachau, H. G. (1971). Half-molecules from phenylalanine tRNA's. *Methods Enzymol.* **XX,** 178–182.

Tsurui, H., Kumazawa, Y., Sanokawa, R., Watanabe, Y., Kuroda, T., Wada, A., Watanabe, K., and Shirai, T. (1994). Batchwise purification of specific tRNAs by solid-phase DNA probe. *Anal. Biochem.* **221,** 166–172.

Uhlenbeck, O. C. (1995). Keeping RNA happy. *RNA* **1,** 4–6.

Urbonavicius, J., Armengaud, J., and Grosjean, H. (2006). Identity elements required for enzymatic formation of N2,N2-dimethylguanosine from monomethyl-derivative and its potential role in avoiding alternative conformations in archaeal tRNAs. *J. Mol. Biol.* **357,** 387–399.

Warner, J. R. (1991). Labeling of RNA and phosphoproteins in *S. cerevisiae. Methods Enzymol.* **194,** 423–428.

Watanabe, K. (1980). Reactions of 2-thioribothymidine and 4-thiouridine with hydrogen peroxide in tRNA from *T. thermophilus* and *Escherichia coli* as studied by circular dichroism. *Biochemistry* **19,** 5542–5549.

Wildenauer, D., Gross, H. J., and Riesner, D. (1974). Enzymatic methylations: III. Cadaverine-induced conformational changes of *E. coli* tRNA^{fMet} as evidenced by the availability of a specific adenosine and specific cytidine residue for methylation. *Nucleic Acids Res.* **1,** 1165–1182.

Wilkinson, K. A., Merino, E. J., and Weeks, K. M. (2006). Selective 2′-hydroxyl acylation analyzed by primer extension (SHAPE): Quantitative RNA structure analysis at single nucleotide resolution. *Nat. Protoc.* **1,** 1610–1616.

Wilkinson, M. L., Crary, S. M., Jackman, J. E., Grayhack, E. J., and Phizicky, E. M. (2007). The 2′-O-methyltransferase responsible for modification of yeast tRNA at position 4. *RNA* **13,** xxxx.

Wintermeyer, W., and Zachau, H. G. (1970). A specific chemical chain scission of tRNA at 7-methylguanosine. *FEBS Lett.* **11,** 160–164.

Wintermeyer, W., and Zachau, H. G. (1974). Replacement of odd bases in tRNA by fluorescent dyes. *Methods Enzymol.* **XX,** 667–673.

Wintermeyer, W., and Zachau, H. G. (1975). Tertiary structure interactions of 7-methylguanosine in yeast tRNA-Phe as studied by borohydride reduction. *FEBS Lett.* **58,** 306–309.

Wittig, B., and Wittig, S. (1978). Reverse transcription of tRNA. *Nucleic Acids Res.* **5,** 1165–1178.

Xue, H., Glasser, A. L., Desgres, J., and Grosjean, H. (1993). Modified nucleotides in *Bacillus subtilis* tRNA^{Trp} hyperexpressed in a *E. coli. Nucleic Acids Res.* **21,** 2479–2486.

Youvan, D. C., and Hearst, J. E. (1979). Reverse transcriptase pauses at N2-methylguanine during *in vitro* transcription of *E. coli* 16S ribosomal RNA. *Proc. Natl. Acad. Sci. USA* **76,** 3751–3754.

Youvan, D. C., and Hearst, J. E. (1981). A sequence from *Drosophila melanogaster* 18S rRNA bearing the conserved hypermodified nucleoside amPsi: Analysis by reverse transcription and high performance liquid chromatography. *Nucleic Acids Res.* **9,** 1723–1741.

Yu, Y. T., and Steitz, J. A. (1997). A new strategy for introducing photoactivable 4-thiouridine (S^4U) into specific positions in a long RNA molecule. *RNA* **3,** 807–810.

Zeevi, M., and Daniel, V. (1976). Aminoacylation and nucleoside modification of *in vitro* synthesized tRNA. *Nature* **260,** 72–74.

Zhao, X., and Yu, Y.-T. (2004). Detection and quantitation of RNA base modifications. *RNA* **10,** 996–1002.

Zimmermann, R. A., Gait, M. J., and Moore, M. J. (1998). Incorporation of modified nucleotides into RNA for studies on RNA structure, function and intermolecular interactions. *In* "Modification and Editing of RNA" (H. Grosjean and R. Benne, eds.), pp. 59–84. ASM Press, Washington, DC.

CHAPTER FOUR

In Vitro Detection of the Enzymatic Activity of Folate-Dependent tRNA (Uracil-54,-C5)-Methyltransferase: Evolutionary Implications

Jaunius Urbonavicius,* Céline Brochier-Armanet,[†]
Stéphane Skouloubris,[‡] Hannu Myllykallio,[‡] and Henri Grosjean[§]

Contents

Abstract

Formation of 5-methyluridine (ribothymidine) at position 54 of the T-ψ loop of tRNA is catalyzed by site-specific tRNA methyltransferases (tRNA[uracil-54,C5]-MTases). In eukaryotes and many bacteria, the methyl donor for this reaction is generally S-adenosyl-L-methionine (S-AdoMet). However, in other bacteria, like

* Laboratoire d'Enzymologie et Biochimie Structurales, Gif-sur-Yvette, France
† Laboratoire de Chimie Bactérienne, Marseille, and Université de Provence, Aix-Marseille I, France
‡ Labortoire de Génomique et Physiologie Microbienne, Université Paris-Sud, Orsay, France
§ Institut de Génétique et Microbiologie, Université Paris-Sud, Orsay, France

Methods in Enzymology, Volume 425
ISSN 0076-6879, DOI: 10.1016/S0076-6879(07)25004-9

Enterococcus faecalis and *Bacillus subtilis*, it was shown that the source of carbon is N^5,N^{10}-methylenetetrahydrofolate ($CH_2=THF$). Recently we have determined that the *Bacillus subtilis gid* gene (later renamed to *trmFO*) encodes the folate-dependent tRNA(uracil-54,C5)-MTase. Here, we describe a procedure for overexpression and purification of this recombinant enzyme, as well as detection of its activity *in vitro*. Inspection of presently available sequenced genomes reveals that *trmFO* gene is present in most Firmicutes, in all α- and δ-Proteobacteria (except Rickettsiales in which the *trmFO* gene is missing), Deinococci, Cyanobacteria, Fusobacteria, Thermotogales, Acidobacteria, and in one Actinobacterium. Interestingly, *trmFO* is never found in genomes containing the gene *trmA* coding for *S*-adenosyl-L-methionine–dependent tRNA (uracil-54,C5)-MTase. The phylogenetic analysis of TrmFO sequences suggests an ancient origin of this enzyme in bacteria.

1. INTRODUCTION

Transfer RNAs (tRNAs) in all cells contain numerous nucleosides that are post-transcriptionally modified (Limbach *et al.*, 1994). One such common modified nucleoside is 5-methyluridine (m^5U or rT for ribothymidine). This C^5-methylated uracil is invariably found at position 54 in T-ψ loop of tRNA of many bacteria and almost all Eukarya (Sprinzl and Vassilenko, 2005). In thermophilic bacteria, such as *Thermus thermophilus* and *Bacillus stearothermophilus* or hyperthermophilic Archaea of the *Thermococcales* order, this m^5U is further modified to a 2-thio-derivative (m^5s^2U or s^2T; Edmonds *et al.*, 1991; Kowalak *et al.*, 1994; Shigi *et al.*, 2002; Watanabe *et al.*, 1976), whereas in tRNAs of metazoan cells, a 2′-O-methyl-derivative is often found (m^5Um). In Archaea, except the *Thermococcales* order (see preceding), other types of U-54 modification are found, such as $m^1ψ$ and Um (see in http://www.uni-bayreuth.de/departments/biochemie/trna/).

Site-specific methylation of uracil-54 in *Escherichia coli* tRNA is catalyzed by tRNA(uracil-54,C5)-methyltransferase (EC.2.1.1.35, Fig. 4.1). This enzyme, initially designated RUMT for RNA uridine methyltransferase, is the first RNA modification enzyme discovered that acts at the polynucleotide level (Fleissner and Borek, 1962; Svensson *et al.*, 1963). This enzyme is also called TrmA (tRNA methyltransferase A), and the corresponding coding gene (*trmA*) was only later identified in *E. coli* (Björk, 1975; Ny and Björk, 1980). In yeast *Saccharomyces cerevisiae*, *S*-AdoMet–dependent Trm2p and the corresponding *TRM2* gene were also identified and characterized (Nordlund *et al.*, 2000), whereas the gene coding for tRNA(uracil-54,C5)-MTase of *Pyrococcus abyssi* was only very recently identified (Urbonavicius *et al.*, submitted).

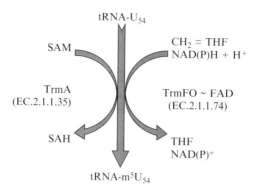

Figure 4.1 Formation of the m^5U_{54} in tRNA is catalyzed either by the SAM-dependent TrmA protein or by the folate-dependent TrmFO flavoprotein.

The TrmA/Trm2p enzymes, as well as most RNA methyltransferases studied so far, use S-adenosyl-L-methionine (S-AdoMet) as the methyl donor (reviewed in Bujnicki *et al.*, 2004). However, previous studies have demonstrated that not all bacterial tRNA(uracil-54,C5)-MTases use S-AdoMet as the methyl donor. For example, in *Enterococcus faecalis* (formerly *Streptococcus faecalis*) and *Bacillus subtilis,* the carbon donor of the methyl group of uracil-54 in tRNA was shown to be N^5, N^{10}-methylenetetrahydrofolate (CH_2=THF; Delk *et al.*, 1980, and references herein). It also requires reduced flavin adenine nucleotide ($FADH_2$, Delk *et al.*, 1979a), thus forming a distinct class of tRNA(uracil-54,C5)-MTases (EC.2.1.1.74; Fig. 4.1). The MTase from *E. faecalis* was purified to almost homogeneity, its molecular weight evaluated from gel electrophoresis, and some biochemical characterization was performed (Delk *et al.*, 1979b), but the corresponding gene still remained to be identified. Benefiting from large-scale microbial sequencing and structural genomics projects, we predicted that the *B. subtilis gid* gene encodes the folate-dependent tRNA(uracil-54,C5)-MTase. This prediction was confirmed through genetic studies and biochemical analyses, and the *gid* gene was renamed *trmFO* (FO for folate, Urbonavicius *et al.*, 2005). Despite their functional similarity, TrmFO (alias Gid) and TrmA/TRM2 are not homologous (i.e., they do not have a common evolutionary origin) and catalyze two distinct methyl transfer reactions. Here we describe a method for purification of recombinant protein and detection of the enzymatic activity of the *B. subtilis* TrmFO protein. We also identify homologs of TrmFO in complete sequenced genomes and discuss their possible evolutionary origin.

2. Overproduction and Purification of *B. subtilis* tRNA (Uracil-54,-C5)-Methyltransferase

2.1. Expression plasmids and strains

The *B. subtilis trmFO* gene was cloned into the pQE80L vector (purchased from Qiagen, cat. No. 32923) to give pQE80L-$_{Bsu}$Gid (Urbonavicius *et al.*, 2005). The production of N-terminal His$_6$-tagged $_{Bsu}$TrmFO protein is under tight control of the powerful phage T5 promoter (recognized by *E. coli* RNA polymerase) and the *lac* operator. Any *E. coli* host strain can be used for the overexpression of this enzyme. *E. coli* Sure® strain (e14⁻[McrA]⁻ Δ[*mcrCB-hsdSMR-mrr*]*171 endA1 supE44 thi-1 gyrA96 relA1 lac recB recJ sbcC umuC::*Tn5 [Kan^r] *uvrC* [F′ *proAB lacI^{q}ZΔM15* Tn*10* [Tet^r]); Stratagene), was used as a host for cloning and overexpressing the *B. subtilis gid trmFO* gene. We also used the *E. coli* GRB113 strain ([*metA, trmA5, zij-90::* Tn*10*], lacking SAM-dependent tRNA(uracil-54,C5)-MTase activity, provided by G. R. Björk, Umeå University, Sweden). We did not see any differences in the activity of the $_{Bsu}$tRNA(uracil-54,C5)-MTase isolated from these two different strains.

2.2. Gene expression

Strains Sure® or GRB113 are transformed with pQE80L-$_{Bsu}$Gid plasmid by use of a standard CaCl$_2$ procedure. Transformants are selected on Luria-Bertani (LB) medium (obtained from Invitrogen) containing 1.5–2% agar and 100 μl/ml ampicillin (Ap, Sigma, cat. no. A0797) or carbenicillin (Cb, Sigma, cat. no. C3416) and purified by restreaking on plates containing the same medium. A single colony is inoculated into 10 ml of LB + Ap (or Cb) and grown overnight at 37°. The preculture is then inoculated (dilution 1:100) into 500 ml of LB + Ap (Cb) and grown until the OD$_{600}$ reaches approximately 0.6. Induction of protein expression is performed by the addition of isopropyl β-D-thiogalactopyranoside (IPTG, purchased from VWR International, cat. No. 03–36–0003–5) to a final concentration of 1 mM. The cultures are grown further for 3 h at 37°, then harvested by centrifugation (4500g for 15 min at 4°), flash-frozen in liquid N$_2$, and stored at −80°.

2.3. Purification of the recombinant enzyme

For enzyme purification, frozen cells are thawed on ice and resuspended in 5 ml of lysis buffer (50 mM sodium phosphate, pH 7.6, 300 mM NaCl, 10% glycerol, and 20 mM imidazole) containing 5 μl Protein Inhibitor Cocktail

(PIC, Sigma, cat. no. P8849) and 1.5 μl β-mercaptoethanol. Cells are broken by two freeze (liquid N_2)/thaw (37°) cycles and ultrasonication (Branson Sonifier S-450A; four bursts of 30 sec performed on ice at 30 W and at a 50% cycle). The lysate is centrifuged for 15 min at 10,000g at $+4°$. The supernatant is transferred to a new tube and then loaded in the cold room onto a small column containing 2 ml of Ni-NTA resin (Qiagen, cat. no. 30410). The resin is washed with 25 ml of cold lysis buffer. The column turns from light blue (normal color of the resin) to light green because of the binding of the yellow TrmFO protein containing tightly bound oxidized flavin cofactor. The enzyme is then eluted with 10 ml of the elution buffer (same as the lysis buffer above, but containing 250 mM imidazole). Fractions, containing the TrmFO protein, are pooled, concentrated with Amicon Ultra-15 devices (Millipore; cat. no. UFC 901024), and dialyzed overnight at 4° (Pierce; Slide-A-Lyzer, cat. no. 66383), against 500 ml of 30 mM HEPES buffer, pH 7.5, containing 200 mM NaCl and 10% glycerol (*NB*: NaCl is absolutely required since we have observed that the enzyme precipitates in its absence; concentrated solution of enzyme is yellowish).

The dialyzed protein is aliquoted into 0.1-ml fractions, flash-frozen in liquid N_2, and stored in hermetically closed Eppendorf tubes at $-80°$ until the enzymatic activity is tested. The aliquot to be used for the determination of the enzymatic activity is usually diluted with glycerol to a final concentration of 50%, stored at $-20°$, and is used as soon as possible, maximum within 2 weeks. Protein concentration is determined with a commercial version of the Bradford assay (Bio-Rad; cat no. 500–0006) that uses bovine serum albumin as a standard. One aliquot of protein ($\sim$5 μg) was heated for 5 min at 90°. The sample was then centrifuged for 15 min at 10,000g. The fluorescence spectrum of the resulting supernatant revealed an emission peak at 520 nm (after excitation at 450 nm) characteristic of flavin nucleotides (Urbonavicius *et al.*, 2005). The yield is approximately 3 mg from 0.5 liter of the growth culture. The purity of the enzyme is >99% as estimated by SDS-PAGE and Coomassie Blue staining.

3. Enzymatic Activity Assay

3.1. Reaction mix

The reaction catalyzed by *B. subtilis* tRNA(uracil-54,C5)-MTase is shown in Fig. 4.2. During the reaction, the methylene group is transferred from CH_2=THF onto the C5 atom of uracil-54 in tRNA. At the same time, the methylene group is reduced to a methyl group by $FADH_2$ that is obtained after the hydride transfer from NADH and/or NADPH. As the substrate, we have used a [α-^{32}P]UTP-labeled yeast tRNA^{Asp} transcript. Its preparation and purification are described elsewhere (Perret *et al.*, 1990; see also

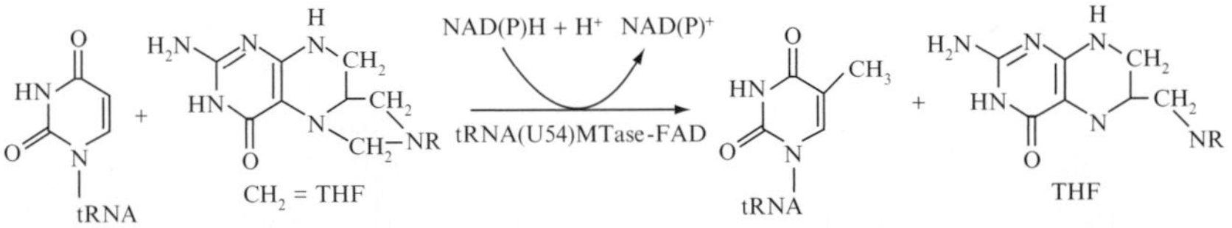

Figure 4.2 Reaction leading to m^5U_{54} formation in tRNA catalyzed by folate-dependent tRNA(uracil-54,C5)-methyltransferase. R means : para-aminobenzoate-glutamate.

Grosjean *et al.*, 2007, this volume). It has been used successfully for identifying the activity of many tRNA modification enzymes *in vitro* and *in vivo* (see for examples: Auxilien *et al.*, 1996; Constantinesco *et al.*, 1999; Grosjean *et al.*, 1996). However, any tRNA transcript containing an intact unmodified T-arm could be used. All the radioactive manipulations are performed with appropriate shielding against β-radiation caused by the radioactive [^{32}P].

The enzyme incubation mix contains in 50 μl (final volume, Eppendorf tubes): 50–100 fmol of radioactive tRNA (5000–10,000 counts per minute, as estimated with a Geiger counter) in 40 mM N-[2-hydroxyethyl] piperazine-N-[2-ethanesulfonic acid]-Na buffer (HEPES-Na, Sigma, cat. no. H8651) at pH 7.0, containing 100 mM ammonium acetate (from a 1 M stock solution), 5% glycerol, 0.25 mM flavin adenine dinucleotide (FAD, disodium salt hydrate, obtained from Sigma, cat. no. F6625, stock solution at 10 mM), 0.5 mM reduced β-nicotinamide adenine dinucleotide (NADH, disodium salt, Sigma, cat. no. N0786, stock solution at 20 mM), 1 mM reduced β-nicotinamide adenine dinucleotide phosphate (NADPH, tetrasodium salt, Sigma, cat. no. N0411, stock solution at 40 mM), 0.25 mM (6R)-N^5,N^{10}-CH$_2$H$_4$PteGlu-Na$_2$ (CH$_2$=THF, provided by Dr. R. Moser, Merck-Eprova AG, Schaffhausen, Switzerland), 5 mM DL-dithiothreitol (DTT, Sigma, cat no. D9779), and 5 μg bovine serum albumin (for molecular biology, RNase-free grade from Roche, cat. no. 711454, stock solution at 1 mg/ml). All cofactor solutions are freshly prepared just before use, especially solutions of CH$_2$=THF, NADH, and NADPH (oxidized NAD$^+$/NADP$^+$ are inhibitory to the methylation reaction) and protected from light. The enzymatic reaction is started by addition of 1 μg (approximately 20 pmol) of the purified enzyme. A blank without enzyme is always included in a series of experiments. The reaction mixture is incubated at 37° for up to 20 min. At the end of the incubation period, 200 μl of 0.3 M Na acetate is added to each tube, and the modified radiolabeled tRNA is extracted with an equal volume (250 μl) of water-saturated phenol/chloroform/ isoamyl alcohol (25:24:1). The tRNA is then precipitated in the cold with pure ethanol and washed twice with 70% ethanol/water (v/v). Ethanol is carefully removed from each tube with a micropipette, and the pellets are dried by incubating the open tubes at 40° for 5–15 min.

3.2. Detection of m⁵U in tRNA

Transfer RNA is completely digested into 5′-monophosphate nucleotides by incubation at 37° overnight with 0.2–0.4 μg/test of nuclease P1 of *Penicillium citrinum* (MP Biomedicals, cat. no. 195352; or Roche, cat. no. 236225) in 10 μl of 50 mM sodium acetate buffer, pH 5.3. Samples are briefly centrifuged and loaded by 0.5-μl portions onto the cellulose plates used for thin layer chromatography (TLC). Usually we use the POLYGRAM-CEL 300–10 (plastic coated, 20 × 20 cm, 0.1-mm thick) purchased from the Macherey-Nagel (cat. No. 808–013) that we cut into 10 × 10 cm sheets for faster analyses (for details see Grosjean *et al.*, in this volume). Thin-layer cellulose plates from Merck (cat. no. 105730–001) are also satisfactory, but the rate of solvent migrations is slower than with the plates from Macherey-Nagel. TLC plates (10 × 10 cm) are run in the chromatography tanks filled by the solvent A: isobutyric acid/concentrated ammonia/water (66:1:33[v:v:v]) for approximately 1.5–2 h. Plates are dried in the hood for approximately 1 h and run in the second direction in buffer B: 100 mM Na-phosphate buffer, pH 6.8/(NH$_4$)$_2$ SO$_4$/n-propanol (100:60:2[v:w:v]), also for approximately 1.5–2 h. Plates are dried again in the hood for 30–60 min and are put into a cassette containing a PhosphoImager screen to be exposed overnight. The next day, TLC plates are scanned with the PhosphoImager screen scanner (Molecular Dynamics) and quantified with the Imagequant program. Figure 4.3 shows typical results with

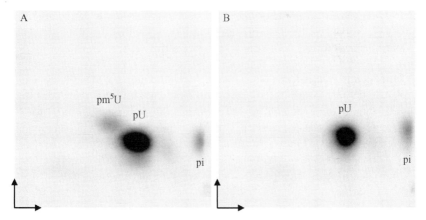

Figure 4.3 Formation of the m⁵U catalyzed by folate-dependent tRNA(uracil-54, C5)-methyltransferase. T7 polymerase transcripts of yeast wild-type tRNA^{Asp}, uniformly labeled with [α-^{32}P]UTP, were used as substrate and incubated for 20 min at 37° with purified recombinant *B. subtilis* TrmFO protein. After the incubation, tRNA was completely digested to monophosphate nucleosides by nuclease P1 and analyzed by 2D-TLC. Radiolabeled compounds were detected and quantified by use of a PhosphoImager detector. (A) In the presence of enzyme; (B) in absence of enzyme (blank). More such data can be found in Urbonavicius *et al.* (2005).

[α-^{32}P]–UTP radiolabeled T7-transcript of yeast tRNA^{Asp} incubated with purified recombinant *B. subtilis* TrmFO (Fig. 4.3A) or in absence of the enzyme (blank, Fig. 4.3B). Typically, yields of 0.3–0.4 mol/mol of tRNA are obtained after 20 min incubation at 37°. Longer incubation is difficult because of oxidation and instability of the substrates in solution. Tests were performed mostly for qualitative evaluation of the reaction products rather than for detailed kinetic analysis of the reaction. No systematic work has been done for optimization of the assay conditions. We have noticed that some enzymatic activity was detected even in the absence of the methyle-netetrahydrofolate, suggesting that some of the folate cofactor copurifies with the TrmFO protein.

4. Phylogenetic Analysis

We have analyzed all TrmFO homologs from 355 completely sequenced archaeal and bacterial genomes retrieved from the NCBI in August 2006 (ftp.ncbi.nih.gov). The "hmmer" package was used to identify known functional domains in the Pfam database and to retrieve the protein sequences from complete genomes (Bateman *et al.*, 2004; Finn *et al.*, 2006). Alignments with the Pfam "hmm" profile domains having an E–value <0.1 were considered significant. The Pfam "hmm" profile PF01134.12 (GIDA, Glucose inhibited division protein A) was identified in TrmFO sequences from *Enterococcus faecalis* (NP_815357) and *Bacillus subtilis* (NP_389495). Next, we retrieved all the protein sequences from complete genomes containing this GIDA domain (i.e., 458 sequences). All the sequences were aligned by ClustalW (Thompson *et al.*, 1994), and the resulting alignment was manually refined using *"ed"* from the "MUST" package (Philippe, 1993). Regions where homology was doubtful were removed from further analysis. A total of 217 amino acid positions were kept for the phylogenetic analysis. Maximum likelihood (ML) phylogenetic tree was inferred by use of PHYML (Guindon and Gascuel, 2003) with the JTT model, an estimated proportion of invariant sites and a gamma correction to take into account the heterogeneity of the evolutionary rates between sites (four categories and an estimated alpha parameter). The robustness of each branch of the ML tree was estimated by nonparametric bootstrap analysis (100 replicates of the original data set) by use of PHYML. The resulting phylogeny shows two distinct well-supported groups corresponding to so-called GidA (304 sequences) and TrmFO (in fact Gid, 131 sequences) sequences, whereas 23 diverging sequences (TrmFO-like/GidA-like sequences, see discussion following) form separated clusters (not shown, but available on http://www.frangun.org/). For further analyses, only the 131 sequences identified as probable TrmFO sequences were considered.

A careful examination of their distribution in available complete genomes shows that they are absent in archaebacteria and eukaryotes but can be found at least in one copy in bacterial genomes from several phyla: in most Firmicutes, in all α-Proteobacteria (except Rickettsiales in which the *trmFO* gene is missing), δ-Proteobacteria, Deinococci, Cyanobacteria, Fusobacteria, Thermotogales, Acidobacteria, and in one Actinobacterium (Table 4.1, detailed table available at http://www.frangun.org/publications. html). Interestingly, *trmFO* are never found in bacterial genomes containing *trmA*. Among the 131 homologous sequences retrieved, 3 were very partial and were not considered in the further phylogenetic analysis. A phylogenetic tree including a subset of 86 TrmFO sequences most representative of the 128 remaining TrmFO sequences was constructed as described previously (Fig. 4.4, a more complete phylogenetic tree including 128 TrmFO, alias Gid sequences is available at http://www.frangun.org/publications.html (Supplementary materials S2 and S3 of the article "Urbonavicius *et al.*, 2007").

5. DISCUSSION

The enzymatic activity of a bacterial folate-dependent tRNA(uracil-54,C5)-MTase was discovered and the corresponding protein of *E. faecalis* was purified approximately 3 decades ago (Delk *et al.*, 1979a). However, no corresponding gene was detected, thus hindering the production of the corresponding protein in large amounts for further characterization. The recent discovery of the gene family (*trmFO*, formerly *gid*) encoding the protein TrmFO allowed the cloning of the *B. subtilis* corresponding gene, its overexpression, and the easy purification of its product (i.e., TrmFO protein) harboring a site-specific methyltransferase activity at C5 of uracil-54 in tRNA (Urbonavicius *et al.*, 2005).

The TrmFO proteins (also called "small GidA," see White *et al.*, 2001) all have a molecular weight of approximately 50 kDa and belong to the cluster of ortholog genes (Tatusov *et al.*, 2003) COG 1206. They are often confused with the longer proteins of the GidA family (glucose-inhibited protein A, belonging to COG 0445), with molecular weight of approximately 70 kDa. These latter proteins are also folate-dependent enzymes that are involved in the enzymatic formation of $cmnm^5U34$ in the anticodon loop of tRNA (Yim *et al.*, 2006, and references therein). Indeed, both TrmFO and GidA share a homologous functional domain called GIDA (Bateman *et al.*, 2004, also see preceding) and belong to the same protein superfamily. However, TrmFO and GidA display significant differences at the sequence level and in size (435 and 628 amino acids for TrmFO and GidA of *B. subtilis*, respectively). Our studies, therefore, reveal that TrmFO and GidA form two distinct families of proteins that probably evolved from

Table 4.1 Taxonomic distribution of the *trmA* and *trmFO* (alias *gid*) genes coding for putative tRNA(uracil-54, C5)-methyltransferases in 355 bacterial and archaeal complete genomes

Genomes	TrmA	TrmFO	Genomes	TrmA	TrmFO
Bacteria			*Bacteria*		
Actinobacteria (22)			*Betaproteobacteria*		
Bifidobacterium longum			*Burkholderiales* (15)		
Actinomycetales (20)			*Thiobacillus denitrificans_ATCC_25259*		
Rubrobacter xylanophilus_DSM_9941		★	*Methylobacillus flagellatus_KT*	★	
Aquificae – Aquifex aeolicus		★	*Chromobacterium violaceum*	★	★
Bacteroidetes (8)			*Neisseria* (3)		
PVC group (12)			*Nitrosomonadales* (2)		
Chloroflexi (2)		★	*Rhodocyclales* (2)	★	
Cyanobacteria (17)		★	*Deltaproteobacteria* (11)		
Deinococci (4)			*Epsilonproteobacteria*		
Firmicutes			*Campylobacter jejuni* (2)	★	
Symbiobacterium thermophilum_IAM14863		★	*Helicobacter acinonychis_Sheeba*	★	
Bacillus (12)		★	*Helicobacter hepaticus*	★	
Geobacillus kaustophilus_HTA426			*Helicobacter pylori* (3)	★	
Oceanobacillus iheyensis			*Thiomicrospira denitrificans_ATCC_33889*		
Listeria (3)		★	*Wolinella succinogenes*	★	
Staphylococcus (13)		★	*Gammaproteobacteria*		
Lactobacillales (27)'		★	*Alteromonadales* (7)		
Clostridium (3)		★	*Nitrosococcus oceani_ATCC_19707*		
Peptococcaceae (2)			*Enterobacteriaceae* (30) '''	★	
'*Thermoanaerobacter tengcongensis*			*Legionellales* (4)		

Moorella thermoacetica_ATCC_39073 ★

Mycoplasmatales (16) – see legend below ★

Fusobacteria – Fusobacterium nucleatum ★

Alphaproteobacteria

Caulobacter crescentus ★

Rhizobiales (18) " ★

Jannaschia CCS1 ★

Rhodobacter sphaeroides_2_4_1 ★

Silicibacter (2) ★

Gluconobacter oxydans_621H ★

Magnetospirillum magneticum_AMB_1 ★

Rhodospirillum rubrum_ATCC_11170 ★

Rickettsiales (16)

Sphingomonadales (4)

Methylococcus capsulatus_Bath ★

Helella dejuensis_KcTC_2396 ★

Pasteurellales (5) ★

Acinetobacter sp_ADP1 ★ ★

Psychrobacter (2) ★

Pseudomonas (8) ★

Francisella tularensis (2) ★

Thiomicrospira crunogena_XCL_2

Baumannia cicadelliniicola_Homalodisca_c

Vibrionales (6)

Xanthomonadales (8)

Spirochaetes (6)

Thermotogae – Thermotoga maritima ★

Acidobacteria – Acidobacteria bacterium_Ellin345 ★

Numbers in brackets indicate the number of complete genomes available for each lineage. Asterisks indicate the presence of putative gene(s) coding for either trmA or trmFO detected in each lineage. Curiously, two putative trmFO genes are found in Mycoplasma capricolum_ATCC_27343, in Mycoplasma mycoides, in Mesoplasma florum_L1, and in Lactobacillus sakey_23K('). Moreover, no trmFO homolog is found in the complete genome of Bartonella quintana_str. toulouse('') whereas its close relative Bartonella henselae str. Houston-1 contains it. Finally, we did not detect trmA homologue in the genomes of three Buchnera aphidicola, of Wigglesworthia brevipalpis and of the two Candidatus blochmania(''). Notice that in genome of many organisms no gene coding for TrmA nor TrmFO are found. However, except for M. capricolum and M. mycoides, no tRNA sequences of these organisms are available, so that we do not know whether m5U at position 54 is present or not in tRNAs of these organisms. More information is available at http://www.frangun.org/publications.html (Supplementary materials S1 of the paper "Urbonavicius et al., 2007").

113

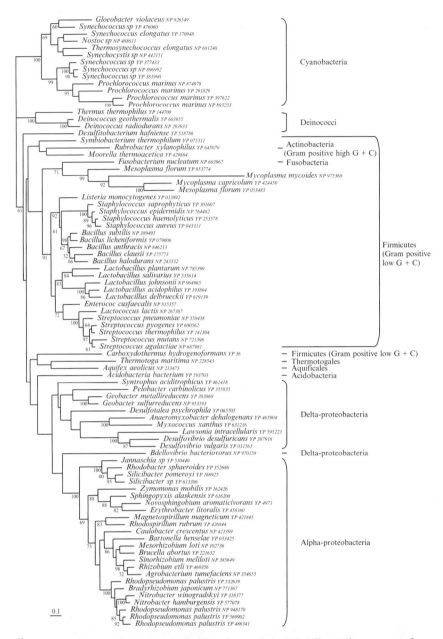

Figure 4.4 Unrooted maximum likelihood tree of TrmFO (alias Gid) sequences from complete genomes (86 sequences, 370 amino acid positions). Numbers at nodes indicate bootstrap values (BV) for 100 replicates of the original data set. For clarity only BV >60% are indicated. The scale bar represents the average number of substitutions per site. More detailed information is available at http://www.frangun.org/publications. html (Supplementary materials S2 and S3 of the article by Urbonavicius *et al.* [2007]).

a common ancestor after a duplication event but acquired different, non-overlapping cellular functions during evolution (Urbonavicius *et al.*, 2005). We have hypothesized that GidA proteins catalyze the first step of $cmnm^5U_{34}$ biosynthesis by introducing the formyl- or methyl-group into U at position 34 of tRNA anticodon. However, we have not detected f^5U_{34} or m^5U_{34} with purified recombinant $_{Bsu}$GidA and several *B. subtilis in vitro*-generated tRNA transcripts as substrates (Urbonavicius, J., Skouloubris, S., Myllykallio, H., and Grosjean, H., unpublished). This suggests that *in vitro*, GidA may be active only together with another protein of the $cmnm^5U$ biosynthetic pathway, most probably MnmE that, indeed, binds strongly to GidA, or that the biosynthetic pathway leading to $cmnm^5U_{34}$ could be different from the one we proposed (see alternative hypothesis in Yim *et al.*, 2006).

The crystal structure of a protein considered to be the "small form of glucose-inhibited protein A" from *Thermus thermophilus* HB8 (accession No. YP_145163 at NCBI; 2CUL at PDB) was recently solved (Iwasaki *et al.*, 2005). However, this protein is shorter (232 aa, molecular weight of approximately 26 kDa) than other TrmFO proteins and displays only 32% and 38% of identity with TrmFO and GidA sequences *of B. subtilis*, respectively. Also, it does not emerge within the TrmFO/GidA family (Fig. 4.4 and http://www.frangun.org/publications.html. Supplementary materials S3 of the article "Urbonavicius *et al.*, 2007"). Finally, its enzymatic activity was not demonstrated *in vitro* (Iwasaki *et al.*, 2005). We suggest that the crystallized protein corresponds not to TrmFO, but to one of the 23 Gid-like/GidA-like proteins that we found distantly related to TrmFO and GidA (see previously), the function of which has still to be determined. Indeed, in addition to the gene coding for a protein YP_145163, the genome of *T. thermophilus* HB8 contains another gene coding for a protein (YP_144708, 443aa, MW approximately 49 kDa) that shows 50% of identity with TrmFO of *B. subtilis*. The sequence of this YP_144708 protein clusters nicely with the other TrmFO sequences in our phylogenetic trees (Fig. 4.4); therefore, it probably has the TrmFO activity for methylating C5 of uridine-54 in *T. thermophilus* tRNA. On the other hand, GidA of *T. thermophilus* HB8 is probably YP_145238 (597 aa, MW 65 kDa) that shares 45% of identity with GidA of *B. subtilis* and belongs within the GidA cluster. Therefore, it is possibly involved in the $cmnm^5U$-34 biosynthesis.

As far as the chemical reaction is concerned (see Fig. 4.2), TrmFO and the FAD-dependent thymidylate synthases of ThyX family (EC.2.1.1.148; Myllykallio *et al.*, 2002) catalyze a very similar, if not identical, type of methylating reaction. This observation raised the possibility that TrmFO and ThyX enzymes methylating very different types of substrates (tRNA and dUMP respectively) could be evolutionary related. However, amino acid sequence comparisons of the two enzymes have clearly indicated that they originate from completely different families of flavoproteins

(Urbonavicius *et al.*, 2005). Note also that a canonical thymidylate synthase ThyA (EC.2.1.1.45) uses yet another way of forming 5-methyluridine in dUMP. In the ThyA reaction, $CH_2=THF$ functions both as the carbon donor and reductant, resulting in production of dihydrofolate during catalysis (Carreras and Santi, 1995). In contrast, ThyX proteins use reduced flavin nucleotides to reduce the methylene group and directly form tetrahydrofolate (Graziani *et al.*, 2006; Myllykallio *et al.*, 2002). Thus, during evolution, enzymes catalyzing the folate-dependent methylation of C5 in uridine of either tRNA or dUMP have been "invented" at least three times.

Surprisingly, the taxonomic distribution of the folate-dependent tRNA (uracil-54,C5)-MTase (TrmFO) seems to be much wider than originally anticipated. Orthologs of the *trmFO* gene are found in most Firmicutes, δ- and α-Proteobacteria, Cyanobacteria, Deinococci, and in representatives of other bacterial phyla (Fusobacteria, Acidobacteria, Thermotogales, and Aquificales [Table 4.1]). It is totally absent in the genomes of Eukarya and Archaea sequenced so far. The ML tree of the putative TrmFO protein sequences is congruent with the bacterial phylogenies based on the ribosomal RNA or on multiple molecular markers (Brochier *et al.*, 2002; Daubin, *et al.*, 2002). In particular, the monophyly of most bacterial phyla is recovered and well supported by high Bootstrap Values (BV) (Cyanobacteria BV = 100%; Deinococci BV = 100%; α-Proteobacteria BV = 100%), whereas most δ-Proteobacteria and most Firmicutes form monophyletic groups (Fig. 4.4). This suggests that TrmFO is a very ancient enzyme, because it was probably present at least in the ancestor of these groups and that it was mainly vertically transmitted during evolution of these bacterial phyla. Moreover, the folate-dependent TrmFO proteins (COG1206) and *S*-AdoMet-dependent TrmA/Trm2p enzymes (COG2265) both acting on tRNA seem to have mutually exclusive taxonomic distributions (Table 4.1). Finally, folate-dependent methylation seems to be restricted to the U_{54} in tRNA and does not occur at few uridines in rRNA as in some cases of the *S*-AdoMet-dependent rRNA(uracil,C5) biosynthetic pathway (Agarwalla *et al.*, 2002; Madsen *et al.*, 2003). Therefore, genes belonging in the TrmFO cluster (Fig. 4.4 and see also http://frangun.org/publications. html) are candidates to encode the folate-dependent formation of the m^5U-54 in exclusively tRNA that should now be studied in more detail to understand the structure and function of this particular methylation enzyme.

ACKNOWLEDGMENTS

We thank Jonatha M. Gott for advice and improvements on the manuscript. The parts of the works of J. U. and H. G. reported in this chapter were supported by research grants from the Centre National de la Recherche Scientifique (CNRS); the Ministère de l'Education

Nationale, la Recherche Scientifique et de la Technologie (Programme Interdépartemental de Géomicrobiologie des Environnements Extrêmes). Only recently it has been supported from funds of University of Orsay to Prof. J. P. Rousset (IGM. Bat 400, Orsay, France), where H. G. has a position of Emeritus scientist. J. U. was the recipient of a FEBS long-term fellowship. He also thanks Prof. G. R. Björk, Department of Molecular Biology, Umeå University, Umeå, Sweden, for support. H. M. and S. S. are supported by CNRS (Programme Microbiologie Fondamentale) and Fondation Bettencourt-Schueller INSERM AVENIR program.

REFERENCES

Agarwalla, S., Kealey, J.T, Santi, D. V., and Stroud, R. M. (2002). Characterization of the 23 S Ribosomal RNA m^5U1939 Methyltransferase from *Escherichia coli. J. Biol. Chem.* **277,** 8835–8840.

Auxilien, S., Crain, P. F., Trewyn, R. W., and Grosjean, H. (1996). Mechanism, specificity and general properties of the yeast enzyme catalyzing the formation of inosine-34 in the anticodon of tRNA. *J. Mol. Biol.* **262,** 437–458.

Bateman, A., Coin, L., Durbin, R., Finn, R. D., Hollich, V., Griffiths-Jones, S., Khanna, A., Marshall, M., Moxon, S., Sonnhammer, E. L. L., Studholme, D. J., Yeats, C., and Eddy, S. R. (2004). The Pfam protein families database. *Nucleic Acids Res.* **32,** D138–D141.

Björk, G. R. (1975). Transductional mapping of gene *trmA* responsible for the production of 5-methyluridine in transfer ribonucleic acid of *Escherichia coli. J. Bacteriol.* **124,** 92–98.

Brochier, C., Bapteste, E., Moreira, D., and Philippe, H. (2002). Eubacterial phylogeny based on translational apparatus proteins. *Trends Genet.* **18,** 1–5.

Bujnicki, J. M., Droogmans, L., Grosjean, H., Purushothaman, S. K., and Lapeyre, B. (2004). Bioinformatics-guided identification and experimental characterization of novel RNA methyltransferases. *In* "Nucleic Acids and Molecular Biology: Practical Bioinformatics Series" (J. M. Bujnicki, ed.), Vol. 15, pp. 139–168. Heidelberg Springer-Verlag, Berlin.

Carreras, C. W., and Santi, D. V. (1995). The catalytic mechanism and structure of thymidylate synthase. *Annu. Rev. Biochem.* **64,** 721–762.

Constantinesco, F., Motorin, Y., and Grosjean, H. (1999). Transfer RNA modification enzymes from *Pyrococcus furiosus*: Detection of the enzymatic activities *in vitro. Nucleic Acids Res.* **27,** 1308–1315.

Daubin, V., Manolo Gouy, M., and Perrière, G. (2002). A phylogenomic approach to bacterial phylogeny: Evidence of a core of genes sharing a common history. *Genome Res.* **12,** 1080–1090.

Delk, A. S., Nagle, D. P., Jr, Rabinowitz, J. C., and Straub, K. M. (1979a). The methylenetetrahydrofolate-mediated biosynthesis of ribothymidine in the transfer-RNA of *Streptococcus faecalis*: Incorporation of hydrogen from solvent into the methyl moiety. *Biochem. Biophys. Res. Commun.* **86,** 244–251.

Delk, A. S., Nagle, D. P., Jr, and Rabinowitz, J. C. (1979b). Purification of methylenetetrahydrofolate-dependent methyltransferase catalyzing biosynthesis of ribothymidine in transfer RNA of *Streptococcus faecalis*. *In* "Chemistry and Biology of Pteridines" (T. L. Kisliuk and G. M. Brown, eds.), pp. 389–394. Elsevier/North Holland Publishing, New York.

Delk, A. S., Nagle, D. P., Jr, and Rabinowitz, J. C. (1980). Methylenetetrahydrofolate-dependent biosynthesis of ribothymidine in transfer RNA of *Streptococcus faecalis*. Evidence for reduction of the 1-carbon unit by FADH2. *J. Biol. Chem.* **255,** 4387–4390.

Edmonds, C. G., Crain, P. F., Gupta, R., Hashizume, T., Hocart, C. H., Kowalak, J. A., Pomerantz, S.C, Stetter, K.O, and McCloskey, J. A. (1991). Posttranscriptional modification of tRNA in thermophilic Archaea. *J. Bacteriol.* **173,** 138–148.

Finn, R. D., Mistry, J., Schuster-Böckler, B., Griffiths-Jones, S., Hollich, V., Lassmann, T., Moxon, S., Marshall, M., Khanna, A., Durbin, R., Sean, R., Eddy, S. R., Sonnhammer, E. L. L., and Bateman, A. (2006). Pfam: Clans, web tools and services. *Nucleic Acids Res.* **34,** D247–D251.

Fleissner, E., and Borek, E. (1962). A new enzyme of RNA synthesis: RNA methylase. *Proc. Natl. Acad. Sci. USA* **48,** 1199–1203.

Graziani, S., Bernauer, J., Skouloubris, S., Graille, M., Zhou, C. Z., Marchand, C., Decottignies, P., van Tilbeurgh, H., Myllykallio, H., and Liebl, U. (2006). Catalytic mechanism and structure of viral flavin-dependent thymidylate synthase ThyX. *J. Biol. Chem.* **281,** 24048–24057.

Grosjean, H., Edqvist, J., Straby, K. B., and Giegé, R. (1996). Enzymatic formation of modified nucleosides in tRNA: Dependence on tRNA architecture. *J. Mol. Biol.* **255,** 67–85.

Grosjean, H., Droogmans, L., Roovers, M., and Gérard, K. (2007). Detection of enzymatic activity of transfer RNA modification enzymes using radiolabelled tRNA substrates. *Methods Enzymol.* **425,** 57–101.

Guindon, S., and Gascuel, O. (2003). A simple, fast, and accurate algorithm to estimate large phylogenies by maximum likelihood. *Syst. Biol.* **52,** 696–704.

Iwasaki, W., Miyatake, H., and Miki, K. (2005). Crystal structure of the small form of glucose-inhibited division protein A from *Thermus thermophilus* HB8. *Proteins* **61,** 1121–1126.

Kowalak, J. A., Dalluge, J. J., McCloskey, J. A., and Stetter, K. O. (1994). The role of posttranscriptional modification in stabilization of tRNA from hyperthermophiles. *Biochemistry* **33,** 7869–7876.

Limbach, P. A., Crain, P. F., and McCloskey, J. A. (1994). Summary: The modified nucleosides of RNA. *Nucleic Acids Res.* **22,** 2183–2196.

Madsen, C. T., Mengel-Jorgensen, J., Kirpekar, F., and Douthwaite, S. (2003). Identifying the methyltransferases for m^5U747 and m^5U1939 in 23S rRNA using MALDI mass spectrometry. *Nucleic Acids Res.* **31,** 4738–4746.

Myllykallio, H., Lipowski, G., Leduc, D., Filee, J., Forterre, P., and Liebl, U. (2002). An alternative flavin-dependent mechanism for thymidylate synthesis. *Science* **297,** 105–107.

Nordlund, M. E., Johansson, J. O. M., von Pawel-Rammingen, U., and Byström, A. S. (2000). Identification of the *TRM2* gene coding the tRNA(m^5U54)methyltransferase of *S. cerevisiae*. *RNA* **6,** 844–860.

Ny, T., and Björk, G. R. (1980). Cloning and restriction mapping of the *trmA* gene coding for transfer ribonucleic acid (5-methyluridine)-methyltransferase in *Escherichia coli* K-12. *J. Bacteriol.* **142,** 371–379.

Perret, V., Garcia, A., Puglisi, J., Grosjean, H., Ebel, J. P., Florentz, C., and Giegé, R. (1990). Conformation in solution of yeast tRNA^{Asp} transcripts deprived of modified nucleotides. *Biochimie* **72,** 735–743.

Philippe, H. (1993). MUST, a computer package of Management Utilities for Sequences and Trees. *Nucleic Acids Res.* **21,** 5264–5272.

Shigi, N., Suzuki, T., Tamakoshi, M., Oshima, T., and Watanabe, K. (2002). Conserved bases in the TPsiC loop of tRNA are determinants for thermophile-specific 2-thiouridylation at position 54. *J. Biol. Chem.* **277,** 39128–39135.

Sprinzl, M., and Vassilenko, K. S. (2005). Compilation of tRNA sequences and sequences of tRNA genes. *Nucleic Acids Res.* **33,** D139–D140.

Svensson, I., Boman, H. G., Eriksson, K. G., and Kjellin, K. (1963). Studies on microbial RNA. I. Transfer of methyl groups from methionine to soluble RNA from *Escherichia coli*. *J. Mol. Biol.* **16,** 254–271.

Tatusov, R. L., Fedorova, N. D., Jackson, J. D., Jacobs, A. R., Kiryutin, B., Koonin, E. V., Krylov, D. M., Mazumder, R., Mekhedov, S. L., Nikolskaya, A. N., et al. (2003). The COG database: an updated version includes eukaryotes. *BMC Bioinformatics* **4,** 41.

Thompson, J. D., Higgins, D. G., and Gibson, T. J. (1994). CLUSTAL W: Improving the sensitivity of progressive multiple sequence alignment through sequence weighting, position-specific gap penalties and weight matrix choice. *Nucleic Acids Res.* **22,** 4673–4680.

Urbonavicius, J., Skouloubris, S., Myllykallio, H., and Grosjean, H. (2005). Identification of a novel gene encoding a flavin-dependent tRNA:m^{5}U methyltransferase in bacteria-evolutionary implications. *Nucleic Acids Res.* **33,** 3955–3964.

Watanabe, K., Shinma, M., Oshima, T., and Nishimura, S. (1976). Heat-induced stability of tRNA from an extreme thermophile, *Thermus thermophilus*. *Biochem. Biophys. Res. Commm.* **72,** 1137–1144.

White, D. J., Merod, R., Thomasson, B., and Hartzell, P. L. (2001). GidA is an FAD-binding protein involved in development of *Myxococcus xanthus Mol. Microbiol.* **42,** 503–517.

Yim, L., Moukadiri, I., Björk, G. R., and Armengod, M. E. (2006). Further insights into the tRNA modification process controlled by proteins MnmE and GidA of *Escherichia coli*. *Nucleic Acids Res.* **34,** 5892–5905.

CHAPTER FIVE

PROBING THE INTERMEDIACY OF COVALENT RNA ENZYME COMPLEXES IN RNA MODIFICATION ENZYMES

Stephanie M. Chervin,* Jeffrey D. Kittendorf,*,† and George A. Garcia*

Contents

Abstract

Within the large and diverse group of RNA-modifying enzymes, a number of enzymes seem to form stable covalent linkages to their respective RNA substrates. A complete understanding of the chemical and kinetic mechanisms of these enzymes, some of which have identified pathological roles, is lacking. As part of our ongoing work studying the posttranscriptional modification of tRNA with queuine, we wish to understand fully the chemical and kinetic mechanisms

* Department of Medicinal Chemistry, College of Pharmacy, University of Michigan, Ann Arbor, Michigan
† Life Sciences Institute, University of Michigan, Ann Arbor, Michigan

Methods in Enzymology, Volume 425
ISSN 0076-6879, DOI: 10.1016/S0076-6879(07)25005-0

involved in this key transglycosylation reaction. In our previous investigations, we have used a gel mobility-shift assay to characterize an apparent covalent enzyme-RNA intermediate believed to be operative in the catalytic pathway. However, the simple observation of a covalent complex is not sufficient to prove intermediacy. To be a true intermediate, the complex must be both chemically and kinetically competent. As a case study for the proof of intermediacy, we report the use of this gel-shift assay under mildly denaturing conditions to probe the kinetic competency of the covalent association between RNA and the tRNA modifying enzyme tRNA-guanine transglycosylase (TGT).

1. INTRODUCTION

Given the abundance and structural diversity of modified nucleosides that have thus far been identified within all types of RNA, it can be hypothesized that they serve important molecular roles in many cellular processes. Although most of these roles remain unknown, it is becoming clear that certain modified nucleosides are involved in the pathogenesis of human disease (Bjork *et al.*, 1999; Jacobs, 2003; Kwak and Kawahara, 2005), whereas others are critical for the virulence of pathogenic organisms such as *Shigella flexneri* (Bjork *et al.*, 1999; Durand and Bjork, 2003). As such, the enzymes responsible for the biosynthesis of modified nucleosides may present unique therapeutic targets for rational drug design.

The characterized catalytic mechanisms of RNA-modifying enzymes are diverse, reflecting the varied structural nature of the modified nucleoside products (Garcia and Goodenough-Lashua, 1998). Within this diverse group, the reaction mechanisms used by a number of enzymes appear to proceed through a stable covalent linkage to their respective RNA substrates (Fig. 5.1). For example, intermediate 1 is postulated to form on the reaction pathway of tRNA U54 methyltransferase (RUMT) (Kealey and Santi, 1995), a member of the m^5U RNA methyltransferase family (similar intermediates have been identified in the related m^5C RNA methyltransferases; Kealey and Santi [1995] and King and Redman [2002]). The covalent intermediate is proposed to form by Michael addition of the cysteine nucleophile to the C6 position of the uracil base. The histidine-linked phosphoamide intermediate 2 has been observed in the action of tyrosyl-DNA phosphodiesterase (TDP) on RNA substrates (Interthal *et al.*, 2005) and the related lysine-linked intermediate 3 has been observed in the action of mRNA-capping enzyme guanylyltransferase (Huang *et al.*, 2005). Pseudouridine synthase is thought to use an aspartate residue to form a covalent complex with its tRNA substrate (Hamilton *et al.*, 2006). In this case, it is not yet clear whether the aspartate nucleophilically attacks the C6 position of the uracil base, resulting in the carbon–oxygen–linked

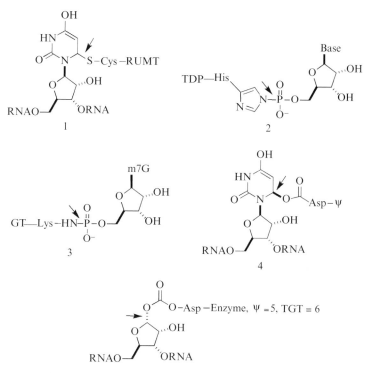

Figure 5.1 Chemical structures of known RNA-enzyme covalent complexes. The shared covalent bond is highlighted with an arrow. Abbreviations: GT, Guanylyl-transferase; Ψ, pseudouridine synthase; RUMT, tRNA U54 methyltransferase; TDP, tyrosyl-DNA-phosphodiesterase (shown working on RNA); TGT, tRNA-guanine transglycosylase.

intermediate 4 or if the aspartate attacks the 1′ position of the ribose yielding intermediate 5. Finally, and most relevant to ongoing investigations in our laboratory, the base-exchange reaction that is catalyzed by the tRNA-modifying enzyme tRNA-guanine transglycosylase proceeds through the ribosyl ester linkage that is present in intermediate 6.

A clear understanding of the mechanisms of RNA-modifying enzymes that may operate by formation of a reaction intermediate such as those described previously requires more than the mere observation of an enzyme-substrate complex. To prove the existence of a reaction intermediate, the following three criteria must be met.

1. Is the intermediate isolable, detectable or trappable? Rigorous characterization of a putative enzyme-RNA covalent intermediate first requires that the intermediate can be isolated. In the best case, the suspected intermediate can be isolated in quantities sufficient for structural analysis (e.g., by X-ray diffraction). In the absence of this, detection of the

intermediate by SDS-PAGE analysis of reaction mixtures or by spectroscopic methods (e.g., NMR, MS) can be suggestive of its existence in the reaction mechanism. However, such analyses can prove difficult to obtain given the technical difficulties associated with analysis of large and highly charged (i.e., RNA phosphodiester backbone) enzyme-RNA complexes. As an alternative approach, the intermediate can be trapped with an appropriately reactive compound, the product of which would provide putative evidence for the intermediate's existence (Hamilton *et al.*, 2006; Redman, 2006).

2. Is the intermediate chemically competent? After the successful isolation or detection of a putative enzyme-RNA intermediate, it is requisite to demonstrate its chemical competence to ensure that the detected intermediate is not an artifact of the experimental conditions. Specifically, the isolated enzyme-RNA intermediate should react to give the same products as does the overall reaction of enzyme and substrate under the identical reaction conditions.

3. Is the intermediate kinetically competent? Finally, and perhaps most experimentally challenging, the kinetic competence of the putative enzyme-RNA intermediate must be demonstrated. No individual step on the reaction pathway can be slower than the overall rate of the reaction (k_{cat}). Therefore, the rates of formation and breakdown of the suspected enzyme-RNA intermediate must be equal to or faster than k_{cat}. The kinetic assessment of the intermediacy of a complex can be complicated by the lack of a suitable assay to quantify the amount of intermediate complex present in the reaction mixture. Here we report the use of a simple gel-shift assay to probe the kinetic competency of the covalent association between RNA and an RNA modifying enzyme.

2. A Case Study for Probing Reaction Intermediacy: The Covalent RNA Complex of tRNA-Guanine Transglycosylase (TGT)

Our laboratory's interest in the catalytic mechanism of tRNA-guanine transglycosylase (TGT) led us to develop and refine methods for the characterization of these stable covalent intermediates to address these criteria for intermediacy. TGT catalyzes the posttranscriptional replacement of guanine with the modified nucleoside queuosine in eukaryotes and the queuosine precursor preQ$_1$ in eubacteria (Noguchi *et al.*, 1982; Okada and Nishimura, 1979; Okada *et al.*, 1979). This base substitution (Fig. 5.2), in which the wobble position guanine is replaced, has been observed in four tRNAs that code for aspartate, asparagine, histidine, and tyrosine. Each of these cognate tRNAs share the common anticodon sequence of $G_{34}U_{35}N_{36}$.

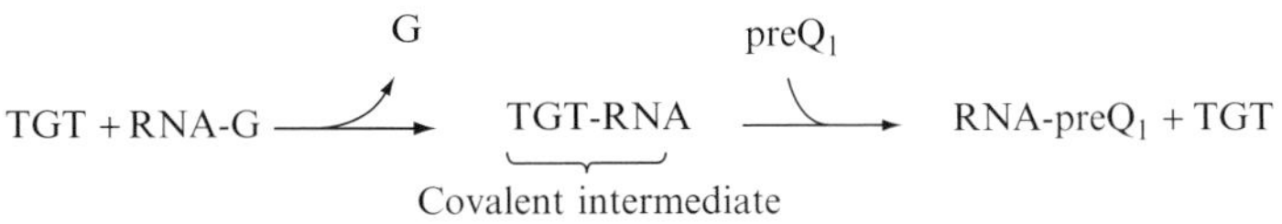

$$\text{TGT} + \text{RNA-G} \xrightarrow{\quad G \quad} \underbrace{\text{TGT-RNA}}_{\text{Covalent intermediate}} \xrightarrow{\quad preQ_1 \quad} \text{RNA-}preQ_1 + \text{TGT}$$

Figure 5.2 The eubacterial TGT-catalyzed reaction exchanging guanine for preQ$_1$ by formation of a covalent enzyme-RNA intermediate.

The TGT reaction proceeds by means of ping–pong kinetics, whereby the tRNA substrate binds first with concomitant loss of guanine leading to the proposed TGT-RNA covalent intermediate (Goodenough-Lashua and Garcia, 2003). The incoming modified base then binds in the active site and nucleophilically displaces the enzyme, resulting in the formation of a new glycosidic bond and the release of the modified RNA product.

The occurrence of an intermediate in catalytic mechanism of TGT was supported by the report of the solution of the X-ray crystal structure of the *Z. mobilis* TGT bound to a minihelical RNA substrate (Xie *et al.*, 2003). Importantly, those results clearly demonstrated that the proposed intermediate was indeed isolable, satisfying criterion 1 (see earlier). The X-ray crystal structure revealed that active site Asp264 (*E. coli* numbering) forms a covalent ribosyl ester linkage with the 1′ ribosyl carbon of the wobble position guanosine of tRNA. From this and other biochemical data (Kittendorf *et al.*, 2001, 2003; Romier *et al.*, 1996a; Romier *et al.*, 1996b), a catalytic mechanism in which Asp264 functions as the enzyme nucleophile leading to the covalent enzyme-RNA intermediate was considered likely (Fig. 5.3).

3. Denaturing Gel Electrophoresis: A Tool to Probe Enzyme-RNA Complexes

In vitro biochemical studies of the TGT-catalyzed reaction have been aided by an extremely facile assay to probe the existence of stable TGT-RNA complexes. Romier and colleagues (1996b) were the first to report the use of a denaturing gel electrophoresis assay to study intermediate components resulting from the incubation of wild-type and mutant *Z. mobilis* TGT with substrate tRNA. The authors observed that denaturing PAGE analysis of equilibrium mixtures contained protein bands of higher apparent molecular weight than monomeric TGT and presumed these bands to be stable *covalent* enzyme-RNA complexes. We routinely use this gel-mobility shift assay to probe the ability of wild-type and mutant *E. coli* TGT to form stable RNA complexes with cognate tRNA (Curnow and Garcia, 1994; Goodenough-Lashua and Garcia, 2003; Kittendorf *et al.*, 2001, 2003), as

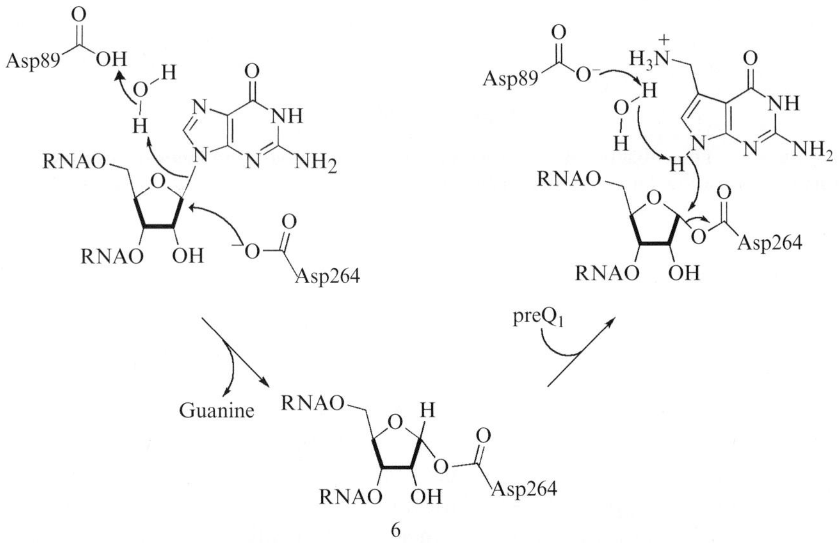

Figure 5.3 Mechanism of the formation of glycosyl ester intermediate 6. Asp264 serves as the catalytic nucleophile that displaces the guanine base. Asp89 is proposed to serve as a general acid/base, possibly through an intermediary water molecule (Kittendorf *et al.*, 2003).

well as with novel RNA structures (Kung *et al.*, 2000; Nonekowski and Garcia, 2001). Although the observation of these bands is not sufficient to argue for a true *covalent* TGT-RNA association, we have observed a correlation between the ability of TGT to form the complexes and enzyme activity. With similar reasoning, Xie and colleagues (2003) used the band-shift assay to investigate the chemical competence of a 9-deazaguanine (9dzG)-stabilized TGT-RNA-9dzG ternary complex, and most recently, Meyer and collaborators (2006) probed the ability of small molecule inhibitors to stabilize a tRNA-TGT intermediate in a ternary complex.

3.1. PAGE band-shift analysis of reaction mixtures

- $5\times$ Bicine reaction buffer: Bicine (250 mM, pH = 7.7), MgCl$_2$ (100 mM), DTT (25 mM).
- Native-PAGE loading buffer: TRIS HCl (60 mM, pH = 7.0), glycerol (50%), bromothymol blue (0.02%).
- Native-PAGE electrophoresis conditions: native buffer strips (Amersham), 5°, 400 V, 10 mA, 2.5 W, 250 AVH.
- SDS-PAGE loading buffer: TRIS HCl (60 mM, pH = 7.0), glycerol (10%), SDS (2%), bromothymol blue (0.02%).
- SDS-PAGE electrophoresis conditions: SDS buffer strips (Amersham), 15°, 250 V, 10 mA, 3 W, 200 AVH.

Equilibrium mixtures were prepared by incubating enzyme TGT (10 μM) with substrate RNA (10 μM) in 1× Bicine reaction buffer at 37° for 20 min.

3.1.1. Native-PAGE

An aliquot (10 μl) of the reaction mixture was combined with native-PAGE loading buffer (10 μl). An aliquot (1 μl) of that combination was loaded onto an 8–25% gradient polyacrylamide gel (PhastGel, Amersham) and electrophoresed under native conditions (see earlier) on a PhastSystem unit.

3.1.2. SDS-PAGE

An aliquot (10 μl) of the reaction mixture was combined with SDS-PAGE loading buffer (10 μl) and allowed to incubate at room temperature for 1 h. An aliquot (1 μl) of that combination was loaded onto an 8–25% gradient polyacrylamide gel (PhastGel, Amersham) and electrophoresed under denaturing conditions (see earlier) on a PhastSystem unit.

Notes: (1) The incubation of the SDS-quenched reaction mixtures on ice or for periods of <1 h resulted in two bands corresponding to TGT-RNA complexes, apparently reflecting different TGT-RNA conformers present in solution. However, after 60 min at room temperature, the upper band coalesced with the lower as shown in Fig. 5.4, suggesting that the band was incompletely denatured complex. (2) This method does not require the use of the PhastSystem electrophoresis equipment. Equivalent results have been obtained with other electrophoresis units and gels (e.g., Bio-Rad).

3.2. Band quantification by fluorescent detection

The plastic gel backing was removed from the polyacrylamide gel with the aid of a gel backing remover wire (Pharmacia) and the gel stained with 1× Spyro Red (Invitrogen) in 7.5% acetic acid for 1 h with shaking. The stained gel was washed with 7.5% acetic acid (2 × 5 ml), followed by water (1 × 5 ml). The bands were visualized by green laser excitation

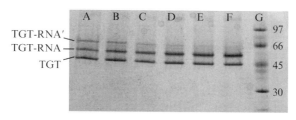

Figure 5.4 Time course of incubation times of TGT-RNA reaction mixtures in SDS loading buffer. Lane A, 5 min; lane B, 15 min; lane C, 30 min; lane D, 45 min; lane E, 60 min; lane F, 75 min; lane G, MW ladder. Protein bands are detected by Spyro Red fluorescent stain.

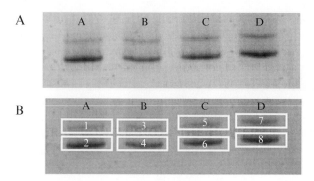

Lane	Rectangle	Volume × 10^6
A	1	4.445
A	2	29.370
B	3	5.073
B	4	17.143
C	5	6.248
C	6	23.864
D	7	8.573
D	8	28.637

Lane	% upper band
A	13.2
B	22.8
C	20.8
D	23.0

Figure 5.5 Representative band volume determination. (A) SDS–PAGE of TGT-RNA reaction mixtures. (B) The same SDS–PAGE of TGT-RNA reaction mixtures with rectangles (white) drawn to encompass gel area for volume calculation. The tables show the individual volumes and derived covalent complex percentages.

(532 nm) and fluorescent detection by use of a Typhoon 9200 gel imaging system (Molecular Dynamics). The band volumes (created by a 3D plot of pixel locations and intensities) were quantified using the ImageQuant software package (Molecular Dynamics). In summary, rectangles of identical size and shape were drawn around bands in the same lane to encompass each band with proportionately equal background space (Fig. 5.5). A report of volumes within each defined rectangle using the "local average background" subtraction method was generated. To calculate the percent of the protein in the upper band of each lane, the volume of the upper rectangle was divided by the sum of the volumes of the upper and lower bands.

Note: For kinetic studies, an enzyme concentration in the final quenched sample of 1.9 μM was sufficient to visualize the resulting bands with the Spyro Red protein stain. Greater final concentrations would be required for Coomassie detection, less for more sensitive detection methods (e.g., radiochemical detection).

3.3. Detection of biotinylated-RNA containing bands by colorimetric assay

- Phosphate-buffered saline-T (PBS-T): NaCl (155 mM), Na$_2$HPO$_4$ (2.97 mM), KH$_2$PO$_4$ (1.06 mM), Tween-20 (0.05%).

After electrophoresis, polyacrylamide gels were blotted onto nitrocellulose membranes (Bio-Rad) according to vendor's protocols. The blots were incubated with horseradish peroxidase–conjugated streptavidin (2 μg/ml) (Pierce) in PBS-T for 30 min and washed with PBS-T (2 × 10 ml). The HRP-conjugated bands were detected with the chromogenic substrate CN/DAB (Pierce) that forms a dark precipitate visible in ambient light.

In our hands, equimolar recombinant *E. coli* TGT (45 kDa) incubated with *E. coli* tRNA^{Tyr} (ECY) or a minihelical analog consisting of the anticodon stem and loop substituted with a 2′-deoxyguanosine located at position 34 (dG$_{34}$ECYMH) (Nonekowski *et al.*, 2002) forms a stable TGT-RNA complex of apparent molecular weight greater than monomeric TGT as assessed by SDS-PAGE analysis (Fig. 5.6A). (*Note*: The 2′-deoxyguanosine modification was designed to probe the role of the 2′ hydroxyl in the TGT reaction.) However, it was unclear whether these higher molecular weight bands represented true covalent complexes or a mixture of covalent and noncovalent (but stable to the denaturing conditions used) associations.

The crystal structure of TGT bound to substrate RNA revealed that under the crystallization conditions, Asp264 could nucleophilically attack the RNA ribose, displacing guanine (Xie *et al.*, 2003). On the assumption that this occurs under normal reaction conditions, we reasoned that a mutation of Asp264 would render TGT unable to form the covalent ribosyl ester linkage; however, the noncovalent association between the enzyme and RNA substrate should be maintained. Moreover, if this noncovalent interaction persisted under mildly denaturing conditions, it should be detected in gel mobility-shift experiments that use the mutant TGT. To probe this hypothesis, aspartate 264 of the *E. coli* TGT was mutated to asparagine (Kittendorf *et al.*, 2003). SDS-PAGE analysis of reaction mixtures containing TGT(D264N) and RNA demonstrated that, even under mild denaturing conditions at decreased temperatures, no TGT(D264N)-RNA complex was formed (Fig. 5.7). By native-PAGE analysis, TGT(D264N) was observed to exist as a mixture of multimeric forms. TGT(D264N) that was incubated with RNA contained a mixture of monomeric TGT(D264N) and a new band of slightly lesser mobility that we have assigned to be the noncovalent TGT-RNA complex (Fig. 5.8A). The suspected TGT-RNA bands that had been observed by SDS-PAGE and native-PAGE analyses were probed with the nucleic acid intercalating agent, ethidium bromide (data not shown); however, no increase in staining was observed relative to the protein bands. Presumably this is due to the cumulative effect of the small size of the RNA mini-helix substrate and of the ability (not widely recognized) of ethidium bromide to stain protein (Csapo *et al.*, 2000; Vincent and Scherre, 1979). To selectively visualize RNA-containing bands, we designed an RNA mini-helix substrate in which the 5′ end contained a biotinylated guanine base, enabling facile detection by Western blot with streptavidin-linked horseradish peroxidase (HRP). After native-PAGE

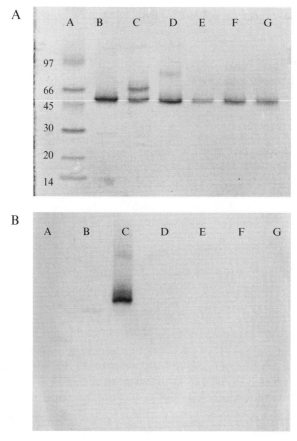

Figure 5.6 SDS-PAGE of the TGT-RNA covalent complex. (A) *E. coli* TGT or TGT (D264N) (10 μM) was incubated with either full-length tRNA (ECY, 10 μM) or biotiny-lated mini-helical substrate (dG$_{34}$ECYMH, 10 μM) in Bicine (pH 7.7) buffer at 37° for 20 min. An aliquot (10 μl) was quenched with SDS loading buffer (10 μl) and incubated at room temperature for 1 h. Lane A, MW ladder; lane B, TGT; lane C, TGT +dG$_{34}$E-CYMH; lane D, TGT + ECY; lane E, TGT(D264N); lane F, TGT(D264N) + dG$_{34}$E-CYMH; lane G, TGT(D264N) + ECY. Protein bands are visualized by Coomassie stain. (B) Western blot of the TGT-RNA covalent complex. SDS-PAGE was performed as in (A). Proteins were blotted onto a nitrocellulose membrane, and RNA-containing bands were identified with a streptavidin-conjugated HRP-based colorimetric stain. Note, uncomplexed RNA substrate runs with the electrophoresis front (not shown). Lanes as indicated for gel A.

separation, Western blot analysis of wild-type and TGT(D264N) reaction mixtures containing the biotinylated RNA revealed that both wild-type and mutant TGT form complexes with RNA (Fig. 5.8B). However, SDS-PAGE separation followed by Western blot analysis indicated that only the wild-type TGT remained associated with RNA under denaturing

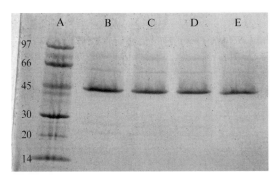

Figure 5.7 SDS-PAGE of TGT(D264N) incubated with RNA. TGT(D264N) (10 μM) was incubated with mini-helical substrate (10 μM) in Bicine buffer (pH 7.7) at 37° for 20 min. An aliquot (10 μl) was quenched with SDS loading buffer (10 μl) and incubated for 1 h at various temperatures. Lane A, MW ladder; lane B, 37°; lane C, 22°; lane D, 10°; lane E, 4°. Protein bands are detected by Spyro Red fluorescent stain.

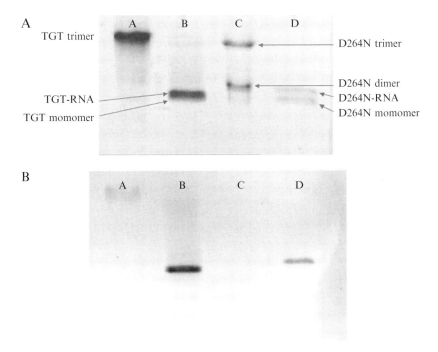

Figure 5.8 Native-PAGE of the TGT-RNA complex. Wild-type TGT and TGT (D264N) (10 μM) were incubated with biotinylated mini-helical substrate (dG_{34} ECYMH, 20 μM) in Bicine (pH 7.7) reaction buffer at 37° for 20 min. An aliquot (10 μl) was treated with native loading buffer (10 μl) and analyzed immediately by native-PAGE. (A) Coomassie-stained native polyacrylamide gel. Lane A, TGT; lane B, TGT + RNA; lane C, TGT(D264N); lane D, TGT(D264N) + RNA. (B) Streptavidin-conjugated HRP-based colorimetric stain of nitrocellulose blot of native polyacrylamide gel to detect RNA-containing bands. Note, uncomplexed RNA substrate runs with the electrophoresis front (not shown). Lanes as indicated for gel A. (See color insert.)

conditions (Fig. 5.6B). Importantly, these experiments provide direct evidence that under native conditions, both noncovalent and covalent TGT-RNA associations are maintained, whereas under denaturing conditions, the noncovalent association is ablated, while the covalent interaction remains intact.

4. ANALYSIS OF THE CHEMICAL COMPETENCY OF COVALENT ENZYME-RNA COMPLEXES

With the successful isolation of the *Z. mobilis* TGT–RNA crystal, Xie and coworkers were able to probe the chemical reactivity of the intermediate with the heterocyclic base substrate preQ$_1$ (Xie *et al.*, 2003). The authors treated the complex with molar excess of preQ$_1$ and found, as assessed by gel-shift analysis, that the complex band disappeared, presumably forming preQ$_1$-substituted RNA. In our hands, a histidine-tagged *E. coli* TGT-RNA complex purified from free RNA components by nickel chelation chromatography similarly reacts with excess preQ$_1$ to yield free TGT (Fig. 5.9). Independently, each of these experiments demonstrates the chemical competence of the proposed enzyme-RNA intermediate.

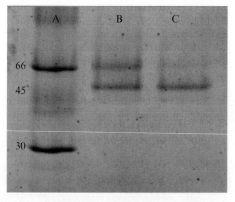

Figure 5.9 SDS-PAGE of TGT-RNA complexes in the absence of and presence of substrate preQ$_1$. Lane A, MW ladder; Lane B, TGT-dG$_{34}$ECYMH and TGT mixture purified from reaction mixture by chelation of histidine-tagged TGT by nickel chromatography; lane C, purified TGT-dG$_{34}$ECYMH intermediate mixture treated with excess preQ$_1$.

5. Analysis of the Rate of Formation of Covalent Enzyme-RNA Complexes

Finally, the kinetic competence of the proposed enzyme-RNA intermediate remained to be substantiated. The denaturing gel mobility-shift assay provides a convenient method for measuring the observed first-order rate constant ($k_{formation}$) for formation of covalent enzyme-RNA intermediates. Such analysis is absolutely necessary for assessing the kinetic competence of any proposed intermediate. As proof of concept, we interrogated the kinetics of formation of the intermediate that is observed on incubation of wild-type TGT with the mini-helical RNA, $dG_{34}ECYMH$. By use of a KinTek rapid quench flow apparatus equipped with a circulating water bath, reaction mixtures containing wild-type TGT and RNA were sampled and quenched over a time course ranging from 0.1 sec to several minutes. After SDS-PAGE separation of the reaction mixtures, the protein bands were stained with a fluorescent protein dye and subsequently visualized by laser excitation, enabling direct protein quantification on the basis of the fluorescence signal (Fig. 5.10).

5.1. Kinetic studies that use rapid quench flow

- Drive buffer: Bicine (50 mM, pH = 7.7).
- SDS quench buffer: TRIS HCl (60 mM, pH = 7.0), glycerol (10%), SDS (2%).

A KinTek RQF-3 rapid quench flow apparatus was charged with drive buffer and SDS quench buffer and equilibrated to 37°. Detailed specifications of the apparatus are available at the KinTek Corp. web site: www.kintek-corp.com. A simplified schematic of the apparatus appears in

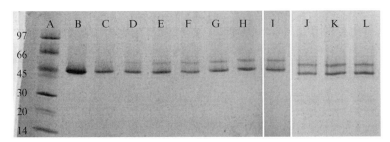

Figure 5.10 SDS-PAGE of the time course of the formation of the TGT-dG_{34}ECYMH covalent intermediate. TGT and dG_{34}ECYMH were incubated at 37° and quenched at intervals from 0.1–120.0 sec with the aid of a rapid-quench device. Protein bands are detected by Spyro Red fluorescent stain. The amount of intermediate formed at each time point was quantified by use of fluorescence detection. Lane A, MW ladder; lane B, TGT; lanes C–L, 0.1, 0.5, 1.0, 3.0, 5.0, 10.0, 20.0, 40.0, 60.0, 120.0 sec, respectively.

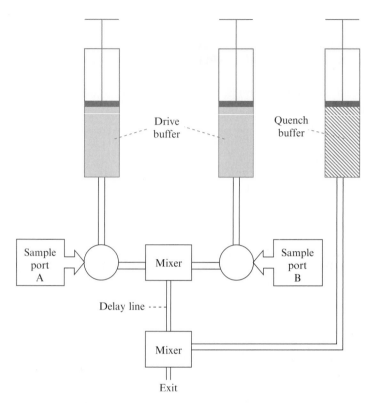

Figure 5.11 Simplified schematic of KinTek RQF-3 Rapid Quench Flow apparatus.

Fig. 5.11 The sample port A was loaded with a solution of enzyme, TGT (15 μM), in Bicine reaction buffer, and sample port B was loaded with a solution of RNA substrate (15 μM) also in Bicine reaction buffer. Aliquots (15 μl) of each reagent were incubated in the mixing chamber for various periods (0.1–120.0 s) and quenched automatically with SDS buffer (90 μl). The mixtures incubated for an additional hour at room temperature before PAGE band-shift analysis.

Note: The quenching agent must be sufficiently denaturing to stop enzymatic function yet sufficiently mild to not disrupt the covalent linkage of the intermediate. Particularly in the case of enzymes that form relatively labile ribosyl ester intermediates, the conventional acidic quench is deleterious to the stability of the covalent bond. A discussion of alternative quenching agents appears in a recent review (Barman *et al.*, 2006).

The ratio of the amount of the TGT-RNA covalent complex to free enzyme was computed by dividing the band volume of the complex band by the total volume of the complex band and free enzyme band. A plot of the percent of complex formed versus time gave a first-order rate of formation of the TGT-RNA covalent complex of 0.684 sec^{-1} (Fig. 5.12).

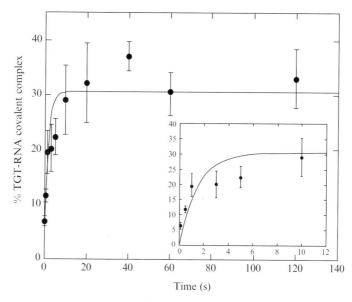

Figure 5.12 Percent covalent complex formation versus time. Each point represents the average of 4–7 independent measurements ($\pm$ SEM). The inset is an expansion of the earliest time points.

It should be noted that the reaction proceeds to approximately 30% completion. Under the conditions used (e.g., 10 μm each, enzyme and RNA) approximately 3 μM free guanine will be generated. The reaction generating the covalent complex is reversible, and the concentration of free guanine generated is approximately $10\times$ K_M (Goodenough-Lashua and Garcia, 2003). Therefore, the reaction actually proceeds to an equilibrium point, which seems to be approximately 30% completion.[1]

6. SUMMARY

A gel-mobility shift assay has been used to probe the occurrence of noncovalent and covalent complexes that form on incubation of RNA and some RNA modifying enzymes. This denaturing band-shift assay is amenable to quantitative analysis and, most importantly, provides data regarding the rate of covalent complex formation. Furthermore, this assay may presumably be extended to determine the rate of covalent complex breakdown, provided that the enzyme-RNA intermediate is stable to isolation.

---

[1] A report on the kinetic competence of the covalent intermediate in the TGT reaction is in preparation.

Given the difficulties associated with obtaining crystallographic data and the hurdles associated with analysis of protein-RNA by mass spectrometry, this method offers a relatively simple assay that uses standard PAGE analyses to probe complex protein–RNA interactions.

ACKNOWLEDGMENTS

We are grateful to Professor Carol Fierke for the use of the rapid quench flow apparatus, to Professor Ronald Woodard for use of the electroblotting apparatus, and to Professor Kyung-Dall Lee for use of the Typhoon imaging system. Work in the Garcia laboratory has been generously supported by grants from NIH (GM065489 & GM45968), NSF (9720139), and the University of Michigan, College of Pharmacy, Vahlteich & Upjohn Research Funds.

REFERENCES

Barman, T. E., Bellamy, S. R. W., Gutfreund, H., Halford, S. E., and Lionne, C. (2006). The identification of chemical intermediates in enzyme catalysis by the rapid quench-flow technique. *Cell. Mol. Life Sci.* **63**, 2571–2583.
Bjork, G. R., Durand, J. M., Hagervall, T. G., Leipuviene, R., Lundgren, H. K., Nilsson, K., Chen, P., Qian, Q., and Urbonavicius, J. (1999). Transfer RNA modification: Influence on translational frameshifting and metabolism. *FEBS Lett.* **452**, 47–52.
Csapo, Z., Gerstner, A., Sasvari-Szekely, M., and Guttman, A. (2000). Automated ultra-thin-layer SDS gel electrophoresis of proteins using noncovalent fluorescent labeling. *Anal. Chem.* **72**, 2519–2525.
Curnow, A. W., and Garcia, G. A. (1994). tRNA-guanine transglycosylase from *Escherichia coli*: Recognition of dimeric, unmodified tRNATyr. *Biochimie* **76**, 1183–1191.
Durand, J. M., and Bjork, G. R. (2003). Putrescine or a combination of methionine and arginine restores virulence gene expression in a tRNA modification-deficient mutant of Shigella flexneri: A possible role in adaptation of virulence. *Mol. Microbiol.* **47**, 519–527.
Garcia, G. A., and Goodenough-Lashua, D. M. (1998). Mechanisms of modifying/editing enzymes. *In* "Modification and Editing of RNA: The Alteration of RNA Structure and Function" (H. Grosjean and R Benne, eds.). ASM Press, Washington, D.C.
Goodenough-Lashua, D. M., and Garcia, G. A. (2003). tRNA-guanine transglycosylase from *E. coli*: A ping-pong kinetic mechanism is consistent with nucleophilic catalysis. *Bioorg. Chem.* **31**, 331–344.
Hamilton, C. S., Greco, T. M., Vizthum, C. A., Ginter, J. M., Johnston, M. V., and Mueller, E. G. (2006). Mechanistic investigations of the pseudouridine synthase RluA using RNA containing 5-fluorouridine. *Biochemistry* **45**, 12029–12038.
Huang, Y.-L., Hsu, Y.-H., Han, Y.-T, and Meng, M (2005). mRNA guanylation catalyzed by the S-adenosylmethionine-dependent guanylyltransferase of Bamboo Mosaic Virus. *J. Biol. Chem.* **280**, 13153–13162.
Interthal, H., Chen, H. J., and Champoux, J. J. (2005). Human Tdp1 cleaves a broad spectrum of substrates, including phosphoamide linkages. *J. Biol. Chem.* **280**, 36518–36528.
Jacobs, H. T. (2003). Disorders of mitochondrial protein synthesis. *Hum. Mol. Genet.* **15**, 293–301.
Kealey, J. T., and Santi, D. V. (1995). Stereochemistry of tRNA(m5U54)-methyltransferase catalysis: 19F NMR spectroscopy of an enzyme-FUraRNA covalent complex. *Biochemistry* **34**, 2441–2446.

King, M. Y., and Redman, K. L. (2002). RNA methyltransferases utilize two cysteine residues in the formation of 5-methylcytosine. *Biochemistry* **41**, 11218–11225.

Kittendorf, J. D., Barcomb, L. M., Nonekowski, S. T., and Garcia, G. A. (2001). tRNA-guanine transglycosylase from *Escherichia coli*: Molecular mechanism and role of aspartate 89. *Biochemistry* **40**, 14123–14133.

Kittendorf, J. D., Sgraja, T., Reuter, K., Klebe, G., and Garcia, G. A. (2003). An essential role for aspartate 264 in catalysis by tRNA-guanine transglycosylase from *Escherichia coli*. *J. Biol. Chem.* **278**, 42369–42376.

Kung, F. L., Nonekowski, S., and Garcia, G. A. (2000). tRNA-guanine transglycosylase from *Escherichia coli*: Recognition of noncognate-cognate chimeric tRNA and discovery of a novel recognition site within the TpsiC arm of tRNA(Phe). *RNA* **6**, 233–244.

Kwak, S., and Kawahara, Y. (2005). Deficient RNA editing of GluR2 and neuronal death in amyotropic lateral sclerosis. *J. Mol. Med.* **83**, 110–120.

Meyer, E. A., Donati, N., Guillot, M., Schweizer, W. B., and Diederich, F. (2006). Synthesis, biological evaluation, and crystallographic studies of extended guanine-based (lin-benzoguanine) inhibitors for tRNA-guanine transglycosylase (TGT). *Helvetica Chim. Acta* **89**, 573–597.

Noguchi, S., Nishimura, Y., Hirota, Y., and Nishimura, S. (1982). Isolation and characterization of an Escherichia coli mutant lacking tRNA-guanine transglycosylase. Function and biosynthesis of queuosine in tRNA. *J. Biol. Chem.* **257**, 6544–6550.

Nonekowski, S. T., and Garcia, G. A. (2001). tRNA recognition by tRNA-guanine transglycosylase from *Escherichia coli*: The role of U33 in U-G-U sequence recognition. *RNA* **7**, 1432–1441.

Nonekowski, S. T., Kung, F.-L, and Garcia, G.A (2002). The *Escherichia coli* tRNA-Guanine transglycosylase can recognize and modify DNA. *J. Biol. Chem.* **277**, 7178–7182.

Okada, N., and Nishimura, S. (1979). Isolation and characterization of a guanine insertion enzyme, a specific tRNA transglycosylase, from *Escherichia coli*. *J. Biol. Chem.* **254**, 3061–3066.

Okada, N., Noguchi, S., Kasai, H., Shindo-Okada, N., Ohgi, T., Goto, T., and Nishimura, S. (1979). Novel mechanism of post-transcriptional modification of tRNA. Insertion of bases of Q precursors into tRNA by a specific tRNA transglycosylase reaction. *J. Biol. Chem.* **254**, 3067–3073.

Redman, K. L. (2006). Assembly of protein-RNA complexes using natural RNA and mutant forms of an RNA cytosine methyltransferase. *Biomacromolecules* **7**, 3321–3326.

Romier, C., Reuter, K., Suck, D., and Ficner, R. (1996a). Crystal structure of tRNA-guanine transglycosylase: RNA modification by base exchange. *EMBO J.* **15**, 2850–2857.

Romier, C., Reuter, K., Suck, D., and Ficner, R. (1996b). Mutagenesis and crystallographic studies of *Zymomonas mobilis* tRNA-guanine transglycosylase reveal aspartate 102 as the active site nucleophile. *Biochemistry* **35**, 15734–15739.

Vincent, A., and Scherre, K. (1979). A rapid and sensitive method for detection of proteins in polyacrylamide SDS gels: Staining with ethidium bromide. *Mol. Biol. Rep.* **5**, 209–214.

Xie, W., Liu, X., and Huang, R. H. (2003). Chemical trapping and crystal structure of a catalytic tRNA guanine transgylcosylase covalent intermediate. *Nat. Struct. Biol.* **10**, 781–788.

CHAPTER SIX

Identification and Characterization of Modification Enzymes by Biochemical Analysis of the Proteome

Jane E. Jackman, Lakmal Kotelawala, Elizabeth J. Grayhack, *and* Eric M. Phizicky

Contents

Abstract

The use of proteomic libraries designed to express the complete set of proteins from an organism has resulted in the identification of many RNA modification enzymes whose function was previously unknown. Here we describe a generalized procedure for the biochemical analysis of a yeast proteomic library for identification of nucleic acid–modifying enzymes, by use of the yeast MORF (*M*oveable *O*pen *R*eading *F*rame) library (Gelperin *et al.*, 2005) as the source of protein activity, and the known yeast tRNA methyltransferase Trm4 as a test case. The procedures outlined in this chapter can be applied to any proteomic expression library from any organism, many of which will become increasingly available as the number of sequenced genomes increases and as genomic cloning techniques improve.

Department of Biochemistry and Biophysics, University of Rochester School of Medicine, Rochester, New York

Methods in Enzymology, Volume 425

ISSN 0076-6879, DOI: 10.1016/S0076-6879(07)25006-2

1. INTRODUCTION

The rapidly expanding number of completed genome sequences provides the potential for identifying the complete set of proteins required for carrying out all biological processes in any given organism. However, although the availability of complete genome sequences can facilitate the identification of the function of many proteins by use of homology to known proteins in other organisms, there are many other proteins for which the molecular function remains unknown, because these proteins represent families with uncharacterized activities, or because of large evolutionary differences between proteins from different organisms, precluding accurate estimation of their activities.

Several nucleic acid modification enzymes have been identified by use of the homology approach, by finding shared sequence motifs between these proteins and a known modification enzyme in the same, or another, organism (Ansmant *et al.*, 2001; Becker *et al.*, 1997; Bjork *et al.*, 2001; Motorin and Grosjean, 1999; Pintard *et al.*, 2002). However, many other modification enzymes have eluded detection with this approach. Instead, for many of these enzymes, the alternative biochemical genomics approach, by use of proteomic expression libraries, has been very successful, yielding the identification of a diverse set of enzymes in recent years, including several methyltransferases (Alexandrov *et al.*, 2002; Jackman *et al.*, 2003; Wilkinson *et al.*, 2007), a highly unusual tRNA^{His} guanylyltransferase (Gu *et al.*, 2003), and an entire family of dihydrouridine synthases (Xing *et al.*, 2002). Of these enzymes, several seem to contain novel recognition and catalytic motifs consistent with the failure to detect these enzymes by use of bioinformatics approaches. The identification of numerous *PUS* (pseudouridine synthase) genes highlights the usefulness of both of these approaches even within the same enzyme family, because many members of this gene family in yeast were implicated in the formation of pseudouridine initially by sequence similarity to known pseudouridine synthase genes (Ansmant *et al.*, 2001; Arluison *et al.*, 1999; Becker *et al.*, 1997; Conrad *et al.*, 1999), whereas Pus7 was identified by parallel biochemical analysis of the yeast proteome (Ma *et al.*, 2003).

Identification of nucleic acid modification enzymes by analysis of the proteome by use of the biochemical genomics approach has been successful largely due to several features inherent to this approach. First, most RNA modification activities can be detected with a high degree of sensitivity. Because the assays routinely contain $\sim 10^{-13}$ moles of each protein (on average), and the RNA species are often labeled with ^{32}P at high specific activity, as few as 10^{-17} moles of reacted substrate can be required to detect activity, corresponding to ~ 0.0001 turnover of protein. Moreover, because the proteins in the pools are highly purified, competing or contaminating background signals are greatly reduced compared with those often observed in a typical cell-free extract, thus enhancing the ability to carry out assays for long times at high protein concentration to yield the most robust signal. Second, the pools of purified proteins are themselves simple and easy to prepare. Because the expression of the protein of interest is controlled by the use of strong inducible promoters in the library expression construct, the size of the culture needed to produce and purify the protein of interest is very small; typically, 2.5 ml of yeast culture is used for each protein in the collection to prepare protein pools for the yeast MORF library. Furthermore, the purification tags allow for rapid and efficient removal of contaminating nucleases and proteases, and the resulting purified preparations are sufficient for ~100–300 assays and are stable for lengthy periods of time. Third, once a biochemical assay has been developed for any desired enzymatic activity, the assignment of that activity to the appropriate gene is very rapid, often taking as little as several days.

2. METHODOLOGY

The process for identification of a yeast gene that catalyzes any nucleic acid modification activity consists of four steps as outlined here.

1. Development of an appropriate biochemical assay for the desired activity and detection of that activity in crude extracts.
2. Parallel biochemical analysis of the yeast proteome for the activity of interest by use of pools of purified proteins derived from the yeast MORF library.
3. Deconvolution of the active pool by preparation and analysis of subpools of proteins from strains derived from the positive pool to identify the individual ORF responsible for the activity of interest.
4. Confirmation of the identity of the ORF by analysis of the individual protein and the corresponding mutant strain.

We describe here the use of these steps to identify the MORF pool containing the known yeast tRNA modification enzyme, Trm4, which has been shown to catalyze the formation of all m^5C in yeast tRNA species (Motorin and Grosjean, 1999).

3. DEVELOPMENT OF AN APPROPRIATE BIOCHEMICAL ASSAY

For any modification activity the most significant impediment to identification of the responsible enzyme may be the ability to develop a biochemical assay for the desired activity. For numerous tRNA base and ribose sugar modification enzymes, successful assays have relied on introduction of a ^{32}P label into the appropriate tRNA substrate, either by standard *in vitro* transcription in the presence of [α-^{32}P]-NTPs or by site-specific labeling of transcribed tRNAs at the position of interest (Yu, 1999). Although the ease of generating uniformly labeled substrates makes the first approach attractive to use as a first attempt, there may be cases where the production of a site-specifically labeled RNA species is desirable, because of the increased signal strength when only a single position is isotopically labeled, and to the elimination of other competing activities that could obscure the visualization of the modification activity of interest. Moreover, use of the site-specifically labeled substrate inherently addresses the question of whether or not the modification occurs at the specific position of interest, which may be an important consideration when the same modification occurs at multiple positions in the substrate tRNA.

For the tRNA 5-methylcytidine methyltransferase activity of Trm4, the activity assay relies on the use of tRNA (in this case, tRNA^{Gly}) uniformly labeled at each cytidine residue by *in vitro* transcription in the presence of [α-^{32}P]-CTP (Fig. 6.1A). After incubation of the substrate tRNA in the presence of yeast crude extract and *S*-adenosylmethionine (SAM), which serves as the methyl group donor, the RNA is purified by phenol extraction followed by ethanol precipitation, and then treated with nuclease P1 to release 5′-monophosphorylated nucleotide species (Fig. 6.1A). Because the cytidine residues are labeled at the 5′-phosphate during *in vitro* transcription, unmodified cytidine residues will be visualized as 5′-p★C, whereas any position that was a substrate for Trm4 activity will be visualized as 5′-p★m^{5}C. These two species are readily resolved from each other by use of thin-layer chromatography (TLC), by spotting the resulting digested reactions onto cellulose plates, and by developing the plates in an isobutyric acid: 0.5 *M* ammonium hydroxide 5:3 (v/v) solvent system (Fig. 6.1B). The TLC migration observed for the products of this reaction is consistent with the published chromatographic mobilities for these nucleotide species. Furthermore, in this test case, because the activity is already known to be the result of Trm4 activity, the identity of the observed reaction products can be confirmed by use of the relevant yeast deletion strain. Thus, the m^{5}C product spot is only observed after incubation with crude extracts derived from a wild-type strain and not with extracts derived from a yeast strain that contains a deletion of *trm4* (Fig. 6.1B).

Several factors must be taken into account during the selection of the appropriate reaction conditions for the desired nucleic acid modification activity. The choice of cofactor to be used in the assay for any given enzymatic activity can influence the success of this approach, because although many required cofactors are available in a typical cell-free extract, these cofactors may not persist in the final purified pools of proteins if they are not tightly bound to the protein of interest and are consequently

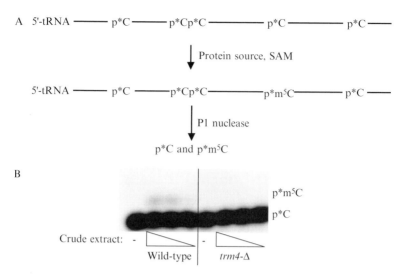

Figure 6.1 Biochemical assay for tRNA m^{5}C methyltransferase activity. (A) Schematic of activity assay with [^{32}P]-labeled tRNA species. After incubation of the labeled tRNA species with the methyl donor cofactor, SAM, and a source of protein, the resulting RNAs are treated with P1 nuclease to release 5'-monophosphorylated nucleotide species that can subsequently be resolved from one another by TLC, by spotting onto cellulose plates and resolving the products in an isobutyric acid/NH$_4$OH:H$_2$O system (66:1:33 v/v/v). (B) Demonstration of Trm4-dependent m^{5}C formation in crude extracts derived from yeast. By use of the assay procedure described previously, cell-free extracts were assayed for m^{5}C methyltransferase activity in a buffer containing 50 mM HEPES, pH 8, 2.5 mM MgCl$_2$, 1 mM dithiothreitol, 50 mM ammonium acetate, 0.05 mM EDTA, 1 mM spermidine, and 0.5 mM SAM for 2 h at 30°. Assays contained decreasing concentrations of crude extracts (~0.1 mg/ml to 1 µg/ml by factors of 10) made from wild-type or *trm4-Δ* yeast strains; -, no extract.

lost during the purification. For example, although many of the known RNA methyltransferases use SAM as a methyl-group donor, other cofactors such as tetrahydrofolate, which is used by some m^{5}U methyltransferases (Urbonavicius *et al.*, 2005), and even cobalamin can be considered if a SAM-dependent activity is not observed. Likewise, the choice of the thin-layer chromatographic system to be used to resolve the reaction components is something to be considered carefully. Several well-known TLC systems have been successfully used for the identification and assay of nucleic acid base and sugar modifications, including the cellulose system described here, which can also be used with other solvent systems (Nishimura and Kuchino, 1983), as well as PEI-cellulose–based assays resolved in either low or high salt buffers, often sodium formate or lithium chloride (Bochner and Ames, 1982; Randerath *et al.*, 1981), and even a silica-TLC–based assay, resolved in an *n*-propanol/H$_2$O/ammonium hydroxide solvent system (55:35:10 v/v/v) that has been successfully used for analysis of modified nucleotide species.

When looking for a modification activity that is catalyzed by an unknown gene product, other steps should be taken to confirm the biochemical assay, such as comigration with standards or further physical analysis of reaction products. After verification of the identity of the observed reaction products and demonstration that the activity can be observed in yeast crude extracts, the analysis of the yeast proteomic library can begin.

4. PARALLEL BIOCHEMICAL ANALYSIS OF THE YEAST PROTEOME WITH THE MORF LIBRARY

Several large-scale collections of cloned yeast genes have been used as the source of whole proteomes for identification of proteins that catalyze specific biochemical activities, bind to particular ligands, or act as substrates for individual posttranslational modifications (Hazbun and Fields, 2002; Martzen *et al.*, 1999; Phizicky and Grayhack, 2006; Ptacek *et al.*, 2005; Ubersax *et al.*, 2003; Zhu *et al.*, 2001). Two early genomic libraries each expressed cloned ORFs (amplified on the basis of the original 1996 annotation of the yeast genome) fused at their N-terminus to a glutathione S-transferase (GST)-tag, and the resulting GST–ORF fusion proteins were purified on glutathione agarose columns (Martzen *et al.*, 1999). The yeast MORF library used here contains 5854 yeast strains, each of which expresses a C-terminally tagged ORF under control of the galactose inducible promoter (Fig. 6.2) (Gelperin *et al.*, 2005). This library is available through Open Biosystems (www.openbiosystems.com/GeneExpression/Yeast/ORF). Each strain in this library contains a plasmid with a single sequence-verified ORF, with an average of 1170 base pairs of completed sequence for each clone, representing a complete gene sequence for more than half of the clones. Moreover, the ORFs in this library derive from a recently updated annotation of the yeast genome and contain a C-terminal tripartite (His_6-HA-3C-ZZ domain of protein A) purification tag that allows for efficient purification by use of IgG Sepharose affinity chromatography based on binding to the ZZ domain, followed by cleavage with 3C protease, with the possibility of an additional (or alternative) purification with immobilized metal ion affinity chromatography, as well as facile detection of the proteins that use the HA epitope (Gelperin *et al.*, 2005). The yeast MORF library expresses proteins to high levels (up to 2 mg or more/liter of culture for the best expressing proteins) and has been used successfully for the identification of new tRNA modification enzymes (Wilkinson *et al.*, 2007), as well as for identification of several previously known tRNA modification activities and a large collection of new N-glycosylated proteins in yeast (Gelperin *et al.*, 2005). Procedures for the growth of this library and preparations of proteins are found elsewhere (Gelperin *et al.*, 2005).

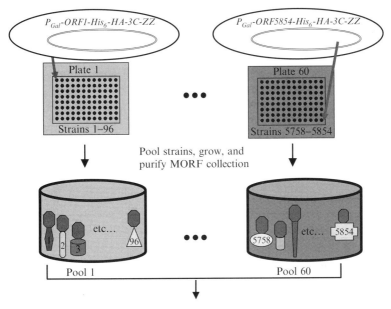

Figure 6.2 Schematic of the biochemical genomics approach by use of the yeast MORF collection. The yeast MORF collection (Gelperin *et al.*, 2005) consists of 5854 yeast strains, each of which contains a single plasmid that can be used to direct the expression of a single-yeast ORF under control of the inducible Gal promoter. The C-terminal His_6-HA-3C-ZZ domain tag fused to each ORF is described in detail in the text. To carry out a proteomic screen for a desired RNA modification activity, the strains are pooled from each 96-well plate such that all of the strains from each plate are grown in a single culture, and this culture is used as the source for purification of all of the MORF proteins in a single tube. The resulting preparation consists of 60 pools of purified proteins, where each pool contains the 96 ORFs that correspond to the 96 strains on the original source plate. Once the pools have been prepared, they can be used as the source of protein to identify the ORF that catalyzes a desired biochemical activity by a parallel screen of all 60 pools. (See color insert.)

By use of the biochemical assay for m^5C modification of tRNA^{Gly} described previously, we carried out a parallel biochemical screen of the MORF library by assaying each of the purified pools of protein (60 total) for the ability to catalyze formation of m^5C on tRNA^{Gly} (Fig. 6.3A). The m^5C modification activity was found in reactions that used pool 22, and thus the gene product responsible for this methylation copurifies with one of the 96 ORFs found in this pool. This result is consistent with the published location of the strain expressing Trm4 in plate 22. We note that a single purification of the pools of proteins from the proteomic library can be used for at least 2 years after preparation, because the pools of purified proteins used for this assay were purified that long ago and have since been stored at $-20°$ in 50% glycerol, yet still retain sufficient activity for easy detection of

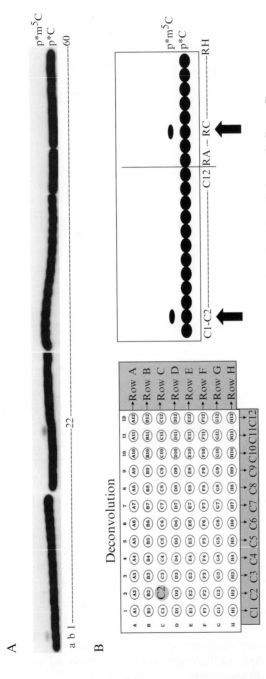

Figure 6.3 Identification of the Trm4 m⁵C methyltransferase with the biochemical genomics approach. (A) Assay of the yeast MORF library for m⁵C methyltransferase activity. By use of the assay described in Fig. 6.1, all 60 pools of the MORF collection were assayed for their ability to form m⁵C on the labeled tRNA^{Gly} substrate. Consistent with the known location of Trm4 on plate 22 of the MORF collection, the only positive signal observed in any of the MORF pools was observed in Pool 22; lane a, negative control; lane b, positive control. (B) Method for identification of the individual protein responsible for a desired RNA modification activity, the plate from which the positive signal was observed is divided into subpools representing either the twelve columns (1–12) or the eight rows (A–H) of the 96-well plate. After growth of these pools and purification of the proteins from each subpool, each can be assayed individually for the ability to catalyze the desired activity. Thus, in the Trm4 example described here, the positive signal from Pool 22 would be recapitulated exclusively in the subpools from column 2 and from row C, implicating the gene at this position in the catalysis of this activity.

the reaction. Furthermore, we have found that several assayed activities are retained in preparations that have been frozen at −70° for several years.

The sensitivity of this method for detection of activities in the purified pools of proteins is also evident from this assay, both in terms of the increase in activity observed over the activity in crude extracts and in the very low background in the negative protein pools. Only 2.7% of the C residues in tRNA^{Gly} were methylated in the reaction with crude extract, (corresponding to an average of 0.5 modifications/tRNA species). The amount of product observed by use of purified pool 22 increased to 5% of the total C residues, representing approximately one m^{5}C modification per tRNA species, and although the extent of modification observed under these conditions is less than the two m^{5}C modifications previously shown to occur in wild-type tRNA^{Gly} (Jackman et al., 2003), this level of activity is easily detected over the background in other pools (∼0.4%).

5. DECONVOLUTION OF THE ACTIVE POOL TO IDENTIFY THE INDIVIDUAL GENE RESPONSIBLE FOR ACTIVITY

The procedure for deconvolution of the positive signal to identify the individual strain required for any given activity involves the purification of a second set of pools of proteins, all derived from the single plate that gave a positive signal in the biochemical assay, for example, plate 22 in the case of Trm4 (Fig. 6.3A) (Martzen et al., 1999; Phizicky and Grayhack, 2006; Wilkinson et al., 2007). Thus, eight pools corresponding to each of the rows (A–H) of the 96-well plate are produced; each of these pools derives from 12 strains and, therefore, contains only 12 purified proteins (Fig. 6.3B). Likewise, 12 pools corresponding to each of the columns (1–12) are prepared in a similar manner, with each pool containing proteins corresponding to the eight ORFs found in that particular column (Fig. 6.3B). The pools of proteins (20 total) are assayed again for the biochemical activity to identify the row and column of the original plate containing the ORF responsible for the desired activity. In the case of Trm4, whose identity is already known, the positive signal would be observed in both the pool of purified proteins from row C and the pool from column 2 (Fig. 6.3B), resulting in identification of the individual gene-specifying activity in a single set of deconvolution assays.

6. CONFIRMATION OF THE IDENTITY OF THE GENE RESPONSIBLE FOR THE MODIFICATION ACTIVITY

The final two steps in the identification of a new enzyme activity are (1) purification and assay of the individual identified protein to determine whether it is both necessary and sufficient for the desired modification activity, and (2) examination of the individual mutant yeast strain to demonstrate that it lacks the corresponding modification, therefore implicating the identified gene in catalysis of the modification *in vivo* as well as *in vitro*. The first step, biochemical characterization of the individual protein, can be readily accomplished by use of the individual MORF strain identified in the biochemical screen (the strain at position C2 on plate 22 for the Trm4 example, previously) both as a source of individually purified protein to verify the identity of the gene responsible for the modification activity and as a source for cloning the yeast ORF into a vector suitable for expression and purification from *E. coli*. If the protein is active when purified from *E. coli*, this indicates that the identified ORF is the active subunit, and no additional yeast proteins are required for modification activity. Although many nucleic acid modifications are catalyzed by single subunit enzymes, a number of known RNA modification enzymes have been demonstrated to act as part of multisubunit enzyme complexes and are, therefore, unable to catalyze their respective modification reactions in the absence of all members of the enzyme complex (Alexandrov *et al.*, 2002; Anderson *et al.*, 1998, 2000; Gerber and Keller, 1999; Ma *et al.*, 2005). Furthermore, if overproduction of the ORF in yeast results in overproduction of the activity in the crude extracts derived from these cells, and the protein purified from *E. coli* has the same specific activity as that purified from yeast, this demonstrates that there are no required posttranslational modifications and likely no required additional subunits for activity. Finally, if the yield of activity is reasonable from crude extracts, this suggests that there are no other required factors lost during the purification. For example, we found the FAD-requirement of Dus1 for synthesis of dihydrouridine (DHU), because we observed a substantial loss of DHU synthase activity during purification of the enzyme (Xing *et al.*, 2004).

The second step in the confirmation of the results of the proteomic assay, analysis of the *in vivo* requirement for the identified modification enzyme, can be accomplished by comparison of nucleosides of appropriate tRNAs derived from wild-type yeast strains with those derived from mutant yeast strains (Gu *et al.*, 2003; Jackman *et al.*, 2003; Wilkinson *et al.*, 2007; Xing *et al.*, 2004). The mutant yeast strains required for this analysis can either be obtained from the yeast deletion strain collection or the yeast conditional knockdown collection or can be constructed by traditional methods.

We note that this proteomic approach can still be used in the case of protein complexes that are responsible for a modification activity. In some cases, multiple positive pools have been observed in the original library screen, reflecting the persistence of the interaction between protein members of the complex throughout the purification of the tagged ORF and, therefore, directly identifying the relevant protein subunits (Alexandrov *et al.*, 2002). Alternately, for some multisubunit complexes, such as the guide-RNA dependent pseudouridylase activity catalyzed by an H/ACA sno-sca RNP at position 42 of U2 snRNA (Ma *et al.*, 2005), a single positive signal was observed in the pool corresponding to one member of the complex. The existence of a multisubunit enzyme complex was confirmed, because the individual positive ORF was inactive when expressed and purified from *E. coli*. In this case, the missing components of the H/ACA sno-sca RNP that were required for activity were identified through the use of chromosomally TAP-tagged strains derived from members of the complex and analysis of the activity from the resulting purifications (Ma *et al.*, 2005). Presumably the other ORFs that are part of this pseudouridine synthase enzyme complex were not present or were not active in the N-terminal GST-ORF library that was tested. An alternate approach to identification of missing partners for a given protein complex would be to rescreen the proteomic library for the ability to enhance the activity of the *E. coli* purified protein.

7. POSSIBLE REASONS FOR LACK OF SUCCESS WITH THIS APPROACH

Although several yeast proteomic libraries have been used successfully to find a number of RNA modification enzymes, there are three features of the library that could prevent detection of activity. First, although the coverage of the yeast proteome, particularly with the MORF library is extremely good, representing >90% of the verified ORFs in SGD (*Saccharomyces* genome database, www.yeastgenome.org), the particular gene of interest may not be present in the library. Second, the identity and position of the tag could adversely affect the activity of the ORF of interest, although the use of alternately tagged versions, such as the previously described N-terminal GST-tagged proteomic library, provides an additional source of protein to be used to circumvent this issue. Third, the ORF may not be present at sufficient concentration in the purified fractions to enable detection of its enzymatic activity. A significant fraction of the individual ORFs in the MORF library are expressed poorly despite the use of the same inducible promoter for expression (Gelperin *et al.*, 2005). Also, an individual ORF may not be soluble or may not survive the purification procedure, and therefore its activity may not be detectable in the purified protein pools.

Also, two features of the assay used for a given biochemical activity may lead to an inability to detect the desired gene. First, as described previously, the absence of the correct cofactor, a required metal ion, or the appropriate buffer conditions during purification of the pools for retention of enzymatic activity would preclude the successful identification of an enzyme in the protein pools. Second, the presence of RNA degradation or proteolytic activities in the same protein pool as the desired activity could, in principle, prevent the detection of an individual gene product in a pool, although this has not been observed to date.

8. Conclusions

Although yeast proteomic expression libraries have been used successfully to identify tRNA and snRNA modification and processing enzymes, the potential applications of proteomic expression libraries are not limited to these enzymes. Any enzymatic activity, or even any DNA or RNA binding activity, could be identified with this approach. Moreover, proteomic expression libraries such as the yeast MORF library have recently been described for other organisms (Dricot *et al.*, 2004; Lamesch *et al.*, 2007; Wei *et al.*, 2005), and the recent advances in cloning and sequencing technology will most certainly result in the construction of similar libraries from other organisms. Therefore, this approach will increasingly be able to be used to identify new and biologically important enzymes from a number of organisms.

REFERENCES

Alexandrov, A., Martzen, M. R., and Phizicky, E. M. (2002). Two proteins that form a complex are required for 7-methylguanosine modification of yeast tRNA. *RNA* **8,** 1253–1266.

Anderson, J., Phan, L., Cuesta, R., Carlson, B. A., Pak, M., Asano, K., Bjork, G. R., Tamame, M., and Hinnebusch, A. G. (1998). The essential Gcd10p-Gcd14p nuclear complex is required for 1-methyladenosine modification and maturation of initiator methionyl-tRNA. *Genes Dev.* **12,** 3650–3662.

Anderson, J., Phan, L., and Hinnebusch, A. G. (2000). The Gcd10p/Gcd14p complex is the essential two-subunit tRNA(1-methyladenosine) methyltransferase of *Saccharomyces cerevisiae*. *Proc. Natl. Acad. Sci. USA* **97,** 5173–5178.

Ansmant, I., Motorin, Y., Massenet, S., Grosjean, H., and Branlant, C. (2001). Identification and characterization of the tRNA:Psi 31-synthase (Pus6p) of *Saccharomyces cerevisiae*. *J. Biol. Chem.* **276,** 34934–34940.

Arluison, V., Batelier, G., Ries-Kautt, M., and Grosjean, H. (1999). RNA: Pseudouridine synthetase Pus1 from *Saccharomyces cerevisiae*: Oligomerization property and stoichiometry of the complex with yeast tRNA(Phe). *Biochimie* **81,** 751–756.

Becker, H. F., Motorin, Y., Planta, R. J., and Grosjean, H. (1997). The yeast gene YNL292w encodes a pseudouridine synthase (Pus4) catalyzing the formation of psi55 in both mitochondrial and cytoplasmic tRNAs. *Nucleic Acids Res.* **25,** 4493–4499.

Bjork, G. R., Jacobsson, K., Nilsson, K., Johansson, M. J., Bystrom, A. S., and Persson, O. P. (2001). A primordial tRNA modification required for the evolution of life? *EMBO J.* **20,** 231–239.

Bochner, B. R., and Ames, B. N. (1982). Complete analysis of cellular nucleotides by two-dimensional thin layer chromatography. *J. Biol. Chem.* **257,** 9759–9769.

Conrad, J., Niu, L., Rudd, K., Lane, B. G., and Ofengand, J. (1999). 16S ribosomal RNA pseudouridine synthase RsuA of *Escherichia coli*: Deletion, mutation of the conserved Asp102 residue, and sequence comparison among all other pseudouridine synthases. *RNA* **5,** 751–763.

Dricot, A., Rual, J. F., Lamesch, P., Bertin, N., Dupuy, D., Hao, T., Lambert, C., Hallez, R., Delroisse, J. M., Vandenhaute, J., Lopez-Goni, I., Moriyon, I., et al. (2004). Generation of the Brucella melitensis ORFeome version 1.1. Genome Res. 14, 2201–2206.

Gelperin, D. M., White, M. A., Wilkinson, M. L., Kon, Y., Kung, L. A., Wise, K. J., Lopez-Hoyo, N., Jiang, L., Piccirillo, S., Yu, H., Gerstein, M., Dumont, M. E., Phizicky, E. M., Snyder, M., and Grayhack, E. J. (2005). Biochemical and genetic analysis of the yeast proteome with a movable ORF collection. Genes Dev. 19, 2816–2826.

Gerber, A. P., and Keller, W. (1999). An adenosine deaminase that generates inosine at the wobble position of tRNAs. Science 286, 1146–1149.

Gu, W., Jackman, J. E., Lohan, A. J., Gray, M. W., and Phizicky, E. M. (2003). tRNAHis maturation: An essential yeast protein catalyzes addition of a guanine nucleotide to the 5′ end of tRNAHis. Genes Dev. 17, 2889–2901.

Hazbun, T. R., and Fields, S. (2002). A genome-wide screen for site-specific DNA-binding proteins. Mol. Cell. Proteomics 1, 538–543.

Jackman, J. E., Montange, R. K., Malik, H. S., and Phizicky, E. M. (2003). Identification of the yeast gene encoding the tRNA m1G methyltransferase responsible for modification at position 9. RNA 9, 574–585.

Lamesch, P., Li, N., Milstein, S., Fan, C., Hao, T., Szabo, G., Hu, Z., Venkatesan, K., Bethel, G., Martin, P., Rogers, J., Lawlor, S., et al. (2007). hORFeome v3.1: A resource of human open reading frames representing over 10,000 human genes. Genomics 89, 307–315.

Ma, X., Yang, C., Alexandrov, A., Grayhack, E. J., Behm-Ansmant, I., and Yu, Y. T. (2005). Pseudouridylation of yeast U2 snRNA is catalyzed by either an RNA-guided or RNA-independent mechanism. EMBO J. 24, 2403–2413.

Ma, X., Zhao, X., and Yu, Y. T. (2003). Pseudouridylation (Psi) of U2 snRNA in S. cerevisiae is catalyzed by an RNA-independent mechanism. EMBO J. 22, 1889–1897.

Martzen, M. R., McCraith, S. M., Spinelli, S. L., Torres, F. M., Fields, S., Grayhack, E. J., and Phizicky, E. M. (1999). A biochemical genomics approach for identifying genes by the activity of their products. Science 286, 1153–1155.

Motorin, Y., and Grosjean, H. (1999). Multisite-specific tRNA:m5C-methyltransferase (Trm4) in yeast Saccharomyces cerevisiae: Identification of the gene and substrate specificity of the enzyme. RNA 5, 1105–1118.

Nishimura, S., and Kuchino, Y. (1983). "Methods of DNA and RNA Sequencing." Praeger, New York.

Phizicky, E. M., and Grayhack, E. J. (2006). Proteome-scale analysis of biochemical activity. Crit. Rev. Biochem. Mol. Biol. 41, 315–327.

Pintard, L., Lecointe, F., Bujnicki, J. M., Bonnerot, C., Grosjean, H., and Lapeyre, B. (2002). Trm7p catalyses the formation of two 2′-O-methylriboses in yeast tRNA anticodon loop. EMBO J. 21, 1811–1820.

Ptacek, J., Devgan, G., Michaud, G., Zhu, H., Zhu, X., Fasolo, J., Guo, H., Jona, G., Breitkreutz, A., Sopko, R., McCartney, R. R., Schmidt, M. C., et al. (2005). Global analysis of protein phosphorylation in yeast. Nature 438, 679–684.

Randerath, K., Reddy, M. V., and Gupta, R. C. (1981). 32P-labeling test for DNA damage. Proc. Natl. Acad. Sci. USA 78, 6126–6129.

Ubersax, J. A., Woodbury, E. L., Quang, P. N., Paraz, M., Blethrow, J. D., Shah, K., Shokat, K. M., and Morgan, D. O. (2003). Targets of the cyclin-dependent kinase Cdk1. Nature 425, 859–864.

Urbonavicius, J., Skouloubris, S., Myllykallio, H., and Grosjean, H. (2005). Identification of a novel gene encoding a flavin-dependent tRNA:m5U methyltransferase in bacteria—evolutionary implications. Nucleic Acids Res. 33, 3955–3964.

Wei, C., Lamesch, P., Arumugam, M., Rosenberg, J., Hu, P., Vidal, M., and Brent, M. R. (2005). Closing in on the C. elegans ORFeome by cloning TWINSCAN predictions. Genome Res. 15, 577–582.

Wilkinson, M., Crary, S., Jackman, J. E., Grayhack, E., and Phizicky, E. (2007). A 2′-*O*-methyltransferase responsible for modification of yeast tRNA at position 4. *RNA* **13,** 404–413.

Xing, F., Hiley, S. L., Hughes, T. R., and Phizicky, E. M. (2004). The specificities of four yeast dihydrouridine synthases for cytoplasmic tRNAs. *J. Biol. Chem.* **279,** 17850–17860.

Xing, F., Martzen, M. R., and Phizicky, E. M. (2002). A conserved family of *Saccharomyces cerevisiae* synthases effects dihydrouridine modification of tRNA. *RNA* **8,** 370–381.

Yu, Y. T. (1999). Construction of 4-thiouridine site-specifically substituted RNAs for cross-linking studies. *Methods* **18,** 13–21.

Zhu, H., Bilgin, M., Bangham, R., Hall, D., Casamayor, A., Bertone, P., Lan, N., Jansen, R., Bidlingmaier, S., Houfek, T., Mitchell, T., Miller, P., *et al.* (2001). Global analysis of protein activities using proteome chips. *Science* **293,** 2101–2105.

CHAPTER SEVEN

IDENTIFICATION OF GENES ENCODING tRNA MODIFICATION ENZYMES BY COMPARATIVE GENOMICS

Valérie de Crécy-Lagard

Contents

Abstract

As the molecular adapters between codons and amino acids, transfer-RNAs are pivotal molecules of the genetic code. The coding properties of a tRNA molecule do not reside only in its primary sequence. Posttranscriptional nucleoside modifications, particularly in the anticodon loop, can modify cognate codon recognition, affect aminoacylation properties, or stabilize the codon-anticodon wobble base pairing to prevent ribosomal frameshifting. Despite a wealth of biophysical and structural knowledge of the tRNA modifications themselves, their pathways of biosynthesis had been until recently only partially character-ized. This discrepancy was mainly due to the lack of obvious phenotypes for tRNA modification–deficient strains and to the difficulty of the biochemical assays used to detect tRNA modifications. However, the availability of hundreds of whole-genome sequences has allowed the identification of many of these missing tRNA-modification genes. This chapter reviews the methods that were used to

Department of Microbiology and Cell Science, University of Florida, Gainesville, Florida

Methods in Enzymology, Volume 425
ISSN 0076-6879, DOI: 10.1016/S0076-6879(07)25007-4

identify these genes with a special emphasis on the comparative genomic approaches. Methods that link gene and function but do not rely on sequence homology will be detailed, with examples taken from the tRNA modification field.

1. INTRODUCTION

The availability of nearly 500 complete genomes (http://www. genomesonline.org/) has changed the manner by which experimental scientists can identify novel enzymes and pathways. Traditionally, linking genes and their functions started with protein purification or mutant isolation steps. Today, the bench scientist can make and validate functional predictions by combining genomic datamining with wet-laboratory experiments (see El Yacoubi *et al.* [2006]; Gerdes *et al.* [2006]; Loh *et al.* [2006]; and Xu *et al.* [2006] for recent examples). No programming skills are needed, because the genomic data and analysis tools are now freely accessible through web-based interfaces.

The sequencing effort of the past decade has revealed that 20–60% of the predicted proteins in any given genome are of unknown function (Osterman and Overbeek, 2003). Experimentalists have in-depth knowledge of specific metabolic and biological areas that most computer scientists lack. If they can harness the genomic data-mining tools, biologists and chemists are uniquely poised to predict the function of the "unknowns" and validate them in the laboratory.

The field of tRNA modification provides a good illustration of the combined power of comparative genomics and experimental validation. Even though most modifications present in tRNA molecules were discovered 20–30 years ago, many tRNA-modification genes were left unidentified (Björk, 1995; Hopper and Phizicky, 2003). The lack of knowledge about the pathways involved in nucleoside modification was largely due to their resistance to traditional biochemical and genetic characterization. Identification and purification of relevant enzyme activities from crude cell-free extracts was complicated by several factors: the difficulty of obtaining appropriate tRNA substrates, the presence of endogenous RNases that degrade the RNA substrates and products, a lack of appropriate assays, and the typically low abundance of the enzymes involved. Likewise, traditional genetic approaches were hindered by the lack of specific phenotypes in most cases. Finally, the unambiguous identification of a gene involved in tRNA modification ultimately depended on determining the presence or absence of the specific modified nucleoside in tRNA—a laborious and technically challenging process when working with large libraries of mutants (Grosjean *et al.*, 2004). As a consequence, the identities of 50% of the tRNA modification genes were still unknown 5 years ago (de Crécy-lagard, 2004; Eastwood Leung *et al.*, 1998). Some, such as the dihydrouridine synthesis genes, were "globally missing," meaning they had not been identified in any organisms. Others, such as the gram-positive m⁵U54 methylase, were "locally missing" and identified only in a subset of organisms. This represented quite a large number of genes given that, in most organisms, ~1% of the genome is dedicated to encoding tRNA modification enzymes (Björk and Kohli, 1990; Hopper and Phizicky, 2003). Clearly, new approaches to identify tRNA modification genes were necessary.

2. METHODS TO IDENTIFY MISSING tRNA MODIFICATION GENES

With the discovery of nearly 50 genes since 2002 (Tables 7.1–7.3), this gap in genetic understanding of tRNA modification has been nearly filled (at least for the model organisms *Escherichia coli* and *Saccharomyces cerevisiae*). Only a handful of these genes were identified by traditional genetic or biochemical methods (Table 7.1), whereas most were found by use of postgenomic experimental platforms (Table 7.1) or bioinformatic tools (Tables 7.2 and 7.3).

The availability of whole-genome sequences has driven large-scale systematic experimental efforts such as structural genomics initiatives, systematic interaction mapping, or systematic gene disruption combined with phenotypic screenings (Huynen *et al.*, 2004; Mittl and Grutter, 2001). For the purpose of identifying missing tRNA modification genes, these approaches have been quite effective. For example, nearly 10 genes (Table 7.1) have been identified by use of "biochemical profiling approaches" (discussed in Chapter 6). In these studies, all the proteins of *S. cerevisiae* (Martzen *et al.*, 1999) and *E. coli* (Kitagawa *et al.*, 2005) have been cloned and expressed and were tested in pools or individually for specific enzyme activities. In other studies, large-scale deletion mutant libraries have been completed for *S. cerevisiae* (Winzeler *et al.*, 1999), *B. subtilis* (Kobayashi *et al.*, 2003), and *E. coli*, and screening these libraries by LC-MS analysis of enzymatic digests of tRNA isolated from individual clones led to the identification of 10 other tRNA-modification gene families (Table 7.1).

Other systematic efforts that could lead to the discovery of missing tRNA-modification genes are the application of microarray technology to detect modifications (Hiley *et al.*, 2005; Peng *et al.*, 2003) and the availability of structural genomics data. To date, 2000 structures have been deposited by structural genomics programs in the Protein Data Bank (http://www.rcsb.org/pdb/), and more than 50,000 of these proteins have been cloned and expressed in the process (http://targetdb.pdb.org/statistics/sites/PSI.html). Structural proteomics has been quite efficient at predicting RNA/protein interactions, a first hint that a protein could be involved in tRNA or rRNA processing (see Yakunin *et al.* [2004] for review).

Table 7.1 tRNA modification genes recently identified by use of experimentally driven approaches

Functional role	Verified in	Protein name	Initial experimental method and reference
Postgenomic systematic approaches			
Dihydrouridine synthases	*S. cerevisiae*	Dus1 Dus2 Dus3 Dus4 Dus5	Biochemical profiling (Xing *et al.*, 2002)
tRNA(His) guanylyltransferase	*S. cerevisiae*	Thg1	Biochemical profiling (Gu *et al.*, 2003)
tRNA m⁷G-methyltransferase	*S. cerevisiae*	Trm8 Trm82	Biochemical profiling (Alexandrov *et al.*, 2002)
2-thiouridine synthesis	*E. coli*	TusA TusB TusC TusD	Systematic mutant analysis (Ikeuchi *et al.*, 2006)
Wybutosine biosynthesis	*S. cerevisiae*	WyeA WyeB WyeC WyeD WyeE	Systematic mutant analysis (Noma *et al.*, 2006)
Classical genetic or biochemical approaches			
tRNA (uridine-5-oxyacetic acid methyl ester) 34 synthase	*E. coli*	CmoA	Genetic screen (Nasvall *et al.*, 2004)
tRNA (5-methoxyuridine) 34 synthase	*E. coli*	CmoB	Genetic screen (Nasvall *et al.*, 2004)
tRNA pseudouridine 13 synthase	*E. coli*	TruD	Protein purification (Kaya and Ofengand, 2003)
tRNA(cytosine32)-2-thiocytidine synthetase	*E. coli*	TtcA	Mapping of a previously identified mutation (Jager *et al.*, 2004)
Queuosine biosynthesis	*E. coli*	QueC	Mutant complementation (Gaur and Varshney, 2005)

Table 7.2 tRNA modification genes recently identified by use of homology-based bioinformatic approaches

Functional role	Verified in	Protein name	Identification method and reference
tRNA-specific adenosine-34 deaminase (EC 3.5.4.-)	E. coli	TadA	BLAST with S. cerevisiae deaminase gene (Wolf et al., 2002)
Selenophosphate-dependent tRNA 2-selenouridine synthase	E. coli	YbbB	Search for proteins containing rhodanese domains (Wolfe et al., 2004)
tRNA pseudouridine synthase (position 55)	P. abyssi	PsuX	Gapped-BLAST with Euglena gracilis Cbf5p (Roovers et al., 2006; Watanabe and Gray, 2000)
tRNA m$^2_2$G10 methyltransferase	S. cerevisiae	Trm112p Trm11p	Protein fold prediction (Purushothaman et al., 2005)
tRNA m^{1}A$_{58}$ methyltransferase	Thermus thermophilus	TrmI	tBlastN with Rv2118c from M. tuberculosis (Droogmans et al., 2003)
tRNA (cytosine32/34–2'-O-)-methyltransferase	S. cerevisiae	Trm7p	BLAST with FtsJ of E. coli (Pintard et al., 2002)
tRNA:Cm$_{32}$/Um$_{32}$ methyltransferase	E. coli	YhfQ = TrmJ	SPOUT domain search (Purta et al., 2006)
tRNA (m^{7}G46) methyltransferase	E. coli	YggH	Protein fold prediction (De Bie et al., 2003)
Wybutosine biosynthesis	S. cerevisiae	WyeC	Methylase Motif searches (Kalhor et al., 2005)
Mitochondrial tRNA-specific 2-thiouridylase 1	Homo sapiens S. cerevisiae	MTU1	BLAST with E. coli MnmA (Umeda et al., 2005)
tRNA (ribose 2'-O-methylase), position cytosine 56	Pyrococus abyssii	PAB1040	SPOUT domain search (Renalier et al., 2005)

Table 7.3 tRNA modification genes identified by use of non-homology–based comparative genomics techniques

Functional role	Verified in	Protein name	Key bioinformatic evidence and reference
5-Methylaminomethyl-2-thiouridine synthase	*E. coli*	MnmC	Clustering/fold recognition (Bujnicki *et al.*, 2004b; de Crécy-lagard, 2004)
Flavin-dependent tRNA:m⁵U methyltransferase	*B. subtilis*	Gid	Occurrence profile (Urbonavicius *et al.*, 2005)
tRNA lysidine synthase	*E. coli*	MesJ	Occurrence profile and essentiality data (Soma *et al.*, 2003)
tRNA Carbamoyl-threonyl-adenosine synthase	*S. cerevisiae*	Sua5	Occurrence profile/structure[a]
Wybutosine biosynthesis	*S. cerevisiae*	WyeA	Occurrence profile (Waas *et al.*, 2005)
Bacterial tRNA dihydrouridine synthase	*E. coli*	DusA DusB DusC	Occurrence profile/operon (Bishop *et al.*, 2002)
Queuosine/archeosine biosynthesis	*Acinetobacter baylyi*	QueE QueC QueD	Occurrence profile/operon (Reader *et al.*, 2004)
PreQ₀ reductase	*E. coli* *B. subtilis*	QueF	Occurrence profile/operon (Reader *et al.*, 2004; Van Lanen *et al.*, 2005)

[a] de Crécy-lagard and collaborators (unpublished results).

These postgenomic methods are still labor intensive and expensive. Starting with protein pools (Martzen *et al.*, 1999) or with mutants carrying large deletions (Ikeuchi *et al.*, 2006), reduces the quantity of assays to manageable numbers, but laboratories that use these systematic approaches to find tRNA-modification genes are still scarce, mainly because of the remaining complexity of the tRNA-modification enzyme assays. However, the availability of these postgenomic resources (clones and mutants) tremendously increases the speed at which bioinformatic-driven predictions can be tested. Hence, most tRNA-modification genes recently identified were found by combining an initial bioinformatic search with an experimental validation step (Tables 7.2 and 7.3). The bioinformatic tools used can be separated into homology based and non-homology based. The homology-based mining tools are known to most experimental scientists and will only be briefly discussed here in the context of tRNA-modification enzymes. The use of the less familiar non-homology–based genomic mining tools is the main focus of this review.

3. HOMOLOGY-BASED GENOMIC DATA MINING METHODS

Functional inferences based on comparative sequence analysis are well-established foundations of genomic annotation. The most significant advances in this field over the past decade are directly related to the dramatic increase in the number of sequenced genomes, as well as to the development of robust and sensitive search algorithms, such as FASTA, BLAST and their modifications (for an overview, see Koonin and Galperin [2003]). Domain analysis and grouping of putative orthologs (such as Cluster of Orthologous Groups or COGs [Tatusov *et al.*, 2001]) play an important role in projection of functional assignments between diverse species. For well-studied gene families, in which the initial annotation has been experimentally verified, these homology-based methods are quite accurate in predicting function (Tian and Skolnick, 2003b). However, factors such as low sequence similarity (Tian and Skolnick, 2003b), multidomain proteins (Hegyi and Gerstein, 2001), gene duplications (Gerlt and Babbitt, 2000; Tian and Skolnick, 2003a), and nonorthologous displacements (Galperin and Koonin, 1998) have all contributed to incorrect or absent annotations. This has been a major problem in the field of tRNA-modification enzymes, because many are members of large paralogous families, and transferring functional annotations with BLAST scores alone can be very dangerous, particularly between kingdoms. Cases where the closest homologs in two genomes do not catalyze the same reaction are numerous in the tRNA-modification field with the added complication of having both tRNA and rRNA as potential substrates (see Jeltsch *et al.* [2006]; Motorin and Grosjean [1999]; Urbonavicius *et al.* [2005]; and Xing *et al.* [2004] for specific examples). That said, the use of sensitive search algorithms such as PSI-BLAST or Gapped-BLAST (Altschul *et al.*, 1997), the development of protein fold–based methods and motifs to differentiate methylase subfamilies (Bujnicki *et al.*, 2004a; Katz *et al.*, 2003), and the identification of RNA binding domains such as THUMP, PUA, or SPOUT (Anantharaman *et al.*, 2002a,b; Aravind and Koonin, 2001; Gustafsson *et al.*, 1996; Kurowski *et al.*, 2003) have led to many of the predictions and validations listed in Table 7.2. These methods are, however, limited to tRNA-modification enzymes that are members of superfamilies such as deaminases, methylases, or pseudouridine synthases. For the other "missing" tRNA-modification genes, the inherent limitations of homology-based approaches (only similar objects can be identified) require the use of non–homology–based comparative genomic methods.

4. Non-Homology–Based Genomic Data Mining Methods

Integrating different types of genomic evidence to identify missing genes or predict the function of unknown genes started in the late 1990s just a few years after the first set of genomes was sequenced (see Bishop *et al.* [2002]; Bobik and Rasche [2001]; Daugherty *et al.* [2001]; Graham *et al.* [2001]; and Heath and Rock [2000] for early examples). Ten years later, the success stories are now plentiful, and several reviews have covered both the techniques and specific examples (Galperin and Koonin, 2000; Huynen *et al.*, 2003; Kharchenko *et al.*, 2006; Makarova and Koonin, 2003). The author recommends starting with the review by Osterman and Overbeek (Osterman and Overbeek, 2003) to grasp the core concepts of this field. These are summarized in Fig. 7.1. In short, analysis of gene clustering on the chromosome, gene fusions events, phylogenetic distribution profiles, interaction data, coexpression data, structural genomics data, phenomics data, and regulatory motifs can lead to non–homology–based predictions that can be then tested experimentally. Comparative genomics platforms in which the experimental scientist would input a gene name or sequence and all the possible functional association would be given as outputs or where one could ask complex questions integrating different types of data and genes answering these criteria would be found automatically are still not available. However, many tools have already been developed and partially integrated. Describing how to make predictions on gene function by use of these tools in a time-efficient manner with just a personal computer and Internet access is the focus of the rest of the review.

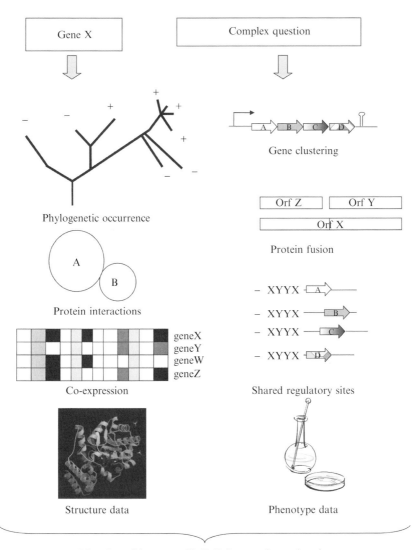

Figure 7.1   Comparative genomic strategies used to make predictions on gene function. (See color insert.)

With so many databases now available (Chen *et al.*, 2007; Field *et al.*, 2005), the "experimental" section cannot be exhaustive and reflects the personal preferences of the author (see Table 7.4 for the list of databases discussed in this review). However, a deliberate choice was made to include only resources that are available through a web interface and that are the most useful to make predictions on gene function. In the limits of the

Table 7.4 Freely available databases and analysis platforms discussed in this review[a]

Name	Location
Integrative databases	
STRING	http://dag.embl-heidelberg.de/newstring_cgi/show_input_page.pl
CMR-genome properties	http://www.tigr.org/tigr-scripts/CMR2/GenomeSlicer.spl
SEED	http://theseed.uchicago.edu/FIG/
NMPDR	http://www.nmpdr.org/
MicrobesOnline	http://www.microbesonline.org/
CoGenT++	http://cgg.ebi.ac.uk/cgg/cpp_sitemap.html
NCBI	http://www.ncbi.nlm.nih.gov/
Protein fusion analysis	
Fusion DB	http://igs-server.cnrs-mrs.fr/FusionDB/
CDART	http://www.ncbi.nlm.nih.gov/Structure/lexington/lexington.cgi
Pfam	http://www.sanger.ac.uk/Software/Pfam/
Phylogenetic distribution analysis	
Protein Link Explorer (Plex)	http://apropos.icmb.utexas.edu/plex/plex.html
Cluster of orthologous groups	http://www.ncbi.nlm.nih.gov/COG/ http://www.ncbi.nlm.nih.gov/COG/old/phylox.html
PhydBac	http://igs-server.cnrs-mrs.fr/phydbac/
MBGD	http://mbgd.genome.ad.jp/
Pathway tools	
GenomeNet and KEGG	http://www.genome.ad.jp/
MetaCyc	http://metacyc.org/
Cytoscape	http://www.cytoscape.org/
Organism-specific databases	
SGD	http://www.yeastgenome.org/
TAIR	http://www.arabidopsis.org/
Array, protein interaction, and phenotype analysis	
Visant	http://visant.bu.edu/
Array prospector	http://www.bork.embl.de/ArrayProspector
Prophecy	http://prophecy.lundberg.gu.se/
DIP	http://dip.doe-mbi.ucla.edu/

[a] See text for references.

allocated space, it was impossible to walk the reader through all the query steps; however, most databases used here are straightforward to navigate. (Readers should consult the original description and/or help sections if they do not find the query processes intuitive.) One exception is the SEED database (Overbeek *et al.*, 2005). To fully take advantage of all the possibilities of this comparative genomic platform requires an initial effort and a few hours of tutorial from a more experienced user. The derived National

Microbial Pathogen Database Resource or NMPDR database (McNeil *et al.*, 2007) is of easier access, and it is recommended to start with that interface before switching to SEED for more elaborate tasks.

4.1. Predictions based on gene clustering on chromosomes

4.1.1. Overview

Genes of a given pathway have a high probability of being physically linked on the chromosome (Overbeek *et al.*, 1999), particularly in prokaryotes. If a gene of unknown function is physically clustered with a gene of known function, a functional relationship can be inferred. The analysis of such clustering relationship is sometimes referred to as functional context analysis (Overbeek *et al.*, 2005). The exponential growth in the number of sequenced genomes increases the chances of making inferences from clustering events at the cost of having to eliminate noninformative clustering information deriving from closely related genomes. Precomputed clustering relationships can be easily accessed through the "Search Tool for the Retrieval of Interacting Proteins" or STRING database (von Mering *et al.*, 2003), the PhydBac database (Enault *et al.*, 2004), or the Regulon tool of MicrobesOnline (Alm *et al.*, 2005). A number of clustering tools are included in SEED and well described in the "functional context section" of the NMPDR tutorial (http://www.nmpdr.org/content/navigate.php). SEED is the only database that differentiates between direct (genes that cluster with a given input gene) and indirect functional coupling (genes that cluster with homologs of an input gene). These different databases will be compared in the case study that follows. The author recommends that readers try to follow the described queries in the different databases when reading the case studies presented in the review.

4.1.2. Case study

The newly discovered 7-aminomethyldeazaguanosine (preQ$_1$) biosynthesis pathway will be used to compare the available clustering analysis platforms. This GTP-derived metabolite is the precursor of the modified base queuosine (Q) found at position 34 of tRNA$_{His,Tyr,Asp,Asn}$ in most bacteria and many eukaryotes (Kersten and Kersten, 1990; Kuchino *et al.*, 1976; Okada *et al.*, 1978). The synthesis of preQ$_1$ most certainly requires several genes, but none had been identified, a typical example of a globally missing pathway. By combining several comparative genomic methods with experimental validation, four new *queCDEF* genes involved in this pathway were identified (Reader *et al.*, 2004). The *B. subtilis queCDEF* genes (*ykvJKLM*) are in an operon, whereas the *E. coli* homologs (*ybaX, ygcM, ygcF,* and *yqcD,* respectively) are scattered around the chromosome

(Reader *et al.*, 2004). Homologs of the four genes are often clustered in phylogenetically diverse genomes as shown in Fig. 7.2.

To test the different platforms, the following questions were asked. Had only one of the four *queECDF* gene families been identified, would the clustering tools allow the identification of the other three? Also, does the choice of the starting gene in a given gene family influence the results? Finally, are the different tools equivalent?

Each of the *queCDEF* genes from *E. coli* and *B. subtilis* (using the organism specific respective names) were used as initial inputs in the STRING, PhydBac, MicrobesOnline, and NMPDR databases to extract the corresponding clustered genes. The results are summarized in Table 7.5. Both the PhydBac and STRING clustering tools found that the four genes were highly clustered independently of the starting input gene. False-positive results were rare in both databases. However, the results were not strictly identical. For example, *yhhQ*, which is predicted to encode a $preQ_1/preQ_0$ transporter (see below), was identified only in PhydBac. SEED and GenomesOnline were both less efficient than STRING and PhydBac at detecting clustering relationships when the input genes were unclustered (as in *E. coli*). On the gene page in NMPDR, a "show functional coupling" link reveals direct clustering events. Clicking on the CL sign near the gene ID will lead to clustering detected with a homolog (indirect clustering events). When starting with the (unclustered) *E. coli* genes, no clustering was detected directly as expected. Only *queE* (*ygcF*) could be detected through the CL tool and only when starting with *queC* (*ybaX*) or *queD* (*ygcM*) not with *queF*

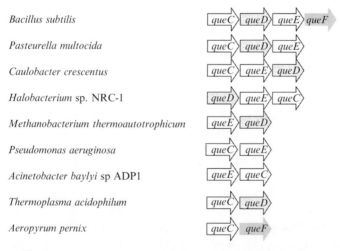

Figure 7.2 Clustering of *queCDEF* genes in several genomes.

Table 7.5 Comparison of the STRING, PhydBac, NMPDR (SEED), and MicrobesOnline platforms to detect clustering events

Input gene	Predicted clustered genes			
	String	PhydBac	NMPDR	Microbes-Online
QueC$_{Ec}$ = *ybaX*	YgcF (0.938)[a] YgcM (0.844) YqcD (0.675)	YgcF YgcM YbgF YqcD Pal	None detected by direct functional coupling. YbaX was detected through the CL tool.	YgcF YgcM
QueD$_{Ec}$ = *ygcM*	YbaX (0.844) YgcF (0.804) YqcD (0.535)	YgcF YbaX YqcD	As above	YbaX
QueE$_{Ec}$ = *ygcF*	YbaX (0.938) YgcM (0.804) YqcD (0.720) PyrG (0.519)	YgcM YqcD YbaX YbgF Pal TolB	As above	YbaX
QueF$_{Ec}$ = *yqcD*	YgcF (0.720) YbaX (0.675) YgcM (0.583) YgdH (0.440)	YgcF YhhQ YgcM YbaX	None detected	None detected
QueC$_{Bs}$ = *ykvJ*	YkvL (0.940) YkvM (0.840) YkvK (0.804)	NA[b]	YkvK (score 12)[a] YkvL (score 12)	YkvJKLM

(continued)

Table 7.5 (*continued*)

| Input gene | Predicted clustered genes | | | |
	String	PhydBac	NMPDR	Microbes–Online
QueD_{Bs} = *ykvK*	YkvL(0.926) YkvJ (0.804) YkvM (0.481)	NA	YkvJ (score 12) YkvL (score 6)	YkvJKLM
QueE_{Bs} = *ykvL*	YkvJ (0.940) YkvK (0.926) YkvM (0.792)	NA	YkvJ (score 12) YkvK (score 6)	YkvJKLM
QueF_{Bs} = *ykvM*	YkvJ (0.840) YkvL (0.792) YkvK (0.481)	NA	None detected by direct functional coupling, YkvJ and YkvL were detected through the CL tool.	YkvJKLM

[a] Database-specific scores.
[b] Not available

(*yqcD*). When starting with the (clustered) *B. subtilis* genes, the clustering of *queCDE (ykvJKL)* was systematically detected (as expected), but the fourth gene of the operon *queF (ykvM)* failed to be identified. Results can be much improved and the clustering of the four genes identified if genes from different organisms are used as inputs (data not shown), confirming what most SEED users know from experience; it is absolutely necessary to check clustering starting from wide range of phylogenetically diverse orthologs.

In terms of visualization tools, all four databases have graphical summaries of the clustering, but they are all precomputed. One exception is the Genome Browser tool of MicrobesOnline that displays the regions surrounding a given gene in different genomes and allows the user to choose the genomes and the size of the regions in a format that can be exported in graphics. This feature can be very useful when preparing figures.

4.1.3. Conclusion

The user should not be faithful to any one database but should try them all when searching for clustering events. As a rule, several members of a given gene family should be used as inputs, particularly in the SEED database, because the strength of SEED lies more in detecting and representing clustering in the context of a subsystem analysis as discussed in the following.

4.2. Detecting protein fusion events

4.2.1. Overview

In a gene fusion event, two separate parent genes are encoded in a single multifunctional polypeptide. These fusions, which have been called Rosetta stone proteins, suggest a high probability of functional interaction between the two proteins (Enright *et al.*, 1999; Pellegrini *et al.*, 1999). As with the inferences driven by the physical clustering analysis described previously, if the function of one of the two genes is known and the other is not, detecting the fusion event can allow strong functional predictions (see Daugherty *et al.* [2002] and Levin *et al.* [2004] for examples). Several web-based platforms have been specifically designed to detect these fusion events, but the author has not found them very effective. One reason is that the best fusion hints often come from comparing eukaryotic and prokaryotic genomes, because fusions events are more frequent in eukaryotic genomes (Veitia, 2002). Unfortunately, the databases that are the most user-friendly and directly integrate the fusion data (FusionDB [Suhre and Claverie, 2004] and STRING) focus mainly on prokaryotic genomes and, therefore, miss many fusion events. Until better specialized databases are available, the author has found that databases that analyze protein domains such as CDART at NCBI (Geer *et al.*, 2002) or P_{fam} at the Sanger Center (Finn *et al.*, 2006) are very effective at detecting protein fusion events. Both cover all known proteins from both prokaryotic and eukaryotic genomes. The output might not be very selective, particularly with protein domains that are ubiquitous; however, both CDART and P_{fam} present the results in graphic summaries that can be analyzed very effectively. Fusion events can also be easily detected in the SEED database with the color coding of the protein similarities table (see [http://theseed.uchicago.edu/FIG/Html/similarity_region_colors.html] for explanations).

4.2.2. Case study

To illustrate both the use of these fusion detection tools and the efficiency of the domain databases with a tRNA-related example, the PAB1506 protein from *Pyrococcus abyssii* that encodes a stand-alone PUA domain was analyzed. The PUA domain is found in many RNA binding proteins (Anantharaman *et al.*, 2002a) and in several tRNA modifying enzymes such as archaeosine tRNA guanine transglycosidase (Tgt) (Ishitani *et al.*, 2002) and pseudouridine synthase (TruB) (Hoang and Ferre-D'Amare, 2001). The function of PAB1506 is unknown. When the PhydBac database is queried with PAB1506, two fusion events can be detected linking PAB1506 to PAB2176 (annotated as an esterase) and to PAB0064 (annotated as a hydrolase). These two fusions were not detected in the STRING database by use of the exact same input protein nor in the CDART, P_{fam} or SEED databases. When analyzed in detail, the PhydBac result is most likely due to sequencing errors. As a rule, fusion events detected in only one genome should be carefully checked.

Neither STRING nor FusionDB detected the PUA fusions to the TGT and TruB domains. This result was expected, because their method is designed to eliminate hits from domains found in many different proteins in the same genome (explained in http://www.igs.cnrs-mrs.fr/FusionDB/methods.html). These were identified with both the CDART and P_{fam} domain analysis tool and the SEED color-coding homology tool. Additional fusions with the metabolic enzymes glutamate-5-kinase and 3'-phosphoadenosine 5'-phosphosulfate sulfotransferase (CysH domain) were also detected in CDART and P_{fam}. All these fusion events had previously been identified in a comprehensive graph-based analysis (Ye and Godzik, 2005).

The fusion with the glutamate 5-kinase is present in nearly all bacterial genomes, and the PUA domain has a role in activation of the enzyme and not in tRNA binding (Perez-Arellano *et al.*, 2005). The fusion of the PUA and CysH domains is limited to the archaeal kingdom. CDART also detected proteins from methanogenic archaea that contained not only the PUA and CysH domains but also additional domains such as cysteine desulfurase domains or ferredoxin domains (data not shown). This observation is interesting, because the sulfur metabolism in methanogenic archaea is not fully understood, and it has been recently proposed that cysteine biosynthesis could occur mainly on the $tRNA_{Cys}$ molecule (Helgadottir *et al.*, 2007; Sauerwald *et al.*, 2005). The archaeal PUA-CysH family could, therefore, be involved in channeling the thiol from the charged tRNA to the target metabolites, and we are currently testing this hypothesis.

4.2.3. Conclusion
As the ultimate in genome clustering events, protein fusions are very powerful prediction tools. However, no database is available that is really efficient in detecting these events yet. The protein domain analysis tools CDART, P_{fam} or the SEED color-coding tools are the best default ones to date, with the caveat that one will retrieve not only true fusion events but also all proteins containing a given domain.

4.3. Searches based on phylogenetic distribution profiles

4.3.1. Overview
Another powerful tool that does not rely on any homology information is to query phylogenetic distribution profiles (Pellegrini *et al.*, 1999). In this application, one needs to compute the proteins that are present in a given set of organisms and absent in another set. The initial COG database was the pioneer for such queries (Tatusov *et al.*, 1997) but became outdated quickly because of its limited set of genomes (43 only). The new version of COG contains 66 genomes (Tatusov *et al.*, 2003), but to the author's dismay, the phylogenetic query tool has disappeared (or is very difficult to find). Protein Links Explorer or PLEX (Date and Marcotte, 2005) has slightly more genomes than the COG platform (88). The strength of COG and PLEX is their speed, because all the phylogenetic patterns are precomputed, but this is also a limitation, because they do not get updated often. STRING and PhydBac have many more genomes and precomputed phylogenetic profiles. These databases are very powerful for identifying genes that follow the same profile as an initial query gene; however, the user cannot extract a list of genes that follow a phylogenetic distribution pattern.

The author is aware of three databases that combine constantly updated genomes with robust phylogenetic distribution query tools. CoGenT++ is part of an extensive computational genomics environment led by the European Bioinformatics Institute and has a phylogenetic profile tool (Goldovsky *et al.*, 2005). The NMPDR database (McNeil *et al.*, 2007) includes a "signature gene tool" with the added possibility of filtering the output list with keywords. Another valuable resource is the orthologous distribution tables of the MicroBial Genome Database for comparative analysis or MBGD (Uchiyama, 2007). A table containing all of the orthologous families in a given set of genomes can be generated and then queried for particular phylogenetic distribution patterns. There again filters can be used to sort the output data. Only 100 genomes can be analyzed, but this is sufficient for most queries.

4.3.2. Case study

To illustrate the power of the use of phylogenetic distribution profiles, we tested the strategy used to identify the gene encoding the missing wybutosine tricyclic guanosine-ring forming enzyme WyeA (Waas *et al.*, 2005) in the different databases. Literature analysis extracted from the tRNA database (Sprinzl *et al.*, 1999) suggested that this gene should be present in archaea, yeast, and *Homo sapiens* and absent in bacteria and fly. An input profile was generated to query the different databases. The gene family should be present in *Methanococcus janaschii, Homo sapiens, S. cerevisiae*, but absent in *E. coli, B. subtilis*, and *Drosophila melanogaster*. To query MBGD that lacks many eukaryotic genomes, *Homo sapiens* was eliminated from the query list and the fly genome was replaced by another insect, *Encephalotazoon cuniculi*. Remarkably the output from the PLEX search showed only the two protein families exemplified by the yeast proteins Ypl207w and Ygl050w. Experimental validation by several groups has shown that both these proteins are, indeed, involved in wybutosine biosynthesis (Kalhor *et al.*, 2005; Noma *et al.*, 2006; Waas *et al.*, 2005). The output from the NMPDR and MBGD, CoGenT++ databases were less selective. Both protein families identified in the PLEX search were identified, but more false-positive hits obscured the result. Thirty-five protein families were extracted by use of MBGD, because fewer genomes were used in the query. Several hundred were found by use of CoGenT++, because the user is not given the choice of a genome for the output list, making the results quite difficult to analyze. The NMPDR analysis gave 435 output proteins, but the NMPDR signature tool also extracts gene families that do not follow exactly the query with perfect matches given the highest score of two.

4.3.3. Conclusion

A well-designed phylogenetic query is very efficient at identifying gene candidates for a given function, and several databases make these queries possible. Success will depend on: (1) the robustness of the initial biological information used to design the profile; (2) imposing the fewest possible query constraints but still being stringent enough so that the output is not too large, which usually means trying different combinations of genome choices as inputs; (3) trying several search databases; and (4) even if the output list of families is large, it can be reduced by combining with other criteria such as keywords or physical clustering as discussed later with the identification of the lysidine synthase gene (*tilS*).

4.4. Mining other types of "Omics" data

Inferences on gene functions can be derived from many types of associations. For example, genes in the same pathways are often regulated by a common protein recognizing a specific DNA sequence or by common riboswitches (Gelfand et al., 2000b). Finding genes that share regulatory sites is, therefore, a powerful method to link genes functionally (see Barrick et al. [2004]; Rodionov et al. [2006]; and Yang et al. [2006] for examples). In an example related to tRNA modification, a riboswitch was identified upstream of the B. subtilis ykvJKLM operon (Barrick et al., 2004). Genes under the control of the same riboswitch in other genomes include yhhQ (Barrick et al., 2004), a predicted transporter protein that also clusters with queuosine pathway genes (see Table 7.5). We are, therefore, currently testing the hypothesis that YhhQ is a $preQ_1/preQ_0$ transporter. Unfortunately, the algorithms to detect conserved DNA motifs over all sequenced genomes such as SignalX are not available yet as web-based applications (Gelfand et al., 2000a). The reader will have to wait before such queries can be performed without the use of specialized programs that require some computer programming skills.

Associations can also be derived from interaction data sets (results of systematic two-hybrid or Tap-Tag experiments), coexpression data sets (results of expression profiling on microarrays), or phenotype arrays. The rapid increase in the volume and quality of functional genomics data is expected to strongly impact functional gene characterization in the near future. Among the growing number of web resources are the Stanford Microarray Database (SMD) for expression data (http://genome-www5.stanford.edu/) and the Database of Interacting Proteins (DIP) for protein–protein interactions (http://dip.doe-mbi.ucla.edu/). For the purpose of using these resources to predict gene function, two main problems remain: (1) the great number of false-positive or noninformative associations; and (2) the difficulty of mining these data at the click of a mouse (particularly for microarray results). However, adding filters and/or combining with other types of information can solve the first problem, as does the STRING database that integrates results from interactions and array experiments with clustering and phylogenetic data (von Mering et al., 2007).

The information that the researcher hunting for a gene's function would like to extract from array data is the list of genes having the same expression profile as the input gene over all expression data available. The Program Array prospector (Jensen et al., 2004) was designed for this purpose, but, unfortunately, the web site did not seem to be working when tested. Organism-specific databases such as Saccharomyces Genome Database (SGD) for yeast (Nash et al., 2007) or TAIR for Arabidopsis thaliana (Rhee et al., 2003) that are constantly integrating all the available "omics" data will allow such queries and are obvious starting points when possible.

Finally, the availability of large-scale mutant libraries allows the implementation of phenomics approaches (consisting of phenotype arrays or multiplexed phenotype tests screening of all mutants). These data are also starting to be mined (Kahraman *et al.*, 2005). One example is the Prophecy database that enables phenotypes of all available *S. cerevisiae* mutants to be accessed through different query formats (Fernandez-Ricaud *et al.*, 2007). However, phenomic information is not yet integrated in comparative genomic databases except for essentiality data. Systematic mutant construction libraries or transposon library mapping (see Osterman and Begley [2006] for descriptions of techniques) allows the prediction of which genes are essential for growth in specific organisms. This information has been integrated in the SEED database.

4.4.1. Case study

One example of the use of essentiality information in the tRNA modification field is the discovery of *tilS* encoding lysidine synthase (Soma *et al.*, 2003). This modification was predicted to be found only in bacteria and to be essential for survival, because in its absence, the minor $tRNA^{Ile}_{CAU}$ would be charged by methionine (Muramatsu *et al.*, 1988). By use of the signature tool of the NMPDR database, the following query can be performed. Which genes are **present** in *Bacillus subtilis* 168, *Buchnera aphidicola* str. APS, *Escherichia coli* K12, *Mycoplasma mycoides* subsp. *mycoides, Wolbachia* sp. endosymbiont of *Drosophila melanogaster* **absent** in *Arabidopsis thaliana, Methanocaldococcus jannaschii, Saccharomyces cerevisiae* and essential for growth in *E. coli*. The output list of 91 genes contains only approximately 10 genes of unknown function; one of them encodes the lysidine synthase.

4.4.2. Conclusion

Both the postgenomic experimental data sets and the platforms to analyze them are constantly improving. We anticipate that if we update this review in a few years, examples of prediction driven by mining postgenomic data will be much more numerous than today.

4.5. Subsystem analysis

4.5.1. Overview

Very early in the genomic era it was apparent that a dramatic enhancement of the quality and utility of genomic annotations can be achieved with metabolic reconstruction technology in which genes encoding metabolic pathways are inventoried in given genome (Galperin and Brenner, 1998; Selkov *et al.*, 1997). By placing genes in the context of metabolic pathways, metabolic reconstruction was a key component of the success of genome sequencing, because the physiology and metabolism of an organism can now be predicted from genomes (Galperin and Brenner, 1998; Overbeek *et al.*, 1999).

Stemming from metabolic reconstruction technology is the possibility of analyzing metabolic pathways across all genomes by computing the presence or absence of pathway genes. The consequence of this type of analysis was the realization that the number of missing genes (both "locally" or "globally" missing) was much larger than expected, reflecting the diversity of metabolic solutions used by life. Many public resources support this approach such as KEGG/GenomeNet (Kanehisa *et al.*, 2006), MetaCyc (Krieger *et al.*, 2004), the CMR-genome properties (Haft *et al.*, 2005), and MicrobesOnline (Alm *et al.*, 2005). In all of these platforms, spreadsheets computing the distribution of the genes of specific pathways in all (or in a subset of) genomes can be generated. MicrobesOnline has also developed very helpful graphical interfaces by use of the KEGG pathway maps as templates. The great limitation of these databases is that all of the pathways that can be profiled are precomputed, and all use the KEGG pathway database as template. This led Ross Overbeek and colleagues to develop the concept of the subsystem, first with the commercial ERGO database (Overbeek, 2003), then with the freely available SEED database (Overbeek *et al.*, 2005). A subsystem is a collection of genes that is built by the user and in which the genes are analyzed as a group. It can consist of genes of a pathway or a complex but is not limited to these. Subsystems can be updated or modified at will. The tools to build and analyze subsystems are at the core of the SEED platform.

4.5.2. Case study

In the case of preQ$_1$ biosynthesis, no pathway was present in KEGG, because the enzymes of the pathway had not been characterized. A queuosine subsystem was created, including the known Q biosynthesis enzymes such as tRNA-guanine transglycolase (TGT) (Noguchi *et al.*, 1982), QueA (Reuter *et al.*, 1991; Slany *et al.*, 1994), and the newly discovered QueC-DEF enzymes. After a step in which all orthologs of the subsystem families are annotated in all genomes by the user (this step should not be performed without adequate SEED training), the subsystem spreadsheet was generated. As shown in Table 7.6, for every genome in the database the presence or absence of the genes of the subsystem is visualized with a link to the corresponding protein. If two genes are physically clustered in a given genome, they will be highlighted in the same color. The clustering of the *queCDEF* genes becomes very apparent with the color coding (Table 7.6). Also rare clustering events such as the *queD-tgt* proximity in *Synechocystis* that had not been detected by any of the clustering tools discussed previously become easy to visualize. This is important, because in some cases, clustering occurring in just a few genomes can give the initial association clue.

Table 7.6 Clustering of the *queCDEF* and *tgt* genes derived from subsystem analysis[a]

Organism	Genome id	QueD	QueC	QueE	QueF	Tgt
Bacteroides fragilis ATCC 25285	272559.3	3666	1310	3665	1311	2272
Porphyromonas gingivalis W83	242619.1	913	1124	914	1154	436
Synechocystis sp. PCC 6803	1148.1	2581	2844	1133	2787	2794
Bacillus subtilis subsp. subtilis str. 168	224308.1	1375	1374	1376	1377	2774
Staphylococcus aureus subsp. aureus COL	93062.4	260	261	259	275	1875
Bradyrhizobium japonicum USDA 110	224911.1	2482	4495	2483	4796	4683
Oceanicaulis alexandrii HTCC2633	314254.3	1577	2212	2211	1787	1713
Rickettsia felis URRWXCal2	315456.3	1093	51	185	539	25
Zymomonas mobilis subsp. mobilis ZM4	264203.3	1550	1861	107	822	859
Bordetella bronchiseptica RB50	257310.1	3059	417	3058	3322	1355
Neisseria meningitidis FAM18	487.2	550	548	552	1595	1794
Nitrosomonas europaea ATCC 19718	228410.1	1457	215	214	2184	1097
Wolinella succinogenes DSM 1740	273121.1	1516	1515	1514	4	1394
Buchnera aphidicola str. Sg (*Schizaphis graminum*)	198804.1	382	435	381	273	122
Escherichia coli K12	83333.1	2721	441	2733	2750	403
Salmonella typhimurium LT2	99287.1	2845	440	2847	2864	391
Yersinia pestis KIM	187410.1	805	1016	804	3097	973
Acinetobacter sp. ADP1	62977.3	2241	2377	2376	2161	128
Psychrobacter sp. 273–4	259536.4	1015	1369	1335	219	1592
Xanthomonas campestris pv. campestris str. 8004	314565.3	4169	1290	1289	3587	2464
Magnetococcus sp. MC–1	156889.1	2790	1234	2789	928	3471

[a] Data extracted from the SEED database. The complete table is found in the "Queuosine and Archaeosine biosynthesis subsystem" MIE Table. Numbers correspond to the FIG identities. Clustered genes are highlighted in identical gray colors.

Table 7.7 Comparison of integrative databases

Properties	Databases					
	Entrez	STRING	SEED/MNPDR	CoGenT++	Microbes-Online	CMR–Genome Properties
Chromosome clustering	No	Yes	Yes	No	Yes	No
Protein fusion analysis	CDART[a]	Yes	Yes[a]	Yes[b]	Yes[a]	Yes[a]
Precomputed phylogenetic profiles	COG	Yes	No	Yes	No	No
Phylogenetic query tool	COG	No	Yes	Yes	No	No
Precomputed pathways analysis tools	No	No	Yes	Not yet	Yes	Yes
User-defined pathways analysis tool	No	No	Yes	No	No	No
Array data	GEO	Yes	No	No	Yes	No
Interaction data	No	Yes	No	No	No	No
Essentiality	No	No	Yes	No	No	No
Domain/family	CDART	No	SEED–FAM	Yes	No	TIGRFAM
Literature integration	PubMed	Yes	No	No	No	No
Elaborate queries	MyNCBI	No	No	No	No	Yes

[a] Through homology tool.
[b] For a small number of genomes only.

175

4.5.3. Conclusion

By focusing on a specific subsystem, the user can identify the globally and locally missing genes, visualize clustering, or see phylogenetic distribution patterns. In this way, SEED soon becomes for the user a virtual laboratory where specific hypotheses can be tested *in silico*.

5. GENERAL CONCLUSION: THE POWER OF INTEGRATION

The possibility of asking complex queries that integrate many types of data is the next bioinformatic challenge. For example, to find the tRNA dihydrouridine synthase genes, one could ask the following question: what protein families are absent in *Pyrococcus* sp. but present in *E. coli, B. subtilis,* and *S. cerevisiae,* and are part of a dehydrogenase family, bind tRNA, and cluster with genes related to translation (Bishop *et al.,* 2002)? The data allowing the correct prediction are available but have not yet been integrated in a database in a queriable form. Several platforms are working toward this goal, and a summary of the integrative capabilities of a few databases is presented in Table 7.7. These tools are constantly improving, and in a few years such complex queries might, indeed, become possible. One example of integration is the query toolbox of CMR-Genome properties that can filter searches by use of different types of characteristic such as MW, pI, or keywords in the annotations. Finally, although the examples in this review were taken from the tRNA modifications field, the techniques discussed here could be applied to any field of metabolism.

ACKNOWLEDGMENTS

This work was supported in part by The National Science foundation (MCB-05169448) and by the National Institutes of Health (R01 GM70641–01). The author thanks Madeline Rasche, Andrew Hanson, Basma El Yacoubi, Andrei Osterman, and Henri Grosjean for helpful discussions and critical reading of the manuscript.

REFERENCES

Alexandrov, A., Martzen, M. R., and Phizicky, E. M. (2002). Two proteins that form a complex are required for 7-methylguanosine modification of yeast tRNA. *RNA* **8,** 1253–1266.

Alm, E. J., Huang, K. H., Price, M. N., Koche, R. P., Keller, K., Dubchak, I. L., and Arkin, A. P. (2005). The MicrobesOnline Web site for comparative genomics. *Genome Res.* **15,** 1015–1022.

Altschul, S. F., Madden, T. L., Schaffer, A. A., Zhang, J., Zhang, Z., Miller, W., and Lipman, D. J. (1997). Gapped BLAST and PSI-BLAST: A new generation of protein database search programs. *Nucl. Acids Res.* **25,** 3389–3402.

Anantharaman, V., Koonin, E. V., and Aravind, L. (2002a). Comparative genomics and evolution of proteins involved in RNA metabolism. *Nucl. Acids Res.* **30,** 1427–1464.

Anantharaman, V., Koonin, E. V., and Aravind, L. (2002b). SPOUT: A class of methyl-transferases that includes *spoU* and *trmD* RNA methylase superfamilies, and novel super-families of predicted prokaryotic RNA methylases. *J. Mol. Microbiol. Biotechnol.* **4,** 71–75.

Aravind, L., and Koonin, E. V. (2001). THUMP—a predicted RNA-binding domain shared by 4-thiouridine, pseudouridine synthases and RNA methylases. *Trends Biochem. Sci.* **26,** 215–217.

Barrick, J. E., Corbino, K. A., Winkler, W. C., Nahvi, A., Mandal, M., Collins, J., Lee, M., Roth, A., Sudarsan, N., Jona, I., Wickiser, J. K., and Breaker, R. R. (2004). New RNA motifs suggest an expanded scope for riboswitches in bacterial genetic control. *Proc. Natl. Acad. Sci. USA* **101,** 6421–6426.

Bishop, A. C., Xu, J., Johnson, R. C., Schimmel, P., and de Crécy-Lagard, V. (2002). Identification of the tRNA-dihydrouridine synthase family. *J. Biol. Chem.* **277,** 25090–25095.

Björk, G. R. (1995). Biosynthesis and Function of Modified Nucleosides. *In* "tRNA: Structure, Biosynthesis, and Function" (U. L. RajBhandary, ed.), pp. 165–206. ASM Press, Washington D. C.

Björk, G. R., and Kohli, J. (1990). Synthesis and Function of Modified Nucleosides in tRNA. *In* "Chromatography and Modification of Nucleosides. Part B. Biological Roles and Function of Modification" (C. Gehrke and K. Kuo, eds.), pp. B13–B67. Elsevier, Amsterdam.

Bobik, T. A., and Rasche, M. E. (2001). Identification of the human methylmalonyl-CoA racemase gene based on the analysis of prokaryotic gene arrangements. Implications for decoding the human genome. *J. Biol. Chem.* **276,** 37194–37198.

Bujnicki, J. M., Droogmans, L., Grosjean, H., Purushothaman, S. K., and Lapeyre, B. (2004a). Bioinformatics-guided identification of novel RNA methyltransferases. "Practical Bioinformatics" (J. M. Bujnicki, ed.),Vol. 15, pp. 139–168. Springer-Verlag, Berlin Heidelberg.

Bujnicki, J. M., Oudjama, Y., Roovers, M., Owczarek, S., Caillet, J., and Droogmans, L. (2004b). Identification of a bifunctional enzyme MnmC involved in the biosynthesis of a hypermodified uridine in the wobble position of tRNA. *RNA* **10,** 1236–1242.

Chen, Y.-B., Chattopadhyay, A., Bergen, P., Gadd, C., and Tannery, N. (2007). The Online Bioinformatics Resources Collection at the University of Pittsburgh Health Sciences Library System—a one-stop gateway to online bioinformatics databases and software tools. *Nucl. Acids Res.* **35,** D780–D785.

Date, S. V., and Marcotte, E. M. (2005). Protein function prediction using the Protein Link EXplorer (PLEX). *Bioinformatics* **21,** 2558–2559.

Daugherty, M., Polanuyer, B., Farrell, M., Scholle, M., Lykidis, A., de Crécy-Lagard, V., and Osterman, A. (2002). Complete reconstitution of the human coenzyme A biosyn-thetic pathway via comparative genomics. *J. Biol. Chem.* **277,** 21431–21439.

Daugherty, M., Vonstein, V., Overbeek, R., and Osterman, A. (2001). Archaeal shikimate kinase, a new member of the GHMP-kinase family. *J. Bacteriol.* **183,** 292–300.

De Bie, L. G., Roovers, M., Oudjama, Y., Wattiez, R., Tricot, C., Stalon, V., Droogmans, L., and Bujnicki, J. M. (2003). The *yggH* gene of *Escherichia coli* encodes a tRNA (m7G46) methyltransferase. *J. Bacteriol.* **185,** 3238–3243.

de Crécy-lagard, V. (2004). Bioinformatics leads the path to the identification of missing tRNA modification genes. "Practical Bioinformatics" (J. M. Bujnicki, ed.), Vol. 15, pp. 169–190. Springer-Verlag, Berlin Heidelberg.

Droogmans, L., Roovers, M., Bujnicki, J. M., Tricot, C., Hartsch, T., Stalon, V., and Grosjean, H. (2003). Cloning and characterization of tRNA (m^{1}A58) methyltransferase (TrmI) from *Thermus thermophilus* HB27, a protein required for cell growth at extreme temperatures. *Nucl. Acids Res.* **31,** 2148–2156.

Eastwood Leung, H.-C., G., H. T., Björk, G. R., and Winkler, M. E. (1998). Genetic locations and database accession numbers of RNA-modifying and -editing enzymes. *In* "Modification and Editing of RNA" (R. Benne, ed.), pp. 561–568. ASM Press, Washington, D.C.

El Yacoubi, B., Bonnett, S., Anderson, J. N., Swairjo, M. A., Iwata-Reuyl, D., and de Crécy-Lagard, V. (2006). Discovery of a new prokaryotic type I GTP cyclohydrolase family. *J. Biol. Chem.* **281,** 37586–37593.

Enault, F., Suhre, K., Poirot, O., Abergel, C., and Claverie, J. M. (2004). Phydbac2: Improved inference of gene function using interactive phylogenomic profiling and chromosomal location analysis. *Nucl. Acids Res.* **32,** W336–W339.

Enright, A. J., Iliopoulos, I., Kyrpides, N. C., and Ouzounis, C. A. (1999). Protein interaction maps for complete genomes based on gene fusion events. *Nature* **402,** 86–90.

Fernandez-Ricaud, L., Warringer, J., Ericson, E., Glaab, K., Davidsson, P., Nilsson, F., Kemp, G. J. L., Nerman, O., and Blomberg, A. (2007). PROPHECY—a yeast phenome database, update 2006. *Nucl. Acids Res.* **35,** D463–D467.

Field, D., Feil, E. J., and Wilson, G. A. (2005). Databases and software for the comparison of prokaryotic genomes. *Microbiology* **151,** 2125–2132.

Finn, R. D., Mistry, J., Schuster-Bockler, B., Griffiths-Jones, S., Hollich, V., Lassmann, T., Moxon, S., Marshall, M., Khanna, A., Durbin, R., Eddy, S. R., Sonnhammer, E. L. L., *et al.* (2006). Pfam: Clans, web tools and services. *Nucl. Acids Res.* **34,** D247–D251.

Galperin, M. Y., and Brenner, S. E. (1998). Using metabolic pathway databases for functional annotation. *Trends Genet.* **14,** 332–333.

Galperin, M. Y., and Koonin, E. V. (1998). Sources of systematic error in functional annotation of genomes: Domain rearrangement, non-orthologous gene displacement and operon disruption. *In Silico Biol.* **1,** 55–67.

Galperin, M. Y., and Koonin, E. V. (2000). Who's your neighbor? New computational approaches for functional genomics. *Nat. Biotechnol.* **18,** 609–613.

Gaur, R., and Varshney, U. (2005). Genetic analysis identifies a function for the *queC* (*ybaX*) gene product at an initial step in the queuosine biosynthetic pathway in *Escherichia coli*. *J. Bacteriol.* **187,** 6893–6901.

Geer, L. Y., Domrachev, M., Lipman, D. J., and Bryant, S. H. (2002). CDART: Protein Homology by Domain Architecture. *Genome Res.* **12,** 1619–1623.

Gelfand, M. S., Koonin, E. V., and Mironov, A. A. (2000a). Prediction of transcription regulatory sites in Archaea by a comparative genomic approach. *Nucl. Acids Res.* **28,** 695–705.

Gelfand, M. S., Novichkov, P. S., Novichkova, E. S., and Mironov, A. A. (2000b). Comparative analysis of regulatory patterns in bacterial genomes. *Brief Bioinform.* **1,** 357–371.

Gerdes, S. Y., Kurnasov, O. V., Shatalin, K., Polanuyer, B., Sloutsky, R., Vonstein, V., Overbeek, R., and Osterman, A. L. (2006). Comparative genomics of NAD biosynthesis in *Cyanobacteria*. *J. Bacteriol.* **188,** 3012–3023.

Gerlt, J. A., and Babbitt, P. C. (2000). Can sequence determine function? *Genome Biol.* **1,** 1–10.

Goldovsky, L., Janssen, P., Ahren, D., Audit, B., Cases, I., Darzentas, N., Enright, A. J., Lopez-Bigas, N., Peregrin-Alvarez, J. M., Smith, M., Tsoka, S., Kunin, V., and Ouzounis, C. A. (2005). CoGenT++: An extensive and extensible data environment for computational genomics. *Bioinformatics* **21,** 3806–3810.

Graham, D. E., Graupner, M., Xu, H., and White, R. H. (2001). Identification of coenzyme M biosynthetic 2-phosphosulfolactate phosphatase. A member of a new class of Mg2+-dependent acid phosphatases. *Eur. J. Biochem.* **268,** 5176–5188.

Grosjean, H., Keith, G., and Droogmans, L. (2004). Detection and quantification of modified nucleotides in RMA using thin-layer chromatography. *In* "Methods in Molecular Biology" (J. Gott, ed.), Vol. 265, pp. 357–391. Humana Press, Totowa, NJ.

Gu, W., Jackman, J. E., Lohan, A. J., Gray, M. W., and Phizicky, E. M. (2003). tRNA^{His} maturation: An essential yeast protein catalyzes addition of a guanine nucleotide to the 5$'$ end of tRNA^{His}. *Genes Dev.* **17,** 2889–2901.

Gustafsson, C., Reid, R., Greene, P. J., and Santi, D. V. (1996). Identification of new RNA modifying enzymes by iterative genome search using known modifying enzymes as probes. *Nucl. Acids Res.* **24,** 3756–3762.

Haft, D. H., Selengut, J. D., Brinkac, L. M., Zafar, N., and White, O. (2005). Genome Properties: A system for the investigation of prokaryotic genetic content for microbiology, genome annotation and comparative genomics. *Bioinformatics* **21,** 293–306.

Heath, R. J., and Rock, C. O. (2000). A triclosan-resistant bacterial enzyme. *Nature* **406,** 145–146.

Hegyi, H., and Gerstein, M. (2001). Annotation transfer for genomics: Measuring functional divergence in multi-domain proteins. *Genome Res.* **11,** 1632–1640.

Helgadottir, S., Rosas-Sandoval, G., Soll, D., and Graham, D. E. (2007). Biosynthesis of phosphoserine in the Methanococcales. *J. Bacteriol.* **189,** 575–582.

Hiley, S. L., Jackman, J., Babak, T., Trochesset, M., Morris, Q. D., Phizicky, E., and Hughes, T. R. (2005). Detection and discovery of RNA modifications using microarrays. *Nucl. Acids Res.* **33,** e2.

Hoang, C., and Ferre-D'Amare, A. R. (2001). Cocrystal structure of a tRNA Psi55 pseudouridine synthase: Nucleotide flipping by an RNA modifying enzyme. *Cell* **107,** 929–939.

Hopper, A. K., and Phizicky, E. M. (2003). tRNA transfers to the limelight. *Genes Dev.* **17,** 162–180.

Huynen, M. A., Snel, B., Mering, C., and Bork, P. (2003). Function prediction and protein networks. *Curr. Opin Cell. Biol.* **15,** 191–198.

Huynen, M. A., Snel, B., and van Noort, V. (2004). Comparative genomics for reliable protein-function prediction from genomic data. *Trends Genet.* **20,** 340–344.

Ikeuchi, Y., Shigi, N., Kato, J., Nishimura, A., and Suzuki, T. (2006). Mechanistic insights into sulfur relay by multiple sulfur mediators involved in thiouridine biosynthesis at tRNA wobble positions. *Mol. Cell.* **21,** 97–108.

Ishitani, R., Nureki, O., Fukai, S., Kijimoto, T., Nameki, N., Watanabe, M., Kondo, H., Sekine, M., Okada, N., Nishimura, S., and Yokoyama, S. (2002). Crystal structure of archaeosine tRNA-guanine transglycosylase. *J. Mol. Biol.* **318,** 665–677.

Jager, G., Leipuviene, R., Pollard, M. G., Qian, Q., and Björk, G. R. (2004). The conserved Cys-X1-X2-Cys motif present in the TtcA protein is required for the thiolation of cytidine in position 32 of tRNA from *Salmonella enterica* serovar *Typhimurium. J. Bacteriol.* **186,** 750–757.

Jeltsch, A., Nellen, W., and Lyko, F. (2006). Two substrates are better than one: Dual specificities for Dnmt2 methyltransferases. *Trends Biochem. Sci.* **31,** 306.

Jensen, L. J., Lagarde, J., von Mering, C., and Bork, P. (2004). ArrayProspector: A web resource of functional associations inferred from microarray expression data. *Nucl. Acids Res.* **32,** W445–W448.

Kahraman, A., Avramov, A., Nashev, L. G., Popov, D., Ternes, R., Pohlenz, H.-D., and Weiss, B. (2005). PhenomicDB: A multi-species genotype/phenotype database for comparative phenomics. *Bioinformatics* **21,** 418–420.

Kalhor, H. R., Penjwini, M., and Clarke, S. (2005). A novel methyltransferase required for the formation of the hypermodified nucleoside wybutosine in eucaryotic tRNA. *Bioch. Bioph. Res. Com.* **334,** 433.

Kanehisa, M., Goto, S., Hattori, M., Aoki-Kinoshita, K. F., Itoh, M., Kawashima, S., Katayama, T., Araki, M., and Hirakawa, M. (2006). From genomics to chemical genomics: New developments in KEGG. *Nucl. Acids Res.* **34,** D354–D357.

Katz, J. E., Dlakic, M., and Clarke, S. (2003). Automated Identification of Putative Methyltransferases from Genomic Open Reading Frames. *Mol. Cell. Proteomics* **2,** 525–540.

Kaya, Y., and Ofengand, J. (2003). A novel unanticipated type of pseudouridine synthase with homologs in bacteria, archaea, and eukarya. *RNA* **9,** 711–721.

Kersten, H., and Kersten, W. (1990). Biosynthesis and Function of Queuine and Queuosine tRNAs. In "Chromatography and Modification of Nucleosides Part B" (K. C. T. Kuo, ed.), pp. B69–B108. Elsevier, Amsterdam.

Kharchenko, P., Chen, L., Freund, Y., Vitkup, D., and Church, G. (2006). Identifying metabolic enzymes with multiple types of association evidence. *BMC Bioinformatics* **7,** 177.

Kitagawa, M., Ara, T., Arifuzzaman, M., Ioka-Nakamichi, T., Inamoto, E., Toyonaga, H., and Mori, H. (2005). Complete set of ORF clones of *Escherichia coli* ASKA library (A Complete Set of *E. coli* K-12 ORF Archive): Unique Resources for Biological Research. *DNA Res.* **12,** 291–299.

Kobayashi, K., Ehrlich, S. D., Albertini, A., Amati, G., Andersen, K. K., Arnaud, M., Asai, K., Ashikaga, S., Aymerich, S., Bessieres, P., Boland, F., Brignell, S. C., *et al.* (2003). Essential Bacillus subtilis genes. *Proc. Natl. Acad. Sci. USA* **100,** 4678–4683.

Koonin, E. V., and Galperin, M. Y. (2003). "SEQUENCE-EVOLUTION-FUNCTION. Computational approaches in comparative genomics." Kluwer Academic Publishers, Dordrecht, Netherlands.

Krieger, C. J., Zhang, P., Mueller, L. A., Wang, A., Paley, S., Arnaud, M., Pick, J., Rhee, S. Y., and Karp, P. D. (2004). MetaCyc: A multiorganism database of metabolic pathways and enzymes. *Nucl. Acids Res.* **32,** D438–D442.

Kuchino, Y., Kasai, H., Nihei, K., and Nishimura, S. (1976). Biosynthesis of the Modified Nucleoside Q in Transfer RNA. *Nucl. Acids Res.* **3,** 393–398.

Kurowski, M., Sasin, J., Feder, M., Debski, J., and Bujnicki, J. (2003). Characterization of the cofactor-binding site in the SPOUT-fold methyltransferases by computational docking of S-adenosylmethionine to three crystal structures. *BMC Bioinformatics* **4,** 9.

Levin, I., Giladi, M., Altman-Price, N., Ortenberg, R., and Mevarech, M. (2004). An alternative pathway for reduced folate biosynthesis in bacteria and halophilic archaea. *Mol. Microbiol.* **54,** 1307–1318.

Loh, K. D., Gyaneshwar, P., Markenscoff Papadimitriou, E., Fong, R., Kim, K. S., Parales, R., Zhou, Z., Inwood, W., and Kustu, S. (2006). A previously undescribed pathway for pyrimidine catabolism. *Proc. Natl. Acad. Sci. USA* **103,** 5114–5119.

Makarova, K., and Koonin, E. (2003). Comparative genomics of archaea: How much have we learned in six years, and what's next? *Genome Biol.* **4,** 115.

Martzen, M. R., McCraith, S. M., Spinelli, S. L., Torres, F. M., Fields, S., Grayhack, E. J., and Phizicky, E. M. (1999). A Biochemical Genomics Approach for Identifying Genes by the Activity of Their Products. *Science* **286,** 1153–1155.

McNeil, L. K., Reich, C., Aziz, R. K., Bartels, D., Cohoon, M., Disz, T., Edwards, R. A., Gerdes, S., Hwang, K., Kubal, M., Margaryan, G. R., Meyer, F., *et al.* (2007). The National Microbial Pathogen Database Resource (NMPDR): A genomics platform based on subsystem annotation. *Nucl. Acids Res.* **35,** D347–D353.

Mittl, P. R., and Grutter, M. G. (2001). Structural genomics: Opportunities and challenges. *Curr. Opin. Chem. Biol.* **5,** 402–408.

Motorin, Y., and Grosjean, H. (1999). Multisite-specific tRNA:m^{5}C-methyltransferase (Trm4) in yeast *Saccharomyces cerevisiae*: Identification of the gene and substrate specificity of the enzyme. *RNA* **5,** 1105–1118.

Muramatsu, T., Yokoyama, S., Horie, N., Matsuda, A., Ueda, T., Yamaizumi, Z., Kuchino, Y., Nishimura, S., and Miyazawa, T. (1988). A novel lysine-substituted nucleoside in the first position of the anticodon of minor isoleucine tRNA from *Escherichia coli*. *J. Biol. Chem.* **263,** 9261–9267.

Nash, R., Weng, S., Hitz, B., Balakrishnan, R., Christie, K. R., Costanzo, M. C., Dwight, S. S., Engel, S. R., Fisk, D. G., Hirschman, J. E., Hong, E. L., Livstone, M. S., *et al.* (2007). Expanded protein information at SGD: New pages and proteome browser. *Nucl. Acids Res.* **35,** D468–D471.

Nasvall, S. J., Chen, P., and Björk, G. R. (2004). The modified wobble nucleoside uridine-5-oxyacetic acid in tRNA^{Pro}(cmo^{5}UGG) promotes reading of all four proline codons *in vivo*. *RNA* **10,** 1662–1673.

Noguchi, S., Nishimura, Y., Hirota, Y., and Nishimura, S. (1982). Isolation and characterization of an *Escherichia coli* mutant lacking tRNA-guanine transglycosylase. Function and biosynthesis of queuosine in tRNA. *J. Biol. Chem.* **257,** 6544–6550.

Noma, A., Kirino, Y., Ikeuchi, Y., and Suzuki, T. (2006). Biosynthesis of wybutosine, a hyper-modified nucleoside in eukaryotic phenylalanine tRNA. *EMBO J.* **25,** 2142–2154.

Okada, N., Noguchi, S., Nishimura, S., Ohgi, T., Goto, T., Crain, P. F., and McCloskey, J. A. (1978). Structure Determination of a Nucleoside Q Precursor Isolated from *E. coli* tRNA: 7-(aminomethyl)-7-deazaguanosine. *Nucl. Acids Res.* **5,** 2289–2296.

Osterman, A., and Begley, T. (2006). A subsystems based approach to the identification of drug targets in bacterial pathogens. *In* "Progress in Drug Research" (H. I. Boshoff and C. E. I. Barrry, eds.), Vol. 64, pp. 133–170. Birkhauser Verlag, Basel (Switzerland).

Osterman, A., and Overbeek, R. (2003). Missing genes in metabolic pathways: A comparative genomics approach. *Curr. Opin. Chem. Biol.* **7,** 238–251.

Overbeek, R., Begley, T., Butler, R. M., Choudhuri, J. V., Chuang, H. Y., Cohoon, M., de Crécy-Lagard, V., Diaz, N., Disz, T., Edwards, R., Fonstein, M., Frank, E. D., *et al.* (2005). The subsystems approach to genome annotation and its use in the project to annotate 1000 genomes. *Nucl. Acids Res.* **33,** 5691–5702.

Overbeek, R., Begley, T., Butler, R. M., Choudhuri, J. V., Chuang, H. Y., Cohoon, M., de Crecy-Lagard, V., Diaz, N., Disz, T., Edwards, R., Fonstein, M., Frank, E. D., *et al.* (2003). The ERGO Genome Analysis and Discovery System. *Nucleic Acids Res.* **31,** 1–8.

Overbeek, R., Fonstein, M., D'Souza, M., Pusch, G. D., and Maltsev, N. (1999). The use of gene clusters to infer functional coupling. *Proc. Natl. Acad. Sci. USA* **96,** 2896–2901.

Pellegrini, M., Marcotte, E. M., Thompson, M. J., Eisenberg, D., and Yeates, T. O. (1999). Assigning protein functions by comparative genome analysis: Protein phylogenetic profiles. *Proc. Natl. Acad. Sci. USA* **96,** 4285–4288.

Peng, W.-T., Robinson, M. D., Mnaimneh, S., Krogan, N. J., Cagney, G., Morris, Q., Davierwala, A. P., Grigull, J., Yang, X., and Zhang, W. (2003). A panoramic view of yeast noncoding RNA processing. *Cell* **113,** 919.

Perez-Arellano, I., Rubio, V., and Cervera, J. (2005). Dissection of *Escherichia coli* glutamate 5-kinase: Functional impact of the deletion of the PUA domain. *FEBS Lett.* **579,** 6903.

Pintard, L., Lecointe, F., Bujnicki, J. M., Bonnerot, C., Grosjean, H., and Lapeyre, B. (2002). Trm7p catalyses the formation of two 2′-O-methylriboses in yeast tRNA anticodon loop. *Embo J.* **21,** 1811–1820.

Purta, E., van Vliet, F., Tkaczuk, K. L., Dunin-Horkawicz, S., Mori, H., Droogmans, L., and Bujnicki, J. M. (2006). The *yfhQ* gene of *Escherichia coli* encodes a tRNA:Cm32/Um32 methyltransferase. *BMC Mol. Biol.* **7,** 23.

Purushothaman, S. K., Bujnicki, J. M., Grosjean, H., and Lapeyre, B. (2005). Trm11p and Trm112p Are both required for the formation of 2-methylguanosine at position 10 in yeast tRNA. *Mol. Cell. Biol.* **25,** 4359–4370.

Reader, J. S., Metzgar, D., Schimmel, P., and de Crécy-Lagard, V. (2004). Identification of four genes necessary for biosynthesis of the modified nucleoside queuosine. *J. Biol. Chem.* **279,** 6280–6285.

Renalier, M.-H., Joseph, N., Gaspin, C., Thebault, P., and Mougin, A. (2005). The Cm56 tRNA modification in archaea is catalyzed either by a specific 2′-O-methylase, or a C/D sRNP. *RNA* **11,** 1051–1063.

Reuter, K., Slany, R., Ullrich, F., and Kersten, H. (1991). Structure and Organization of *E. coli* Genes Involved in Biosynthesis of the Deazaguanine Derivative Queuine, a Nutrient Factor for Eukaryotes. *J. Bacteriol.* **173,** 2256–2264.

Rhee, S. Y., Beavis, W., Berardini, T. Z., Chen, G., Dixon, D., Doyle, A., Garcia-Hernandez, M., Huala, E., Lander, G., Montoya, M., Miller, N., Mueller, L. A., et al. (2003). The *Arabidopsis* Information Resource (TAIR): A model organism database providing a centralized, curated gateway to *Arabidopsis* biology, research materials and community. *Nucl. Acids Res.* **31**, 224–228.

Rodionov, D. A., Hebbeln, P., Gelfand, M. S., and Eitinger, T. (2006). Comparative and functional genomic analysis of prokaryotic nickel and cobalt uptake transporters: Evidence for a novel group of ATP-binding cassette transporters. *J. Bacteriol.* **188**, 317–327.

Roovers, M., Hale, C., Tricot, C., Terns, M. P., Terns, R. M., Grosjean, H., and Droogmans, L. (2006). Formation of the conserved pseudouridine at position 55 in archaeal tRNA. *Nucl. Acids Res.* **34**, 4293–4301.

Sauerwald, A., Zhu, W., Major, T. A., Roy, H., Palioura, S., Jahn, D., Whitman, W. B., Yates, J. R., 3rd, Ibba, M., and Söll, D. (2005). RNA-dependent cysteine biosynthesis in Archaea. *Science* **307**, 1969–1972.

Selkov, E., Maltsev, N., Olsen, G. J., Overbeek, R., and Whitman, W. B. (1997). A reconstruction of the metabolism of *Methanococcus jannaschii* from sequence data. *Gene* **197**, GC11–GC26.

Slany, R. K., Bosl, M., and Kersten, H. (1994). Transfer and isomerization of the ribose moiety of AdoMet during the biosynthesis of queuosine tRNAs, a new unique reaction catalyzed by the QueA protein from *Escherichia coli*. *Biochimie* **76**, 389–393.

Soma, A., Ikeuchi, Y., Kanemasa, S., Kobayashi, K., Ogasawara, N., Ote, T., Kato, J., Watanabe, K., Sekine, Y., Suzuki, T., Muramatsu, T., Nishikawa, K., et al. (2003). An RNA-modifying enzyme that governs both the codon and amino acid specificities of isoleucine tRNA. *Mol. Cell* **12**, 689–698.

Sprinzl, M., Vassilenko, K. S., Emmerich, J., and Bauer, F. (1999). tRNA Compilation 2000 http://www.uni-bayreuth.de/departments/biochemie/trna/.

Suhre, K., and Claverie, J.-M. (2004). FusionDB: A database for in-depth analysis of prokaryotic gene fusion events. *Nucl. Acids Res.* **32**, D273–D276.

Tatusov, R., Fedorova, N., Jackson, J., Jacobs, A., Kiryutin, B., Koonin, E., Krylov, D., Mazumder, R., Mekhedov, S., Nikolskaya, A., Rao, B. S., Smirnov, S., et al. (2003). The COG database: An updated version includes eukaryotes. *BMC Bioinformatics* **4**, 41.

Tatusov, R. L., Koonin, E. V., and Lipman, D. J. (1997). A genomic perspective on protein families. *Science* **278**, 631–637.

Tatusov, R. L., Natale, D. A., Garkavtsev, I. V., Tatusova, T. A., Shankavaram, U. T., Rao, B. S., Kiryutin, B., Galperin, M. Y., Fedorova, N. D., and Koonin, E. V. (2001). The COG database: New developments in phylogenetic classification of proteins from complete genomes. *Nucl. Acids Res.* **29**, 22–28.

Tian, W., and Skolnick, J. (2003a). How well is enzyme function conserved as a function of pairwise sequence identity? *J. Mol. Biol.* **333**, 863–882.

Tian, W., and Skolnick, J. (2003b). How well is enzyme function conserved as a function of pairwise sequence identity? *J. Mol. Biol.* **333**, 863–882.

Uchiyama, I. (2007). MBGD: A platform for microbial comparative genomics based on the automated construction of orthologous groups. *Nucl. Acids Res.* **35**, D343–D346.

Umeda, N., Suzuki, T., Yukawa, M., Ohya, Y., Shindo, H., Watanabe, K., and Suzuki, T. (2005). Mitochondria-specific RNA-modifying enzymes responsible for the biosynthesis of the wobble base in mitochondrial tRNAs: Implications for the molecular pathogenesis of human mitochondrial diseases. *J. Biol. Chem.* **280**, 1613–1624.

Urbonavicius, J., Skouloubris, S., Myllykallio, H., and Grosjean, H. (2005). Identification of a novel gene encoding a flavin-dependent tRNA:m^{5}U methyltransferase in bacteria—evolutionary implications. *Nucl. Acids Res.* **33**, 3955–3964.

Van Lanen, S. G., Reader, J. S., Swairjo, M. A., de Crécy-Lagard, V., Lee, B., and Iwata-Reuyl, D. (2005). From cyclohydrolase to oxidoreductase: Discovery of nitrile reductase activity in a common fold. *Proc. Natl. Acad. Sci. USA* **102,** 4264–4269.

Veitia, R. (2002). Rosetta Stone proteins: "Chance and necessity"? *Genome Biology* **1001,** 1–3.

von Mering, C., Huynen, M., Jaeggi, D., Schmidt, S., Bork, P., and Snel, B. (2003). STRING: A database of predicted functional associations between proteins. *Nucleic Acids Res.* **31,** 258–261.

von Mering, C., Jensen, L. J., Kuhn, M., Chaffron, S., Doerks, T., Kruger, B., Snel, B., and Bork, P. (2007). STRING 7—recent developments in the integration and prediction of protein interactions. *Nucl. Acids Res.* **35,** D358–D362.

Waas, W. F., Crécy-Lagard., d., and Schimmel, P. (2005). Discovery of a gene family critical to wyosine base formation in a subset of phenylalanine-specific transfer RNAs. *J. Biol. Chem.* **280,** 37616–37622.

Watanabe, Y.-i., and Gray, M. W. (2000). Evolutionary appearance of genes encoding proteins associated with box H/ACA snoRNAs: Cbf5p in *Euglena gracilis*, an early diverging eukaryote, and candidate Gar1p and Nop10p homologs in archaebacteria. *Nucl. Acids Res.* **28,** 2342–2352.

Winzeler, E. A., Shoemaker, D. D., Astromoff, A., Liang, H., Anderson, K., Andre, B., Bangham, R., Benito, R., Boeke, J. D., Bussey, H., Chu, A. M., Connelly, C., *et al.* (1999). Functional characterization of the *S. cerevisiae* genome by gene deletion and parallel analysis. *Science* **285,** 901–906.

Wolf, J., Gerber, A. P., and Keller, W. (2002). *tadA*, an essential tRNA-specific adenosine deaminase from *Escherichia coli*. *EMBO J.* **21,** 3841–3851.

Wolfe, M. D., Ahmed, F., Lacourciere, G. M., Lauhon, C. T., Stadtman, T. C., and Larson, T. J. (2004). Functional diversity of the rhodanese homology domain: The *Escherichia coli ybbB* gene encodes a selenophosphate-dependent tRNA 2-selenouridine synthase. *J. Biol. Chem.* **279,** 1801–1809.

Xing, F., Hiley, S. L., Hughes, T. R., and Phizicky, E. M. (2004). The specificities of four yeast dihydrouridine synthases for cytoplasmic tRNAs. *J. Biol. Chem.* **279,** 17850–17860. Epub 2004 Feb 16.

Xing, F., Martzen, M. R., and Phizicky, E. M. (2002). A conserved family of *Saccharomyces cerevisiae* synthases effects dihydrouridine modification of tRNA. *RNA* **8,** 370–381.

Xu, X. M., Carlson, B. A., Mix, H., Zhang, Y., Saira, K., Glass, R. S., Berry, M. J., Gladyshev, V. N., and Hatfield, D. L. (2006). Biosynthesis of Selenocysteine on Its tRNA in Eukaryotes. *PLoS Biol.* **5,** e4.

Yakunin, A. F., Yee, A. A., Savchenko, A., Edwards, A. M., and Arrowsmith, C. H. (2004). Structural proteomics: A tool for genome annotation. *Curr. Opin. Chem. Biol.* **8,** 42–48.

Yang, C., Rodionov, D. A., Li, X., Laikova, O. N., Gelfand, M. S., Zagnitko, O. P., Romine, M. F., Obraztsova, A. Y., Nealson, K. H., and Osterman, A. L. (2006). Comparative genomics and experimental characterization of N-acetylglucosamine utilization pathway of Shewanella oneidensis. *J. Biol. Chem.* **281,** 29872–29885.

Ye, Y., and Godzik, A. (2005). Multiple flexible structure alignment using partial order graphs. *Bioinformatics* **21,** 2362–2369.

CHAPTER EIGHT

Identification and Characterization of Archaeal and Fungal tRNA Methyltransferases

David E. Graham[*,†] *and* Gisela Kramer[*]

Contents

Abstract

All organisms modify their tRNAs by use of evolutionarily conserved enzymes. Members of the Archaea contain an extensive set of modified nucleotides that were early evidence of the fundamental evolutionary divergence of the Archaea from Bacteria and Eucarya. However, the enzymes responsible for these

[*] Department of Chemistry and Biochemistry, The University of Texas at Austin, Austin, Texas
[†] Institute for Cellular and Molecular Biology, The University of Texas at Austin, Austin, Texas

Methods in Enzymology, Volume 425	© 2007 Elsevier Inc.
ISSN 0076-6879, DOI: 10.1016/S0076-6879(07)25008-6	All rights reserved.

posttranscriptional modifications were largely unknown before the advent of genome sequencing. This chapter explains methods to identify tRNA methyltransferases in genome sequences, emphasizing the identification and characterization of six enzymes from the hyperthermophilic archaeon *Methanocaldococcus jannaschii*. We describe methods to express these proteins, purify or synthesize tRNA substrates, measure methyltransferase activity, and map tRNA modifications. Comparison of the archaeal methyltransferases with their yeast homologs suggests that the common ancestor of the archaeal and eucaryal organismal lineages already had extensive tRNA modifications.

1. INTRODUCTION

Mutations in genes encoding modification enzymes that act on the tRNA anticodon loop often produce dramatic phenotypes because of their direct effects on translation. Mutations in the tRNA 1-methylguansoine (37) methyltransferase genes in *Escherichia coli* and *Saccharomyces cerevisiae* severely impair growth (Björk *et al.*, 1989, 2001). In contrast, changes in enzymes modifying nucleotides in the rest of the tRNA often show no discernible phenotype and require more complicated genetic screens for identification. Although the functions of many modifications are not known, the modifying enzymes are highly conserved throughout the tree of life.

Methylation is one of the simplest and most common modifications of tRNAs. The RNA modification database (Rozenski *et al.*, 1999) and the MODOMICS database (Dunin-Horkawicz *et al.*, 2006) list numerous posttranscriptionally methylated nucleosides. Yet far fewer methyltransferases are known to produce these modifications. Cells often express tRNA methyltransferases at low levels. The enzymes have low substrate turnover, and they usually modify only a single position on certain tRNA species. Therefore, it is not surprising that few tRNA methyltransferases were biochemically identified by conventional protein purification from cell extracts (Kline and Söll, 1982). Modification enzymes that were characterized came from model organisms (mainly *E. coli* or *S. cerevisiae*), where genetics implicated specific genes in translational fidelity. However, the advent of genome sequencing and high-level protein expression systems opened the door to characterizing new tRNA modifying enzymes from diverse organisms.

Methanocaldococcus jannaschii is a hyperthermophilic euryarchaeon whose complete genome sequence was among the first published (Bult *et al.*, 1996). tRNAs from *M. jannaschii* have at least 24 modified nucleosides, including 19 methylated nucleosides (McCloskey *et al.*, 2001). Because only one tRNA methyltransferase was recognized in the original genome annotation, this organism was a model system for the use of bioinformatics to identify candidate tRNA methyltransferase genes. Many of the methods described in this chapter were developed in a doctoral dissertation project to identify and biochemically characterize tRNA methyltransferases from *M. jannaschii* (Graham, 2000). Results from this project (and other groups working independently) demonstrated that archaea and eucarya share many tRNA modification systems; thus, results from studies of the archaeal enzymes often apply to fungal and higher eukaryotic systems as well.

This chapter, along with others in this volume, supplements previous reviews of methods used to characterize tRNA methyltransferases (Grosjean *et al.*, 1998). Here, we first discuss the logic and bioinformatic strategies used to identify putative tRNA methyltransferases in complete genome sequences. The specific challenges of working with *S*-adenosylmethionine and tRNA substrates are then addressed, followed by methods to assay tRNA methyltransferase activity and to identify sites of tRNA modification.

2. IDENTIFICATION OF tRNA METHYLTRANSFERASES BY USE OF GENOME SEQUENCES

Except for the folate-dependent m^5U methyltransferase (described in Chapter 14 of this volume), RNA methyltransferases catalyze the transfer of a methyl group from *S*-adenosyl-L-methionine (AdoMet or SAM) to a nucleophile in an S_N2 reaction (Kealey *et al.*, 1994). These enzymes bind AdoMet and the methyl acceptor substrate; then they promote group transfer through covalent catalysis or acid/base chemistry (Takusagawa *et al.*, 1998). Crystal structure models of several RNA methyltransferases show these proteins have α/β structures. However, the only recognizable sequence similarity among these methyltransferases with different specificities is an AdoMet-binding motif that is similar to the NAD(P)-binding Rossmann fold (Fauman *et al.*, 1999). The canonical AdoMet-dependent methyltransferase fold contains alternating β-strands and α-helices with an AdoMet binding site formed by the central β-sheet (Martin and McMillan, 2002). Amino acid residues in this binding site form three characteristic clusters, although there is considerable variation in sequence of the clusters in different tRNA methyltransferases (Fig. 8.1A). Each cluster lies at the carboxy terminus of a β-strand: a glycine-containing loop (Motif I in Fig. 8.1A); adjacent acidic and hydrophobic residues (Motif II); and a proline-containing loop (Motif III). In addition to these three motifs, each tRNA methyltransferase family forms other side chain and main chain hydrogen bonds to AdoMet, which are not universally conserved.

Another group of tRNA methyltransferases belongs to the SPOUT family of AdoMet-dependent enzymes (Koonin and Rudd, 1993). Structural models of guanosine 2′-*O*-ribose methyltransferase (TrmH) and 1-methylguanosine methyltransferase (TrmD) show that members of this

A

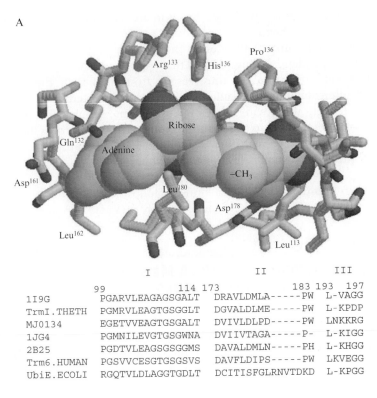

	I	II	III
	99 114	173 183 193	197
1I9G	PGARVLEAGAGSGALT	DRAVLDMLA-----PW	L-VAGG
TrmI.THETH	PGMRVLEAGTGSGGLT	DGVALDLME-----PW	L-KPDP
MJ0134	EGETVVEAGTGSGALT	DVIVLDLPD-----PW	LNKKRG
1JG4	PGMNILEVGTGSGWNA	DVIIVTAGA-----P-	L-KIGG
2B25	PGDTVLEAGSGSGGMS	DAVALDMLN-----PH	L-KHGG
Trm6.HUMAN	PGSVVCESGTGSGSVS	DAVFLDIPS-----PW	LKVEGG
UbiE.ECOLI	RGQTVLDLAGGTGDLT	DCITISFGLRNVTDKD	L-KPGG

B

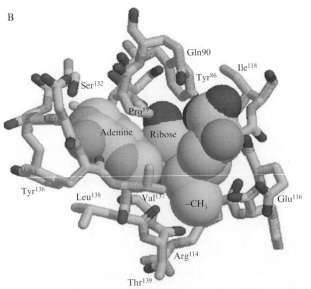

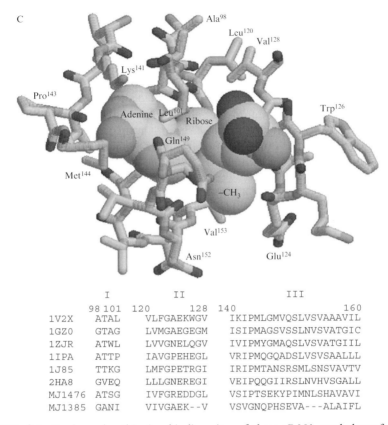

```
              I              II                 III
           98 101    120          128  140                      160
    1V2X   ATAL      VLFGAEKWGV        IKIPMLGMVQSLVSVAAAVIL
    1GZ0   GTAG      LVMGAEGEGM        ISISPMAGSVSSLNVSVATGIC
    1ZJR   ATWL      LVVGNELQGV        IVIPMYGMAQSLVSVATGIIL
    1IPA   ATTP      IAVGPEHEGL        VRIPMQGQADSLVSVSAALLL
    1J85   TTKG      LMFGPETRGI        IRIPMTANSRSMLSNSVAVTV
    2HA8   GVEQ      LLLGNEREGI        VEIPQQGIIRSLNVHVSGALL
    MJ1476 ATSG      IVFGREDDGL        VSISPTSEKYPIMNLSHAVAVI
    MJ1385 GANI      VIVGAEK--V        VSVGNQPHSEVA---ALAIFL
```

Figure 8.1 *S*-adenosylmethionine binding sites of three tRNA methyltransferase classes. AdoMet is shown in space-filling style, whereas protein amino acids are shown as sticks. Light gray atoms are carbon, medium gray atoms are nitrogen, and dark gray atoms are oxygen. Amino acid names and positions are indicated next to key residues. (A) The tRNA(m^{1}A) methyltransferase from *Mycobacterium tuberculosis* (TrmI, pdb|1I9G) binds AdoMet in an extended conformation (Gupta *et al.*, 2001). Three canonical methyltransferase motifs are conserved in this SAM-MT fold, although individual amino acid positions can vary significantly. Alignments of these motifs in homologous proteins are shown for *Thermus thermophilus* TrmI, MJ0134—a tRNA(m^{1}A) methyltransferase from *M. jannaschii, Pyrococcus furiosus* L-isoaspartyl O-methyltransferase (pdb|1JG4), a human tRNA(m^{1}A) methyltransferase (pdb|2B25), human tRNA (m^{1}A58) methyltransferase (Trm6), and the *E. coli* ubiquinone methyltransferase (UbiE). (B) The tRNA(m^{1}G37) methyltransferase from *Haemophilus influenzae* (TrmD, pdb|1UAK) binds AdoMet in a bent, L-shaped conformation (Ahn *et al.*, 2003). The amino acids shown in the AdoMet binding site are highly conserved in TrmD homologs. (C) The tRNA(Gm18) methyltransferase from *T. thermophilus* (TrmH, pdb|1V2X) shares a similar fold and AdoMet binding conformation with the TrmD proteins (Nureki *et al.*, 2004). Alignments of three AdoMet binding motifs are shown for TrmH, *E. coli* large subunit rRNA(Gm2251) methyltransferase (RimB, pdb|1GZ0), *Aquifex aeolicus* TrmH (pdb|1ZJR), *T. thermophilus* RNA 2′O-methyltransferase (RrmA, pdb|1IPA), *H. influenzae* tRNA(Gm) methyltransferase (YibK, pdb|1J85), HIV-1 TAR RNA binding protein 1 (pdb|2HA8), *M. jannaschii* hypothetical 2′-O-methyltransferase (MJ1476) and *M. jannaschii* tRNA(Cm) methyltransferase (aTrm56, MJ1385).

family have unusual dimeric structures with trefoil knots (Ahn *et al.*, 2003). These proteins are not related to the canonical methyltransferases. Members of the SPOUT family bind AdoMet in a bent, L-shaped conformation (Fig. 8.1B,C). However, the AdoMet binding sites of TrmD and TrmH share no conserved amino acid residues or motifs. The highly conserved TrmD homologs probably all function as tRNA(m^{1}G37) methyltransferases, whereas TrmH homologs methylate a broad range of substrates (Fig. 8.1C).

A third class of RNA methyltransferases relies on guide RNAs (C/D box snoRNAs) to direct a complex of methylation proteins to a specific RNA sequence (Bortolin *et al.*, 2003). This RNA-directed system probably accounts for most 2′-O-methyl modifications in rRNA (Lowe and Eddy, 1999), and it produces 2′-O-methylcytosine(56) in some archaeal tRNAs (Renalier *et al.*, 2005). The fibrillarin protein, a member of the methylase complex, shares a similar fold with canonical AdoMet-dependent methyltransferases (Aittaleb *et al.*, 2003).

In addition to these three known classes of tRNA methyltransferases, there are examples of methyltransferases with unique AdoMet binding sites. These include the bacterial enzymes methionine synthase (MetH), cobalt-precorrin-4 methyltransferase (CbiF) and diphthine synthase (Dixon *et al.*, 1999). This diversity suggests that AdoMet binding sites can arise readily through convergent evolution, and there may be more examples of unrecognized tRNA methyltransferases that remain to be discovered.

Variations in AdoMet-binding motif sequence and location complicate searches for new methyltransferases. Basic pattern-matching algorithms cannot recognize short motifs separated by variable numbers of amino acids. Local alignment algorithms may fail to recognize the motif when highly variable regions separate conserved residues. Alternately, local alignments can identify related, but misleading, nucleotide binding domains (Rossmann folds and ATP-binding motifs). Therefore, the most successful strategies to identify methyltransferase sequences have been transitive searching and profiling.

Transitive searching involves querying a database by use of a previously identified methyltransferase sequence, and then repeating the search with the most diverged homolog as a query sequence (Park *et al.*, 1997). This technique was previously used to identify putative tRNA modifying enzymes in the *E. coli* genome (Gustafsson *et al.*, 1996). To identify Ado-Met-binding domains in the set of all predicted open reading frames (ORFs) from *M. jannaschii*, we used ungapped BLAST searches (Altschul *et al.*, 1990) with the previously characterized AdoMet-dependent methyltransferases Trm1p (sp|P15565), ErmC (sp|P13956), and HhaI (sp|P05102) as query sequences. Searches were iterated until no new significant matches could be identified. This search does not detect members of the SPOUT family, so the BLAST search must be repeated with representatives of each methyltransferase family.

Alternately, profile searching uses probabilistic models to identify members of a protein family. Programs construct profiles by aligning recognized homologs, modeling domains conserved in the alignment, searching a database with that profile, and then iterating the process until no additional homologs can be identified. Several software tools have automated this process: HMMER constructs a hidden Markov model of specified sequences (Eddy, 1998); Probe uses Gibbs sampling to construct an alignment of conserved sequence regions (Neuwald et al., 1997); and PSI-BLAST similarly builds a profile from local alignments (Altschul et al., 1997). These methods complement results from ungapped searches and frequently identify distantly related homologs.

Searches of the *M. jannaschii* ORFs identified 51 putative AdoMet-binding proteins in *M. jannaschii*. To restrict the search for tRNA methyltransferases, apparent orthologs of previously characterized rRNA-, DNA-, protein-, porphyrin-, and quinone-modifying methyltransferases were not further considered. To confirm the α/β structure of putative methyltransferases, secondary structure predictions can be made for each protein family by use of the GOR V algorithm (Kloczkowski et al., 2002) or the PredictProtein server (Rost et al., 2004).

The remaining candidate genes are evaluated for the presence or absence of closely related homologs in other archaeal genome sequences. tRNA hydrolysates from many archaea share a similar set of methylated nucleosides (m^{1}A, m^{5}C, m^{2}G, m$^2_2$G, m^{1}G, G$_m$, m^{1}I, C$_m$, and m$^1\Psi$) (Gupta and Woese, 1980). Therefore, we assumed that the respective methyltransferase genes would have similar phylogenetic distributions. A molecular version of Koch's postulates, these guidelines require that each archaeal tRNA methyltransferase be found only in organisms containing the corresponding tRNA modification and that the enzyme be found in all organisms with the equivalent modification. The large number of complete archaeal genome sequences (currently 34) makes this comparison a powerful tool. However, assumptions about phylogenetic distributions can be misleading when modifications are found at multiple tRNA positions or are independently introduced by separate modification enzymes.

The binding site for nucleophilic substrates is poorly conserved except for a catalytic cysteine residue that enhances the C-5 atom's nucleophilicity in 5-methylpyrimidine DNA and RNA methyltransferases (Wu and Santi, 1987). The cysteine thiol acts as a nucleophile, forming a covalent enzyme-substrate complex to activate the nucleotide substrate for methylation (Ivanetich and Santi, 1992). 5-Methylpyrimidine methyltransferases of both RNA and DNA use this mechanism. Combined with the three motifs that are characteristic of AdoMet-binding domains, this conserved cysteine provides a signature for identifying m⁵C methyltransferases. Nevertheless, a previous computational search for m⁵C RNA methyltransferases did not identify the MJ1653 protein described later (Reid *et al.*, 1999), probably because of its close relationship to m⁵U RNA methyltransferases.

SnoRNAs have sequences complementary to their RNA targets. Therefore, modifications directed by these guide RNAs can be accurately predicted. The Snoscan and SnoGPS web servers automate the detection of conserved motifs in these small RNAs (Lowe and Eddy, 1999).

2.1. Cloning, expression, and protein purification

Because of their low natural abundance and the difficulty of purifying native tRNA methyltransferases, these enzymes are most efficiently expressed in heterologous systems fused to an affinity tag. *M. jannaschii* genes encoding six putative methyltransferases were amplified by PCR and cloned in *E. coli* expression vector pET-19b (Novagen) for high-level expression from the T7 RNA polymerase promoter. These proteins were solubly expressed with an aminoterminal decahistidine tag in *E. coli*. Cell-free extracts were heated at 70° to denature native proteins. The residual soluble proteins were loaded onto a nickel affinity column, and the target methyltransferases were eluted with imidazole.

However, not all tRNA methyltransferases can be expressed in *E. coli* to produce high levels of soluble protein. The human and yeast Trm5p proteins are expressed mostly in insoluble and inactive forms in *E. coli*, probably as inclusion bodies (Brulé *et al.*, 2004; G. K., unpublished data). Low temperature expression and fusion to a large, soluble protein may facilitate expression and folding of these difficult proteins.

3. Substrates for Methyltransferase Assays

3.1. S-adenosylmethionine substrate

The ubiquitous enzyme methionine adenosyltransferase catalyzes the biosynthesis of (*S,S*)-S-adenosylmethionine from ATP and L-methionine. However, this biologically active isomer of AdoMet is chirally and covalently unstable. At neutral pH conditions, AdoMet spontaneously hydrolyzes to produce 5′-methylthioadenosine (MTA) and homoserine lactone (HSL) (Fig. 8.2). This reaction proceeds with a first-order rate constant $4.5–6 \times 10^{-6}$ s^{-1} (at 37°, pH 7.5) (Hoffman, 1986; Wu *et al.*, 1983). Alternately, AdoMet depurinates in acid or base-catalyzed reactions to produce adenine and pentosyl methionine with a first-order rate constant 3×10^{-6} s^{-1} (at 37°, pH 7.5) (Hoffman, 1986). Finally, the chiral sulfonium center of AdoMet undergoes racemization to produce the inactive

(R,S)-adenosylmethionine isomer with a first-order rate constant $2 \times 10^{-6}\,\mathrm{s}^{-1}$ (at 37°, pH 7.5) (Hoffman, 1986). The rate of epimerization changes little over a broad pH range, but has a large temperature dependence (Matos and Wong, 1987). However, the equilibrium ratio of (S,S) to (R,S)-AdoMet is approximately 7:3 at 23° (Matos and Wong, 1987). At 70°, the temperature optimum for most *M. jannaschii* methyltransferases, AdoMet decomposes with a half-life of less than 25 min at neutral pH. For comparison, apparent rates of enzyme-catalyzed methyltransferase reactions range from 5×10^{-6} to $5 \times 10^{-3}\,\mathrm{s}^{-1}$.

Because of these undesirable side reactions, AdoMet should be stored desiccated below −20°. The *p*-toluene sulfonate salt of AdoMet (CAS 485–80–3) is commercially available from Sigma-Aldrich or MP Biomedicals, and it seems to be more stable than the chloride salt (Matos and Wong, 1987). For the standard methyltransferase assays described later, radioisotopic compounds [methyl-^{3}H]-AdoMet and [methyl-^{14}C]-AdoMet are available from several sources, including GE Healthcare, MP Biomedicals, and American Radiolabeled Chemicals. Stock solutions should be freshly prepared in 1–10 mM sulfuric acid (pH 3–4) and stored at −20°. Aliquots should be thawed and used promptly for each experiment.

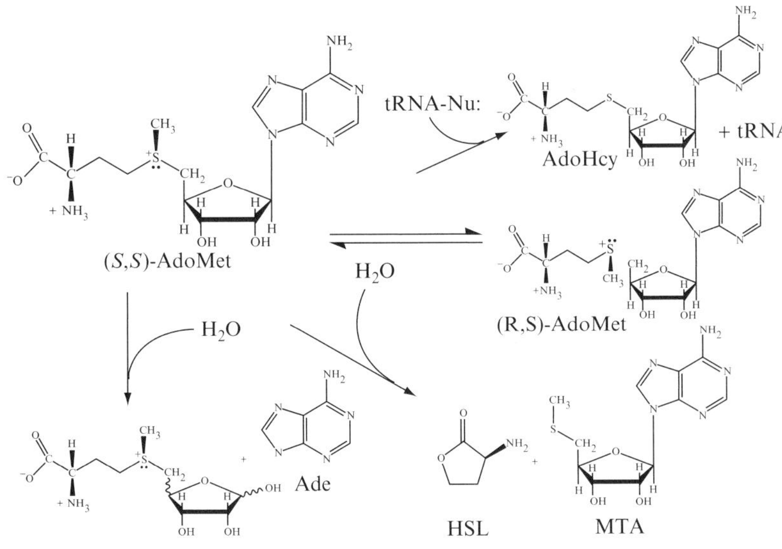

Figure 8.2 Reactions of (S,S)-S-adenosylmethionine (S,S-AdoMet) (modified from Wu *et al.* [1983]). Methyltransferase enzymes catalyze the S_N2 reaction of a nucleophilic group on the tRNA with *(S,S)*-AdoMet. Spontaneous racemization of the sulfonium center produces the biologically inactive isomer (R,S)-AdoMet. Nonenzymatic hydrolysis produces homoserine lactone (HSL) and 5′-deoxy-5′-methylthioadenosine (MTA). Depurination releases adenine and pentosyl methionine.

3.2. tRNA substrates

The order of tRNA modification events inside the cell is poorly understood, and most tRNA modifying enzymes are assumed to act independently of each other, despite evidence for methyltransferase complexes in eukaryotes (Purushothaman *et al.*, 2005). The ideal substrate for *in vitro* methyltransferase assays is a single folded tRNA species that is fully modified at each position except for the target position. The second-best substrates are undermethylated tRNAs purified from a mutant strain missing the target modifying enzyme.

An insertion mutation in the *Methanococcus maripaludis* JJ *trm1* gene was constructed to compare the archaeal 2-dimethylguanosine(26) methyltransferase to the characterized yeast homolog (Graham, 2000). As observed for the yeast *trm1* mutant (Ellis *et al.*, 1986), the *trm1*$^-$ genotype does not substantially affect the growth of *M. maripaludis*. The generation time for *trm1*$^-$ cells was 1.9 h vs. 1.4 h for wild-type cells. tRNAs purified from *M. maripaludis* JJ wild-type and *trm1*$^-$ cells were enzymatically hydrolyzed, and modified nucleosides were analyzed by liquid chromatography–mass spectrometry (S. Zhou and J. A. McCloskey, personal communication). The chromatograms are indistinguishable except for a significant decrease in levels of m$^2_2$G in the *trm1*$^-$ hydrolysate. Relative to the UV absorbance of 2′-O-methylcytidine (Cm) used as a standard, the amount of m$^2_2$G from *trm1*$^-$ tRNA was 26% of the wild-type level. This residual m$^2_2$G probably comes from m$^2_2$G-10, which is formed by a separate tRNA methyltransferase. In contrast, levels of m^{2}G were essentially unchanged in the *trm1* tRNAs. Therefore, tRNAs from this *trm1*$^-$ strain should be ideal substrates for methylation *in vitro*. Purified *M. jannaschii* Trm1 protein transferred sixfold more methyl groups from AdoMet to *trm1*$^-$ tRNAs than to wild-type tRNAs (56 pmol A$_{260}$$^{-1}$ vs 9 pmol A$_{260}$$^{-1}$, respectively). tRNAs from *trm1*$^-$ strains were also methylated twice as much as heterologous *E. coli* tRNAs.

However, making directed mutations in most organisms is still laborious, and null alleles of some tRNA methyltransferases appear lethal. For example, Δ*trm5* mutants of *S. cerevisiae* grow very slowly (Björk *et al.*, 2001), reflecting the direct importance of the m^{1}G37 modification on translational fidelity. tRNAs purified from haploid Δtrm5 mutants are excellent substrates for Trm5p, however (G. K., unpublished data). Fortunately organisms have naturally evolved a variety of tRNA modification schemes. tRNAs purified from cells that lack a specific modification enzyme can be used as substrates to characterize heterologous enzymes. The large number of publicly available genome sequences makes it simple to identify candidate organisms. Purified native tRNAs from *S. cerevisiae*, *E. coli*, bovine liver, and wheat germ are commercially available (Sigma-Aldrich). Total RNA from other sources can be readily purified by acid guanidinium phenol chloroform extraction (Chomczynski and Sacchi, 1987) or by TRI reagent

(Chomczynski, 1993). tRNAs can be further purified by rapid anion exchange chromatography (described later), or single tRNA species can be purified by preparative scale reversed-phase chromatography (Kelmers *et al.*, 1971). Purified tRNAs should be stored at $-20°$ in RNase-free buffer (10 mM ammonium acetate, pH 5).

Mitochondrial tRNAs are often undermodified compared with cytoplasmic tRNAs (Martin, 1995; Sprinzl and Vassilenko, 2005). Mitochondria can be prepared from *S. cerevisiae* cells following the protocol of Glick and Pon (1995) with minor modifications. The washed cells from 4 liters yeast culture are incubated with Lyticase (Sigma) for 80 min at 30° to prepare spheroplasts. After lysis of the spheroplasts, differential centrifugation is used to prepare pellets of crude mitochondria, which are extracted with 2 ml TRI Reagent according to the manufacturer's instructions (Molecular Research Center). One-quarter volume of high salt precipitation solution (containing 0.8 M sodium citrate and 1.2 M NaCl) is added to the RNA extract in TRI Reagent. After 5–10 min incubation at room temperature, the mixture is centrifuged at 12,000g for 8 min to precipitate ribosomal and high molecular weight RNA. tRNA in the supernatant is precipitated with isopropanol. Approximately 50 μg tRNA is obtained from this preparation.

3.3. Protocol for small RNA purification from total RNA

Small RNAs, including tRNA and 5S rRNA, can be purified from total RNA by weak anion exchange chromatography with a short gravity flow column (Gupta, 1995; Tanner, 1989). We have used this method to purify tRNAs from *M. maripaludis, M. jannaschii,* and *E. coli.* tRNAs can be further fractionated by use of large anion exchange columns with gradient elution as described by Nishimura (1971).

1. Prepare a weak anion exchange column by placing sterile glass wool in the bottom of a 5-ml syringe barrel with an attached Luer-Lok stopcock.
2. Equilibrate DEAE Sephadex A-25 in ribonuclease-free binding buffer (100 mM NaCl, 10 mM TRIS-HCl, pH 7) and load 2–3 ml resin in the syringe barrel. Use the stopcock to adjust the flow rate to 0.2 ml/min^{-1} and wash the column with 2–3 bed volumes of binding buffer.
3. Isolate total RNA from 2–3 g cells. Suspend the RNA in 0.5–1 ml binding buffer and load the sample on the DEAE column.
4. If the sample is highly viscous because of contamination by significant amounts of carbohydrate, pass the sample several times through a syringe fitted with a 22-gauge needle, centrifuge the sample at 16,000g for 5 min, and apply the soluble portion to the column. Wash the column with several bed volumes of water before proceeding.

5. Wash the column with 4 bed volumes of binding buffer followed by 4 bed volumes of wash buffer (250 mM NaCl, 10 mM TRIS-HCl, pH 7).

6. Elute small RNAs with 4 bed volumes of elution buffer (1 M NaCl, 10 mM TRIS-HCl, pH 7), collecting 0.5-ml fractions in microcentrifuge tubes. Fractions containing tRNAs can be rapidly identified by UV absorbance spectroscopy or by analyzing an aliquot by urea–polyacrylamide gel electrophoresis (urea-PAGE) with UV shadowing or ethidium bromide staining. 1 A$_{260}$ unit is approximately 50 μgml^{-1} tRNA.

7. Add two volumes of cold ethanol to tubes containing RNA and chill at $-20°$ for at least 2 h. Collect RNA by centrifugation (18,000g, 10 min at 4°) and wash the pellet with cold 80% ethanol.

8. Dry the pellet thoroughly and then resuspend the RNA in water or 10 mM ammonium acetate (pH 5).

9. DEAE columns may be discarded or regenerated by washing with 2 bed volumes of 2 M NaCl, followed by 2 bed volumes of 0.1 N NaOH. Wash the columns with water and then store them in 3 bed volumes of binding buffer with 20% (v/v) ethanol.

3.4. *In vitro* transcripts

Transcripts of tRNAs prepared *in vitro* can be used to determine the modification positions and the identity elements that enzymes use to recognize their tRNA substrates (Milligan and Uhlenbeck, 1989). Although unmodified transcripts can fold into native tRNA structures with lower melting temperatures than the corresponding fully modified native tRNAs, their homogeneity makes them valuable substrates. Constantinesco *et al.* (1999) used transcripts of yeast and halophilic archaeal tRNAs to detect tRNA modifying enzymes in the euryarchaeon *Pyrococcus furiosus*.

T7 RNA polymerase is commonly used to produce these transcripts; it is available in high activity commercial preparations, or can be readily purified from *E. coli* expression systems. Runoff transcripts from PCR products or linearized plasmids can be used as templates for high-yield RNA synthesis. There are several technical problems inherent in this system. First, high rates of transcription initiation are required to produce high yields of tRNA. Initiation is often the rate-limiting step, so the commercially available MEGAshortscript kit (Ambion) has been designed specifically for this purpose. Alternately, high levels of T7 RNA polymerase can be used to improve yield. The high concentrations of template required for efficient initiation may be produced by PCR amplification, although this method can introduce additional nucleotides at the 3′ end of the transcript because of nontemplated nucleotide addition by DNA polymerases. Second, T7 RNA polymerase requires that at least the first two positions of the coding strand be G for efficient transcription initiation. Because of base pair complementarity in tRNA acceptor stems, this constraint affects several positions on the tRNA. Thirty-five years after the synthesis of the first tRNA gene, the low cost of oligonucleotide primer synthesis makes it easy to engineer tRNA templates with appropriate initiation sites (Khorana *et al.*, 1972). However, if the tRNA transcript must be identical to a native tRNA without a 5′-guanosine (for example, the acceptor stem may be an identity element), an alternative strategy must be used, such as the hammerhead ribozyme described later.

To characterize *S. cerevisiae* Trm5p, an unmodified transcript of the mitochondrial initiator tRNA (ymt-tRNA$^{Met}_f$) was required (G.K., unpublished data). The 5′ terminal nucleotide of this tRNA is unfavorable for T7 transcription. This problem was overcome by introducing a hammerhead ribozyme sequence between a strong T7 RNA polymerase promoter and the gene for ymt-tRNA$^{Met}_f$ (Fechter *et al.*, 1998). In high Mg^{2+} conditions, the ribozyme cleaves the transcript generating the desired ribonucleotide sequence (Spencer *et al.*, 2004). The template containing T7 promoter, hammerhead ribozyme, and tRNA gene was produced from oligonucleotides by standard methods, including 5′-phosphorylation, ligation, and PCR amplification by use of terminal primers. The purified product was cloned into pUC18 in the unique *Hind*III and *Bam*HI sites. For efficient production of tRNA$^{Met}_f$, this plasmid was used as template in a PCR reaction with a standard upstream primer and a 2′-O-methyl group on the terminal ribose of the reverse primer. This modification may prevent unwanted extension of the PCR product beyond the CCA end (Sherlin *et al.*, 2001). The PCR product was gel-purified and then used as a template for transcription.

3.5. Protocol for *in vitro* transcription and cleavage of the hammerhead ribozyme

The transcription reactions are carried out as suggested in Fechter *et al.* (1998) with minor modifications.

1. Assemble reactions (100 μl volume) containing 200–300 ng purified DNA template in 10 mM TRIS-HCl (pH 8), 10 μl transcription buffer (Ambion), 8 μl of a 25 mM nucleoside triphosphate solution, 3.5 μl of 0.2 M magnesium acetate, 20 units T7 RNA polymerase Plus (Ambion) and 1 μl SUPERase-In (Ambion).
2. Incubate reactions for 3 h at 37°. Then add 4 units ribonuclease-free DNase (Ambion) and continue the incubation for 1 h at 37°.
3. Dilute the reaction to 500 μl with a solution containing 40 mM TRIS-HCl (pH 7.5), 10 mM DTT, and 31.5 mM magnesium acetate. Incubate for 1 h at 55° to enhance ribozyme cleavage.
4. Add 1 ml cold ethanol and precipitate the RNA at −20° for at least 4 h. Collect the RNA by centrifugation at 12,000g for 10 min at 4°.
5. Dissolve the pellet in 100 μl sterile 10 mM sodium acetate (pH 5), and purify the RNA using a nucleotide removal kit (QIAquick Nucleotide

Removal Kit, Qiagen). Elute the RNA from the membrane with 100 μl elution buffer (10 mM TRIS-HCl, pH 8), but add 1 μl of 2 M sterile potassium acetate (pH 5) immediately after the elution to reduce the pH. Determine the absorbance at 260 nm using a 3 μl aliquot in 0.7 μl H$_2$O. Typical yields are 10 –15 pmol μl^{-1} RNA (25 ng ≈1 pmol).

6. If necessary, the ribozyme moiety can be separated from the cleaved tRNA by urea-PAGE, followed by elution of the tRNA band.

4. METHYLTRANSFERASE ACTIVITY ASSAYS

Standard methyltransferase assays measure the incorporation of tritium-labeled methyl groups from [methyl-^{3}H]-S-adenosylmethionine into TCA precipitable product (Greenberg and Dudock, 1979). Compared with most characterized enzymes, macromolecular methyltransferases demonstrate low activities and turnover numbers *in vitro*. Although this discrepancy may be partially attributable to artifacts of heterologous tRNA substrates and *in vitro* conditions, these enzymes are relatively proficient considering their high selectivity (Takusagawa *et al.*, 1998). Spontaneous nucleoside methylation rates are negligible, and the reverse reaction, demethylation, is very unfavorable, so background levels of activity are low with appropriate washing. S-Adenosylhomocysteine (AdoHcy), a reaction product, inhibits many methyltransferases. Because AdoMet is usually present in a large excess over tRNA, it is not clear whether AdoHcy production is sufficient to cause significant product inhibition in these assays.

A continuous assay for methyltransferase activity has been proposed on the basis of coupling methyltransferase activity with enzymes to produce homocysteine that can be detected with Ellman's reagent (Hendricks *et al.*, 2004). This assay was recently adapted to measure protein arginine methyltransferase activity with peptide substrates (Dorgan *et al.*, 2006). However, the continuous assay currently has a relatively high limit of detection (~4 μM) and has not been tested with RNA methyltransferases.

4.1. tRNA methyltransferase activity protocol

Methyltransferase activity is measured in enzyme-limiting concentrations.

1. Assemble reactions in 0.5-ml microcentrifuge tubes in a total volume of 30 μl. Reactions should include enzyme, tRNA (~800 pmol total tRNA or 25 pmol synthetic tRNA) and buffer. Reaction buffer consists of 100 mM HEPES/NaOH (pH 8.0), 200 mM KCl, 5 mM MgCl$_2$, 1 mM DTT, and 0.01% Nonidet P-40 (or Igepal CA-630). Some enzymes may be more active in lower salt buffers (100 mM KCl).

2. For assays of thermophilic methyltransferases, preincubate mixtures without AdoMet for 10 min at the desired reaction temperature to equilibrate

the temperature. If hyperthermophilic enzymes are assayed at $70°$, cover reactions with 15 μl mineral oil (Sigma) to reduce evaporation. Alternately, reactions can be prepared in thin-wall PCR tubes and incubated in a thermal cycler with a heated lid.

3. Add 3 μl 10× AdoMet stock, containing 0.5 mM S-adenosyl-L-[methyl-^{3}H]methionine (320 Ci mol^{-1}). This stock solution can be prepared by diluting 30 μl S-adenosyl-L-[methyl-^{3}H] methionine (0.55 μCi μl^{-1}, 11.2 Ci mmole^{-1}; Perkin Elmer), 2.5 μl of 20 mM AdoMet, and 10 μl of 50 mM H$_2$SO$_4$ to 100 μl, with water.

4. Incubate tubes for 12–30 min and then stop the reactions with one of the two procedures described below.

4.1.1. Quantitation option 1

1. Transfer 25 μl of the completed reaction mixture to a glass tube (10 × 70 mm) that has been cooled on ice and that contains 4 μl of 10 mgml^{-1} bovine serum albumin. Add 2 ml of cold 10% trichloroacetic acid (TCA) and vortex. Let the tubes sit on ice for 10 min.

2. Connect a Millipore 1225 sampling manifold to a pump or vacuum aspirator. Put a glass fiber filter (Schleicher & Schuell #34, 27 mm diameter) over as many holes as assay tubes were incubated. Cover the extra holes with rubber stoppers. Rinse the glass fiber filters with 10% TCA before filtration.

3. Filter the reaction mixtures, rinse each tube, and filter three times with 3 ml 10% TCA.

4. Dry the filters in a ventilated drying oven at $150°$ for 15 min and place the filters in counting vials containing 5 ml scintillation fluid. Determine radioactivity in a liquid scintillation counter. It is important that the glass fiber filters dry completely; otherwise the counting efficiency is much lower and may vary from one sample to the next.

4.1.2. Quantitation option 2

1. Dilute 20 μl of the reaction product into 380 μl cold 5% TCA.

2. Precipitate the tRNA on ice with TCA for 20 min and then centrifuge at 16,000g for 15 min. Wash the pellet with 500 μl 5% TCA and then dissolve the pellet in 100 μl of 10 mM NaOH.

3. Measure tritium incorporation by liquid scintillation counting. Counting efficiencies (typically 40%) can be calculated by measuring tritium in 5 μl of the remaining reaction.

Specific activities are reported as nmol of methyl groups incorporated from AdoMet per min per mg protein. For reactions carried out under tRNA-limiting conditions, tRNA modification results are reported as nmol of methyl groups incorporated from AdoMet per A$_{260}$ unit of tRNA.

5. IDENTIFYING MODIFICATION PRODUCTS

tRNA substrates are incubated with an excess of enzyme and AdoMet to fully modify the target sites; 0.5 A_{260} units of tRNA are incubated in a 50 μl reaction volume with 0.15 μCi S-adenosyl-L-[methyl-^{14}C]methionine (54.0 mCi/mmol) (Amersham), excess enzyme (usually 0.56–5.2 μg protein, depending on purity and activity), and reaction buffer for 30 min at 60°. Precipitate the tRNAs at −20° with 0.3 M sodium acetate (pH 5.2) and 2 volumes of ethanol. Product is recovered by centrifugation at 16,000g for 15 min and then washed with 80% ethanol. Dry the tRNA pellet under vacuum and then dissolve in ribonuclease-free water. Solutions containing tRNA are diluted to 1–3 μg/μl^{-1} in water and then enzymatically hydrolyzed to ribonucleosides as described by Crain (1990).

The preferred method for identifying modified nucleosides in RNA hydrolysate is HPLC separation on a reversed-phase column with detection by UV visible absorbance or electrospray mass spectrometry (Pomerantz and McCloskey, 1990). The Supelcosil LC-18-S is an analytical end-capped octadecylsilane reversed-phase column (Supelco) that was specifically designed for nucleoside separation. However, we have obtained adequate separations of modified nucleosides from yeast tRNA hydrolysates by use of other C_{18} columns as well. Standard HPLC methods that use gradient elution programs were described by Gehrke and Kuo (1990) and Pomerantz and McCloskey (1990). Although few modified nucleosides are commercially available, a nucleoside test mixture is available (Supelco), and standards can be prepared by hydrolyzing purified tRNAs (Sigma). Other modified nucleosides such as N^1-methylguanosine can be synthesized as described (Broom et al., 1964). The RNA modification database lists references for nucleoside synthesis and properties (Rozenski et al., 1999). In general, modified purines have longer retention times in reversed-phase HPLC than modified pyrimidines and can be differentiated and measured by their UV absorbance spectra (Dunn and Hall, 1970). For rapid and routine analysis, detection by UV absorbance and correlation with the retention times of standards is most convenient.

Thin-layer chromatography (TLC) offers a rapid and inexpensive alternative to HPLC analysis for processing many samples. [^{14}C]Methyl incorporation can be measured on TLC plates by autoradiography. Radioisotopic methods can be used to distinguish modifications made in vitro from similar modifications at other tRNA positions that are introduced by different enzymes. We used this method to characterize the m^2G and m^2_2G reaction products of the M. jannaschii 2-methylguanosine(37) tRNA methyltransferase (Fig. 8.3).

Celluose-coated TLC plates with fluorescent indicator (Sigma) should be prechromatographed in solvent. tRNA hydrolysates are spotted on top

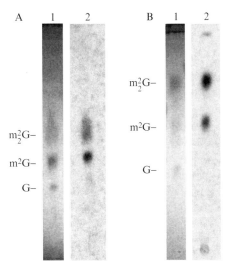

Figure 8.3 Products of MjTrm1. (A) TLC developed in solvent system A: Lane 1, UV shadowed m^2G, m^2_2G, and G standards. Lane 2, Autoradiogram of Trm1-methylated tRNA hydrolysate. (B) TLC developed in solvent system F. Lanes are the same as in (A).

of nucleoside standards (20 nmol each). Dry the spots completely under a stream of nitrogen or dry air. Two useful solvent systems from Rogg *et al.* (1976) are solvent A, *n*-butanol/isobutyric acid/ammonium hydroxide/water (53.6:26.8:1.8:17.9) and solvent F, acetonitrile/ethyl acetate/*n*-butanol/isopropanol/ammonium formate (40 mM, pH 7.6) (51.1:14.6:7.3:7.3:19.7). Reversed-phase C_{18}-coated TLC plates with fluorescent indicator (E. Merck) are developed in KH_2PO_4 (10 mM, pH 5.0)/methanol (86:14). After development, plates are dried, nucleoside standards are visualized by UV absorbance, and autoradiography is performed with BioMAX film (Kodak) at $-80°$ for 2 days. Alternately, a PhosphorImager cassette can be used according to the manufacturer's instructions.

6. IDENTIFYING MODIFICATION SITES

The next step after identifying the modified nucleoside product is mapping that modification to a specific position on a tRNA. Often the site of modification can be predicted on the basis of known tRNA sequences (Sprinzl and Vassilenko, 2005). By comparing methylation of a wild-type tRNA transcript with methylation of a variant differing in only one site, it may be possible to deduce the site of modification. However, folding problems and unknown identity elements may confuse these results, so direct sequencing of the modified product is usually preferred. Most experiments use a single synthetic tRNA transcript as a template, modify the transcript *in vitro*, and then localize the modification by enzymatic or chemical cleavage. Methods for postlabeling RNAs with ^{32}P, digesting them chemically or enzymatically, and separating the products by chromatography to create a "fingerprint" of the RNA have been reviewed by many authors since their development by Sanger and colleagues (Barrell, 1971; Kuchino *et al.*, 1987). However, newer methods that use liquid chromatography-mass spectrometry (LC/MS) to identify modifications in ribonuclease T_1 cleavage products show great promise (Kowalak *et al.*, 1993). The nearest neighbor method of RNA sequencing by LC/MS that was described by Rozenski and McCloskey (1999) complements the analysis of larger cleavage products. As high-resolution Fourier transform mass spectrometers proliferate, top-down RNA sequencing methods may compete with current bottom-up approaches (Kellersberger *et al.*, 2004). Compared with classical chromatographic and electrophoretic sequencing methods, the mass spectral methods require less tRNA and provide more structural information about modifications.

6.1. Mapping modifications by primer extension

Nucleoside base modifications that interfere with base pairing or polymerase processivity can be mapped by primer extension analysis. Reverse transcriptase extends a ^{32}P-labeled primer complementary to a region of the tRNA template downstream from the modification site (Maden *et al.*, 1995). Under low deoxyribonucleoside triphosphate (dNTP) concentrations, reverse transcriptase pauses and frequently dissociates from the RNA-DNA hybrid one nucleotide before the modification. For comparison, an RNA ladder can be synthesized by use of primer, RNA template, and dNTPs mixed with a 10-fold excess of a single dideoxyribonucleotide triphosphate (ddNTP) terminator. The extended products can be separated by urea-PAGE, analyzed by autoradiography, and the modification can be mapped relative to ladder. We used this method to map modifications at position 37 by yeast Trm5p on ymt-tRNA$^{Met}_f$ with a mixed pool of tRNAs (Fig. 8.4). This method is readily scalable—multiple reactions can be processed in parallel.

6.2. Primer extension protocol

This procedure is modified from one described by Hahn *et al.* (1989).

1. Design and purchase an oligodeoxynucleotide primer complementary to the tRNA template. For primer extension on yeast mitochondrial tRNA$^{Met}_f$, the DNA sequence complementary to nucleotides 71–52 (conventional tRNA numbering) was used (5'-GCAATAATAC-GATTTGAACG).

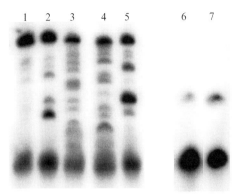

Figure 8.4 PhosphorImage of primer extension reaction products separated by urea-PAGE. The primer was complementary to *S. cerevisiae* mitochondrial tRNA^{Met}f. Lane 1, Unmodified T7 RNA polymerase transcript of yeast mitochondrial tRNA^{Met} with M-MLV reverse transcriptase and deoxyribonucleotides. Lanes 2–5, reverse transcriptase reaction products including ddGTP (lane 2), ddATP (lane 3), ddTTP (lane 4), and ddCTP (lane 5). Lane 6, reverse transcriptase product from purified yeast mitochondrial tRNA with deoxyribonucleotides. Lane 7, reverse transcriptase product from total yeast tRNA (Sigma-Aldrich). The images were edited using Photoshop software (Adobe) to increase contrast, rotate, and crop the image.

2. Phosphorylate the oligonucleotide primer with T4 polynucleotide kinase (NEB). In a 30-μl reaction, mix 135 pmol primer with reaction buffer (NEB), and 7.8 M [γ-^{32}P]ATP (429 Ci/mmol; Perkin-Elmer).

3. Incubate the reaction for 40 min at 37°. Inactivate the enzyme by incubation at 70° for 10 min. Purify the primer by use of a Biogel P6 column (Bio-Rad) following the manufacturer's instructions. In a typical reaction, 72% of the primer is labeled, as determined by liquid scintillation counting.

4. In a 10-μl reaction, anneal 4–6 pmol labeled primer to the tRNA template. The reaction buffer contains 50 mM TRIS-HCl (pH 8.3), 30 mM NaCl, and 10 mM DTT. Typical amounts of tRNA template used were 4 pmol synthetic ymt-tRNA^{Met}f, 100 pmol purified mitochondrial tRNA, or 400 pmol total yeast tRNA.

5. Incubate the annealing reaction for 5 min at 95°, and then cool the reaction to room temperature.

6. Mix 5 μl of the primer-template product with 5 units of M-MLV reverse transcriptase (Ambion) in the accompanying buffer (50 mM TRIS-HCl [pH 8.3], 50 mM KCl, 3 mM MgCl$_2$, 5 mM DTT), in the presence of 0.25 mM of each dNTP in a 10-μl reaction.

7. Incubate the reverse transcriptase reaction for 1 h at 42°. Terminate the reaction by adding 10 mM EDTA. Store the extension product at −20° for analysis by urea-PAGE.

Table 8.1 Identified and predicted tRNA methyltransferases in *Methanocaldococcus jannaschii*[a]

ID	Specific activity[b] (nmol min^{-1} mg^{-1})	Total modification[c] (pmol A$_{260}$$^{-1}$)	Products	Position[d]	*S. cerevisiae* homolog
M.j. Extract	0.064 (±0.005)	162 (±25)	Misc.	nd[e]	—
MJ0438	7.1 (±0.3)	58 (±3.2)	m^{2}G		YJL125C (Gcd14)
MJ0134	3.0 (±0.2)	40 (±7.1)	m^{1}A	57, 58	YOL124C (Trm11)
MJ0710	0.16 (±0.05)	28 (±3.6)	m^{2}G	10	YDR120C (Trm1)
MJ0946	0.35 (±0.04)	25 (±2.2)	m^{2}G, m$^2_2$G	26	YHR070 (Trm5)
MJ0883	0.013 (±0.005)	3.7 (±0.79)	m^{1}G	37	
MJ1385	nd	nd	Cm	56	
MJ1557	nd	nd	imG-14	37	YML005W (Trm12)
MJ1653	0.0070 (±0.002)	3.1 (±1.3)	m^{5}C	nd	—
MJ1649	nd	nd	m^{5}C	nd	—

[a] Data from Graham (2000) unless otherwise referenced.

[b] Methyl groups incorporated from AdoMet at 70° using *E. coli* tRNAs.

[c] Methyl groups incorporated into *E. coli* tRNA under tRNA-limiting conditions at 65°.

[d] Modification position data based on *S. cerevisiae* homolog, except for Trm1.

[e] nd, not determined.

7. tRNA Methyltransferases from *M. jannaschii*

Complete genome sequences of several archaea led to the identification of six new tRNA methyltransferases (Graham, 2000). The original annotation of the *M. jannaschii* genome sequence identified one tRNA methyltransferase: a homolog of the yeast Trm1p enzyme, which forms m^2G and m^2_2G at position 37 (Ellis *et al.*, 1986). Subsequently, the archaeal *P. furiosus* homolog was identified (Constantinesco *et al.*, 1998), and we confirmed that the *M. jannaschii* homolog (MJ0946) catalyzes the same reaction (Table 8.1). A *trm1⁻* mutant of *M. maripaludis* has reduced levels of m^2_2G in its tRNAs, but still has m^2G, probably because of the activities of MJ0438 and MJ0710 methyltransferase homologs. We demonstrated that the MJ0710 enzyme also produces 2-methylguanosine, and its yeast homolog, Trm11p, was recently identified and shown to methylate guanosine-10 (Purushothaman *et al.*, 2005). The MJ0134 methyltransferase is homologous to the yeast Gcd14 protein: both produce 1-methyladenosine. The *Pyrococcus abyssi* homolog was recently shown to modify tRNA positions 57 and 58 (Roovers *et al.*, 2004). The MJ0883 enzyme was the first identified 1-methylguanosine methyltransferase in archaea or eucarya; curiously, it is unrelated to the bacterial TrmD $m^1G(37)$ methyltransferases, as described previously. The human Trm5p protein was subsequently identified (Brulé *et al.*, 2004), and the *M. jannaschii* homolog has been independently identified (Christian *et al.*, 2004). The MJ1557 protein is implicated in the formation of a highly modified wyosine modification on the basis of the identification of homologous *S. cerevisiae* Trm12 (Kalhor *et al.*, 2005). *M. jannaschii* tRNAs contain a monomethyl wyosine derivative (imG-14) (McCloskey *et al.*, 2001; Zhou *et al.*, 2004). Two putative 5-methylcytidine methyltransferases (MJ1653 and MJ1649) seem to have evolved from a family of 5-methyluridine methyltransferases. Finally, the Trm56 protein from *P. abyssi* was recently shown to form $2'$-O-methylcytidine(56) (Renalier *et al.*, 2005). This member of the SPOUT family of methyltransferases was not detected in the original screen for AdoMet binding proteins because it belongs to an unrelated methyltransferase family.

ACKNOWLEDGMENTS

This work was partially supported by grant F-1576 from the Welch Foundation. We thank Dean Appling, Pam Crain, Ramesh Gupta, Jim McCloskey, Gary Olsen, Claudia Reich, Dieter Söll, and Carl Woese for helpful discussions about archaeal tRNA methyltransferases and RNA methodology.

REFERENCES

Ahn, H. J., Kim, H.-W., Yoon, H.-J., Lee, B.-I., Suh, S. W., and Yang, J. K. (2003). Crystal structure of tRNA(m^1G37)methyltransferase: Insights into tRNA recognition. *EMBO J.* **22,** 2593–2603.

Aittaleb, M., Rashid, R., Chen, Q., Palmer, J. R., Daniels, C. J., and Li, H. (2003). Structure and function of archaeal box C/D sRNP core proteins. *Nat. Struct. Biol.* **10**, 256–263.

Altschul, S. F., Gish, W., Miller, W., Myers, E. W., and Lipman, D. J. (1990). Basic local alignment search tool. *J. Mol. Biol.* **215**, 403–410.

Altschul, S. F., Madden, T. L., Schaffer, A. A., Zhang, J., Zhang, Z., Miller, W., and Lipman, D. J. (1997). Gapped BLAST and PSI-BLAST: A new generation of protein database search programs. *Nucleic Acids Res.* **25**, 3389–3402.

Barrell, B. G. (1971). Fractionation and sequence analysis of radioactive nucleotides. *In* "Procedures in Nucleic Acids Research" (G. L. Cantoni and D. R. Davies, eds.), Vol. 2, pp. 751–779. Harper & Row, New York.

Björk, G. R., Jacobsson, K., Nilsson, K., Johansson, M. J. O., Byström, A. S., and Persson, O. P. (2001). A primordial tRNA modification required for the evolution of life? *EMBO J.* **20**, 231–239.

Björk, G. R., Wikström, P. M., and Byström, A. S. (1989). Prevention of translational frameshifting by the modified nucleoside 1-methylguanosine. *Science* **244**, 986–989.

Bortolin, M.-L., Bachellerie, J.-P., and Clouet-d'Orval, B. (2003). *In vitro* RNP assembly and methylation guide activity of an unusual box C/D RNA, cis-acting archaeal pre-tRNATrp. *Nucleic Acids Res.* **31**, 6524–6535.

Broom, A. D., Townsend, L. B., Jones, J. W., and Robins, R. K. (1964). Purine Nucleosides. VI. Further Methylation Studies of Naturally Occurring Purine Nucleosides. *Biochemistry* **3**, 494–500.

Brulé, H., Elliott, M., Redlak, M., Zehner, Z. E., and Holmes, W. M. (2004). Isolation and characterization of the human tRNA-(N^1G37) methyltransferase (TRM5) and comparison to the *Escherichia coli* TrmD protein. *Biochemistry* **43**, 9243–9255.

Bult, C. J., White, O., Olsen, G. J., Zhou, L., Fleischmann, R. D., Sutton, G. G., Blake, J. A., FitzGerald, L. M., Clayton, R. A., Gocayne, J. D., Kerlavage, A. R., Dougherty, B. A., *et al.* (1996). Complete genome sequence of the methanogenic archaeon, *Methanococcus jannaschii*. *Science* **273**, 1017–1140.

Chomczynski, P. (1993). A reagent for the single-step simultaneous isolation of RNA, DNA and proteins from cell and tissue samples. *BioTechniques* **15**, 532–536.

Chomczynski, P., and Sacchi, N. (1987). Single-step method of RNA isolation by acid guanidinium thiocyanate-phenol-chloroform extraction. *Anal. Biochem.* **162**, 156–169.

Christian, T., Evilia, C., Williams, S., and Hou, Y. M. (2004). Distinct origins of tRNA (m1G37) methyltransferase. *J. Mol. Biol.* **339**, 707–719.

Constantinesco, F., Benachenhou, N., Motorin, Y., and Grosjean, H. (1998). The tRNA (guanine-26, N^2-N^2) methyltransferase (Trm1) from the hyperthermophilic archaeon *Pyrococcus furiosus*: Cloning, sequencing of the gene and its expression in *Escherichia coli*. *Nucleic Acids Res.* **26**, 3753–3761.

Constantinesco, F., Motorin, Y., and Grosjean, H. (1999). Transfer RNA modification enzymes from *Pyrococcus furiosus*: Detection of the enzymatic activities *in vitro*. *Nucleic Acids Res.* **27**, 1308–1315.

Crain, P. F. (1990). Preparation and enzymatic hydrolysis of DNA and RNA for mass spectrometry. *Methods Enzymol.* **193**, 782–790.

Dixon, M. M., Fauman, E. B., and Ludwig, M. L. (1999). The black sheep of the family: AdoMet-dependent methyltransferases that do not fit the consensus structural fold. *In* "S-Adenosylmethionine-Dependent Methyltransferases: Structures and Functions" (X. Cheng and R. M. Blumenthal, eds.), pp. 39–54. World Scientific Publishing, New York.

Dorgan, K. M., Wooderchak, W. L., Wynn, D. P., Karschner, E. L., Alfaro, J. F., Cui, Y., Zhou, Z. S., and Hevel, J. M. (2006). An enzyme-coupled continuous spectrophotometric assay for S-adenosylmethionine-dependent methyltransferases. *Anal. Biochem.* **350**, 249–255.

Dunin-Horkawicz, S., Czerwoniec, A., Gajda, M. J., Feder, M., Grosjean, H., and Bujnicki, J. M. (2006). MODOMICS: A database of RNA modification pathways. *Nucleic Acids Res.* **34,** D145–D149.

Dunn, D. B., and Hall, R. H. (1970). Purines, pyrimidines, nucleosides and nucleotides: Physical constants and spectral properties. *In* "Handbook of Biochemistry: Selected Data for Molecular Biology" (H. A. Sober, ed.), pp. G3–G238. Chemical Rubber Co., Cleveland, OH.

Eddy, S. R. (1998). Profile hidden Markov models. *Bioinformatics* **14,** 755–763.

Ellis, S. R., Morales, M. J., Li, J.-M., Hopper, A. K., and Martin, N. C. (1986). Isolation and characterization of the *TRM1* locus, a gene essential for the N^2,N^2-dimethylguanosine modification of both mitochondrial and cytoplasmic tRNA in *Saccharomyces cerevisiae*. *J. Biol. Chem.* **216,** 9703–9709.

Fauman, E. B., Blumenthal, R. M., and Cheng, X. (1999). Structure and evolution of AdoMet-dependent methyltransferases. *In* "S-Adenosylmethionine-Dependent Methyltransferases: Structures and Functions" (X. Cheng and R. M. Blumenthal, eds.), pp. 1–38. World Scientific Publishing, New York.

Fechter, P., Rudinger, J., Giegé, R., and Théobald-Dietrich, A. (1998). Ribozyme processed tRNA transcripts with unfriendly internal promoter for T7 RNA polymerase: Production and activity. *FEBS Lett.* **436,** 99–103.

Gehrke, C. W., and Kuo, K. C. (1990). Ribonucleoside analysis by reversed-phase high performance liquid chromatography. *In* "Chromatography and Modification of Nucleosides: Part a" (C. W. Gherke and K. C. T. Kuo, eds.), Vol. 45A, pp. A3–A71. Elsevier, Amsterdam.

Glick, B. S., and Pon, L. A. (1995). Isolation of highly purified mitochondria from Saccharomyces cerevisiae. *Methods Enzymol.* **260,** 213–223.

Graham, D. E. (2000). Archaeal gene identification Ph.D. Thesis, University of Illinois at Urbana-Champaign.

Greenberg, R., and Dudock, B. S. (1979). Bacterial tRNA methyltransferases. *Methods Enzymol.* **59,** 190–203.

Grosjean, H., Motorin, Y., and Morin, A. (1998). RNA-modifying and RNA-editing enzymes: Methods for their identification. *In* "Modification and Editing of RNA" (H. Grosjean and R. Benne, eds.), pp. 21–46. ASM Press, Washington, DC.

Gupta, A., Kumar, P. H., Dineshkumar, T. K., Varshney, U., and Subramanya, H. S. (2001). Crystal structure of Rv2118c: An AdoMet-dependent methyltransferase from Mycobacterium tuberculosis H37Rv. *J. Mol. Biol.* **312,** 381–391.

Gupta, R. (1995). Preparation of transfer RNA, aminoacyl-tRNA synthetases, and tRNAs specific for an amino acid from extreme halophiles. *In* "Archaea: A laboratory manual (Halophiles)" (S. DasSarma and E. M. Fleischmann, eds.), pp. 119–131. Cold Spring Harbor Laboratory Press, Plainview, NY.

Gupta, R., and Woese, C. R. (1980). Unusual modification patterns in the transfer ribonucleic acids of archaebacteria. *Curr. Microbiol.* **4,** 245–249.

Gustafsson, C., Reid, R., Greene, P. J., and Santi, D. V. (1996). Identification of new RNA modifying enzymes by iterative genome search using known modifying enzymes as probes. *Nucleic Acids Res.* **24,** 3756–3762.

Hahn, C. S., Strauss, E. G., and Strauss, J. H. (1989). Dideoxy sequencing of RNA using reverse transcriptase. *Methods Enzymol.* **180,** 121–130.

Hendricks, C. L., Ross, J. R., Pichersky, E., Noel, J. P., and Zhou, Z. S. (2004). An enzyme-coupled colorimetric assay for S-adenosylmethionine-dependent methyltransferases. *Anal. Biochem.* **326,** 100–105.

Hoffman, J. L. (1986). Chromatographic analysis of the chiral and covalent instability of S-adenosyl-L-methionine. *Biochemistry* **25,** 4444–4449.

Ivanetich, K. M., and Santi, D. V. (1992). 5,6-dihydropyrimidine adducts in the reactions and interactions of pyrimidines with proteins. *Prog. Nucl. Acid Res. Mol. Biol.* **42,** 127–156.

Kalhor, H. R., Penjwini, M., and Clarke, S. (2005). A novel methyltransferase required for the formation of the hypermodified nucleoside wybutosine in eucaryotic tRNA. *Biochem. Biophys. Res. Commun.* **334,** 433–440.

Kealey, J. T., Gu, X., and Santi, D. V. (1994). Enzymatic mechanism of tRNA (m5U54) methyltransferase. *Biochimie* **76,** 1133–1142.

Kellersberger, K. A., Yu, E., Kruppa, G. H., Young, M. M., and Fabris, D. (2004). Top-down characterization of nucleic acids modified by structural probes using high-resolution tandem mass spectrometry and automated data interpretation. *Anal. Chem.* **76,** 2438–2445.

Kelmers, A. D., Weeren, H. O., Weiss, J. F., Pearson, R. L., Stulberg, M. P., and Novelli, G. D. (1971). Reversed-phase chromatography systems for transfer ribonucleic acids–preparatory-scale methods. *Methods Enzymol.* **20,** 9–34.

Khorana, H. G., Agarwal, K. L., Buchi, H., Caruthers, M. H., Gupta, N. K., Kleppe, K., Kumar, A., Otsuka, E., RajBhandary, U. L., Van de Sande, J. H., Sgaramella, V., Terao, T., *et al.* (1972). Studies on polynucleotides. 103. Total synthesis of the structural gene for an alanine transfer ribonucleic acid from yeast. *J. Mol. Biol.* **72,** 209–217.

Kline, L. K., and Söll, D. (1982). Nucleotide modification in RNA. *In* "The Enzymes" (P. D. Boyer, ed.), Vol. XV, pp. 567–582. Academic Press, New York.

Kloczkowski, A., Ting, K. L., Jernigan, R. L., and Garnier, J. (2002). Combining the GOR V algorithm with evolutionary information for protein secondary structure prediction from amino acid sequence. *Proteins* **49,** 154–166.

Koonin, E. V., and Rudd, K. E. (1993). SpoU protein of *Escherichia coli* belongs to a new family of putative rRNA methylases. *Nucleic Acids Res.* **21,** 5519.

Kowalak, J. A., Pomerantz, S. C., Crain, P. F., and McCloskey, J. A. (1993). A novel method for the determination of post-transcriptional modification in RNA by mass spectrometry. *Nucleic Acids Res.* **21,** 4577–4585.

Kuchino, Y., Hanyu, N., and Nishimura, S. (1987). Analysis of modified nucleosides and nucleotide sequence of tRNA. *Methods Enzymol.* **155,** 379–396.

Lowe, T. M., and Eddy, S. R. (1999). A computational screen for methylation guide snoRNAs in yeast. *Science* **283,** 1168–1171.

Maden, B. E., Corbett, M. E., Heeney, P. A., Pugh, K., and Ajuh, P. M. (1995). Classical and novel approaches to the detection and localization of the numerous modified nucleotides in eukaryotic ribosomal RNA. *Biochimie* **77,** 22–29.

Martin, J. L., and McMillan, F. M. (2002). SAM (dependent) I AM: The S-adenosylmethionine-dependent methyltransferase fold. *Curr. Opin. Struct. Biol.* **12,** 783–793.

Martin, N. C. (1995). Organellar tRNAs: Biosynthesis and function. *In* "tRNA: Structure, Biosynthesis and Function" (D. Söll and U. RajBhandary, eds.), pp. 127–140. American Society for Microbiology, Washington, DC.

Matos, J. R., and Wong, C.-H. (1987). S-Adenosylmethionine: Stability and stabilization. *Bioorg. Chem.* **15,** 71–80.

McCloskey, J. A., Graham, D. E., Zhou, S., Crain, P. F., Ibba, M., Konisky, J., Söll, D., and Olsen, G. J. (2001). Post-transcriptional modification in archaeal tRNAs: Identities and phylogenetic relations of nucleotides from mesophilic and hyperthermophilic *Methanococcales*. *Nucleic Acids Res.* **29,** 4699–4706.

Milligan, J. F., and Uhlenbeck, O. C. (1989). Synthesis of small RNAs using T7 RNA polymerase. *Methods Enzymol.* **180,** 51–62.

Neuwald, A. F., Liu, J. S., Lipman, D. J., and Lawrence, C. E. (1997). Extracting protein alignment models from the sequence database. *Nucleic Acids Res.* **25,** 1665–1677.

Nishimura, S. (1971). Fractionation of transfer RNA by DEAE-Sephadex A-50 column chromatography. In "Procedures in Nucleic Acid Research" (G. L. Cantoni and D. R. Davies, eds.), Vol. 2, pp. 542–564. Harper & Row, New York.

Nureki, O., Watanabe, K., Fukai, S., Ishii, R., Endo, Y., Hori, H., and Yokoyama, S. (2004). Deep knot structure for construction of active site and cofactor binding site of tRNA modification enzyme. *Structure* **12,** 593–602.

Park, J., Teichmann, S. A., Hubbard, T., and Chothia, C. (1997). Intermediate sequences increase the detection of homology between sequences. *J. Mol. Biol.* **273,** 349–354.

Pomerantz, S. C., and McCloskey, J. A. (1990). Analysis of RNA hydrolyzates by liquid chromatography-mass spectrometry. *Methods Enzymol.* **193,** 796–824.

Purushothaman, S. K., Bujnicki, J. M., Grosjean, H., and Lapeyre, B. (2005). Trm11p and Trm112p are both required for the formation of 2-methylguanosine at position 10 in yeast tRNA. *Mol. Cell. Biol.* **25,** 4359–4370.

Reid, R., Greene, P. J., and Santi, D. V. (1999). Exposition of a family of RNA m⁵C methyltransferases from searching genomic and proteomic sequences. *Nucleic Acids Res.* **27,** 3138–3145.

Renalier, M.-H., Joseph, N., Gaspin, C., Thebault, P., and Mougin, A. (2005). The Cm56 tRNA modification in archaea is catalyzed either by a specific 2'-O-methylase, or a C/D sRNP. *RNA* **11,** 1051–1063.

Rogg, H., Brambilla, R., Keith, G., and Staehelin, M. (1976). An improved method for the separation and quantitation of the modified nucleosides of transfer RNA. *Nucleic Acids Res.* **3,** 285–295.

Roovers, M., Wouters, J., Bujnicki, J. M., Tricot, C., Stalon, V., Grosjean, H., and Droogmans, L. (2004). A primordial RNA modification enzyme: The case of tRNA (m1A) methyltransferase. *Nucleic Acids Res.* **32,** 465–476.

Rost, B., Yachdav, G., and Liu, J. (2004). The PredictProtein server. *Nucleic Acids Res.* **32,** W321–W326.

Rozenski, J., Crain, P. F., and McCloskey, J. A. (1999). The RNA Modification Database: 1999 update. *Nucleic Acids Res.* **27,** 196–197.

Rozenski, J., and McCloskey, J. A. (1999). Determination of nearest neighbors in nucleic acids by mass spectrometry. *Anal. Chem.* **71,** 1454–1459.

Sherlin, L. D., Bullock, T. L., Nissan, T. A., Perona, J. J., Lariviere, F. J., Uhlenbeck, O. C., and Scaringe, S. A. (2001). Chemical and enzymatic synthesis of tRNAs for high-throughput crystallization. *RNA* **7,** 1671–1678.

Spencer, A. C., Heck, A., Takeuchi, N., Watanabe, K., and Spremulli, L. L. (2004). Characterization of the human mitochondrial methionyl-tRNA synthetase. *Biochemistry* **43,** 9743–9754.

Sprinzl, M., and Vassilenko, K. S. (2005). Compilation of tRNA sequences and sequences of tRNA genes. *Nucleic Acids Res.* **33,** D139–D140.

Takusagawa, F., Fujioka, M., Spies, A., and Schowen, R. L. (1998). S-adenosylmethionine (AdoMet)-dependent methyltransferase. In "Comprehensive Biological Catalysis" (M. Sinnott, ed.), Vol. 1, pp. 1–30. Academic Press, New York.

Tanner, N. K. (1989). Purifying RNA by column chromatography. *Methods Enzymol.* **180,** 25–41.

Wu, J. C., and Santi, D. V. (1987). Kinetic and catalytic mechanism of HhaI methyltransferase. *J. Biol. Chem.* **262,** 4778–4786.

Wu, S.-E., Huskey, W. P., Borchardt, R. T., and Schowen, R. L. (1983). Chiral instability at sulfur of S-adenosylmethionine. *Biochemistry* **22,** 2828–2832.

Zhou, S., Sitaramaiah, D., Noon, K. R., Guymon, R., Hashizume, T., and McCloskey, J. A. (2004). Structures of two new "minimalist" modified nucleosides from archaeal tRNA. *Bioorg. Chem.* **32,** 82–91.

CHAPTER NINE

MASS SPECTROMETRIC IDENTIFICATION AND CHARACTERIZATION OF RNA-MODIFYING ENZYMES

Tsutomu Suzuki, Yoshiho Ikeuchi, Akiko Noma, Takeo Suzuki, *and* Yuriko Sakaguchi

Contents

Abstract

Posttranscriptional modifications are characteristic structural features of RNA molecules. To study the functional roles played by RNA modifications, it is necessary to identify the genes and enzymes that are responsible for their biosynthesis. Many uncharacterized genes for RNA modifications still remain buried in the genomes of model organisms. We describe here a systematic genomewide screening method that uses a reverse genetic approach combined with mass spectrometry, which we have named "ribonucleome analysis," to identify uncharacterized genes that are involved in generating RNA modifications.

Department of Chemistry and Biotechnology, Graduate School of Engineering, The University of Tokyo, Bunkyo-ku, Tokyo, Japan

Methods in Enzymology, Volume 425
ISSN 0076-6879, DOI: 10.1016/S0076-6879(07)25009-8

1. INTRODUCTION

Noncoding RNAs have emerged as regulatory elements of gene expression and are involved in various biological processes. RNA molecules mature through various posttranscriptional processing events in a spatiotemporal manner. Posttranscriptional modifications (or RNA editing) are characteristic structural features of RNA molecules and are required for their proper functioning. To date, more than 100 different RNA modifications have been reported (Rozenski *et al.*, 1999). Most of these modifications have been found in abundant RNA molecules, such as tRNAs, rRNAs, and/or UsnRNAs. Modified bases in tRNAs participate in stabilizing the tertiary structure of the tRNA, modulating tRNA–protein interactions, and deciphering the genetic code on the ribosome (Grosjean, 2005; Suzuki, 2005). It has recently been reported that even small RNAs are modified. In mammals, certain populations of miRNAs (~6%) contain inosine (I) (Blow *et al.*, 2006; Luciano *et al.*, 2004). In plants, the 3′-termini of miRNAs are modified by 2′-O-methylation during the maturation step (Yu *et al.*, 2005), and this modification is required for normal maturation of the miRNAs (Li *et al.*, 2005). We recently reported that mouse Piwi-interacting RNAs (piRNAs) also have 3′-terminal 2′-O-methylations (Ohara *et al.*, 2007).

To investigate the functional roles played by RNA modifications, it is necessary to identify the genes and enzymes that are responsible for these chemical modifications. There still remain many RNA modification genes buried in uncharacterized genes in the genomes that have been sequenced. We describe here a systematic genomewide screening method that uses a reverse genetic approach combined with mass spectrometry, which we have named "ribonucleome analysis" (Suzuki, 2005), to identify uncharacterized genes that are involved in generating RNA modifications (Fig. 9.1). In addition, we illustrate our mass spectrometric analysis of RNA modifications in individual RNAs to characterize RNA-modifying enzymes.

2. RIBONUCLEOME ANALYSIS: QUEST FOR RNA-MODIFYING GENES

Ribonucleome analysis uses a series of knockout strains of *E. coli* or yeast. Each knockout strain is cultured in parallel in a deep-well plate and then transferred to 96-well plates. Thereafter, total RNA is extracted and digested into nucleosides and automatically analyzed by LC/MS by use of an ion trap mass spectrometer. This analysis allows us to determine whether a particular gene deletion results in the absence of a specific modified base and, thus, permits us to identify the enzyme or protein responsible for this

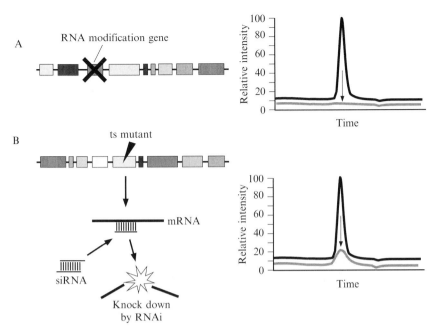

Figure 9.1 Ribonucleome analysis: a systematic genomewide screen for genes responsible for RNA modifications by reverse genetics combined with mass spectrometry. (A) Mass spectrometric identification of a nonessential RNA modification by screening knockout strains for knockouts that cause a particular modified nucleoside to be absent from the mass chromatogram. (B) To identify genes involved in essential RNA modifications, temperature sensitive (ts) strains or expression-controlled strains are available for screening essential genes responsible for RNA modifications. In mammals, including humans, RNA interference with siRNA is available for knocking down specific candidate genes. An essential gene responsible for RNA modification can be identified by searching for reduced amounts of the target RNA modification in the mass chromatogram.

RNA modification (Fig. 9.1). By use of this approach, we identified five new genes (*tusA, B, C, D,* and *E*) working as sulfur mediators responsible for 2-thiouridine formation in the 5-methylaminomethyl-2-thiouridine (mnm^{5}s^{2}U) of *E. coli* tRNAs (Ikeuchi *et al.*, 2006), and four new genes (*TYW1, 2, 3,* and *4*) responsible for synthesis of wybutosine (yW) in *Saccharomyces cerevisiae* tRNA^{Phe} (Noma *et al.*, 2006).

Because it is not possible to obtain deletion strains of essential genes that are involved in RNA modifications, temperature-sensitive (ts) strains are available to identify essential genes mediating RNA modifications. The ts strains can be cultured at the nonpermissive temperature, and reductions in the amounts of specific RNA modifications can be searched for in the chromatogram. In fact, by use of this approach, we identified an essential gene (*tilS*) responsible for lysidine formation in *E. coli* tRNA^{Ile} (Soma *et al.*, 2003).

Alternately, expression-controlled strains are also available. In this case, the deletion of an essential gene in the genome and its rescue by expressing the same gene under the control of an inducible promoter on a plasmid will not result in a reduction in an RNA modification unless the cells are grown without induction of the essential gene. By use of this approach, we found the essential yeast gene *Nfs1* to be involved in 2-thiouridine formation at the wobble position of both mitochondrial and cytoplasmic tRNAs (Nakai *et al.*, 2004). Furthermore, RNA interference is now a powerful tool that can be used to explore the genes that are responsible for both essential and nonessential RNA modifications in mammals. We have, in fact, used an siRNA knockdown technique to identify a human mitochondrial RNA-modifying enzyme (MTU1) (Umeda *et al.*, 2005).

3. STRAINS USED FOR THE RIBONUCLEOME ANALYSIS

E. coli strains used in this study are available from the NIG collection (National Institute of Genetics, Japan, http://shigen.lab.nig.ac.jp/ecoli/strain/nbrp/resource.jsp) and the CGSC (the *E. coli* Genetic Stock Center, Yale University http://cgsc2.biology.yale.edu/index.php). To reduce the number of candidate genes for screening, large-scale chromosomal deletion mutants were first analyzed. Each of the strains lacks approximately 20 kbp of the *E. coli* chromosome, which would code for approximately 20 genes (Hashimoto *et al.*, 2005). If a strain lacks a particular RNA modification, a candidate gene responsible for the modification must be localized in the deleted region. A series of disruptant strains, in which a single gene in the deleted region was knocked out, was obtained from the Keio collection (Baba *et al.*, 2006). Alternately, the absence of the RNA modification can be rescued by introducing a plasmid expressing the gene deleted in the chromosome region. For this purpose, we used a genomewide set of mobility plasmids (Saka *et al.*, 2005), each of which bears one *E. coli* open-reading frame.

A *S. cerevisiae* strain BY4742 (*Matγ; his3Δ1; leu2Δ0; lys2Δ0; ura3Δ0*) and a series of single-gene disruptant strains, were obtained from EUROSCARF (http://web.uni-frankfurt.de/fb15/mikro/euroscarf/). In this study, we used two strains: Y16650 (BY4742, *YOL141w::kanmx4*) and Y17332 (BY4742, *YOR274w::kanmx4*).

4. PARALLEL PREPARATION OF TOTAL RNAs FROM *E. COLI* OR YEAST STRAINS

An outline for parallel preparation of total RNAs is shown in Fig. 9.2. The wild-type *E. coli* A19 strain and a series of mutant strains were grown in 5 ml of LB medium (1% tryptone, 1% NaCl, and 0.5% yeast extract) in a

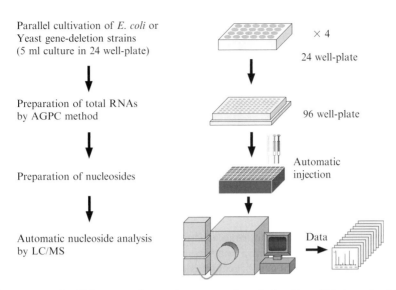

Figure 9.2 Parallel preparation and automated analysis of modified nucleosides. A series of knockout strains of *E. coli* or yeast were cultured in a 24-deep-well plate and then transferred to 96-well plates, where total RNA was extracted, and the samples were digested into nucleosides. The samples were analyzed by LC/MS by use of an ion trap mass spectrometer to detect loss of a specific modified nucleoside caused by the gene deletion. This system allows us to analyze 500 strains per month.

24 deep-well plate at 37° overnight. Ninety-six–well plates with silicon rubber lids were used to prepare total RNA from the harvested cells by the acid-guanidinium thiocyanate-phenol-chloroform (AGPC) method (Chomczynski and Sacchi, 1987) with 700 μl of the ISOGEN reagent (Nippon Gene, Japan) per well, according to the manufacturer's instructions. The extracted supernatants (400 μl), containing *E. coli* total RNA, were carefully removed from the 96-well plates by use of a 12-channel pipetter and transferred to a clean 96-well plate. To avoid protein and phenol contamination, only approximately one half to two thirds of the supernatant was removed. An equal volume of 2-propanol was added to each supernatant and mixed thoroughly to precipitate the total RNA, which was then centrifuged at 5800g (6000 rpm) for 30 min at 4° with a Sigma 4K15C centrifuge (QIAGEN). By use of UV transparent 96-well plates (Costar #3235, Corning), the amount of total RNA was quantified with a SpectraMax190 plate reader (Molecular Devices, Inc.). To prepare an enriched fraction of rRNAs, we used the RNeasy Mini Kit (QIAGEN) according to the manufacturer's instructions.

Yeast strains were grown in 5 ml of YPD medium (1% yeast extract, 2% Bactopeptone, and 2% glucose) in a 24 deep-well plate at 30° for 36 h, and cells were harvested during late-log phase growth (OD_{660} ~1.5–2.0). Cell pellets were resuspended in 500 μl of lysis buffer (20 mM Tris-HCl [pH 7.5], 10 mM MgCl$_2$) and transferred to a 96-well plate. Total RNA was extracted with 100 μl of phenol/chloroform saturated with a neutral pH buffer by shaking for 3 h at room temperature. Total RNA was then recovered by ethanol precipitation from the aqueous phase. The RNA pellets were rinsed with 70% ethanol, dissolved in 100 μl of ddH$_2$O, and stored at $-20°$. The amount of purified RNA was quantified as described previously.

5. ISOLATION AND PURIFICATION OF INDIVIDUAL tRNAS FROM YEAST

Rapid and simple isolation of individual tRNAs is essential to be able to characterize a modification intermediate in a specific tRNA obtained from cells that lack an RNA-modifying enzyme. We describe here our standard procedure for tRNA isolation and mass spectrometry analysis of hypomodified yeast tRNA^{Phe}. Yeast strains were grown in 2 liter of YPD medium and harvested during late-log phase growth (OD$_{660}$ $\sim$1.5–2.0). Total RNA was extracted as described previously. Approximately 700 A$_{260}$ units of total RNA was obtained from a 2-liter culture. To isolate individual tRNAs with the highest efficiency, we have devised, and successfully improved, a novel solid-phase DNA probe method, which has been named "chaplet column chromatography" (Kaneko *et al.*, 2003; Suzuki, 2005; Suzuki and Suzuki, 2007). The detailed procedure is described in Chapter 10 in this volume (Suzuki and Suzuki, 2007). A 3'-biotinylated DNA probe, 5'-tgcgaattctgtggatcgaacacaggacct-3', complementary to yeast tRNA^{Phe}, was immobilized on avidin Sepharose (Amersham-Pharmacia) packed in a mini-column (GL science, Japan) with a 200-μl bed volume. The crude total RNA (600 A$_{260}$ units) dissolved in 4.5 ml of binding buffer (1.2 M NaCl, 30 mM HEPES-KOH [pH 7.5], 15 mM EDTA) was circulated through the column by use of a peristaltic pump at a temperature of 70° to entrap the target tRNA. Nonspecific tRNAs were washed out with wash buffer (0.6 M NaCl, 15 mM HEPES-KOH [pH 7.5], 7.5 mM EDTA) until the UV absorbance fell below 0.01 A$_{260}$ units, and tRNA^{Phe} was then eluted from the column with a low-salt buffer (20 mM NaCl, 0.5 mM HEPES-KOH [pH 7.5], 0.25 mM EDTA) at 70°. Approximately 0.5 A$_{260}$ units of tRNA^{Phe} were isolated with good purity. As appropriate, the isolated tRNA was further purified by denaturing polyacrylamide gel electrophoresis. Recently, we reported another technology termed "reciprocal circulating chromatography (RCC)" for the isolation of individual RNAs, which enables parallel, automated purification of different RNA species (Miyauchi *et al.*, 2007).

6. NUCLEOSIDE PREPARATION

To analyze the nucleoside composition of the RNA, total RNA (or purified tRNA) obtained from each strain was digested to nucleosides with nuclease P1 (SEIKAGAKU CORPORATION) and bacterial alkaline phosphatase (BAP) derived from *E. coli* strain C75 (BAP C75, Takara) (Ikeuchi *et al.*, 2006; Noma *et al.*, 2006; Suzuki *et al.*, 2002). Nuclease P1, which is supplied as a lyophilized powder, is dissolved in 50 mM ammonium acetate (pH 5.3) to a concentration of 0.5 units/μl and stored at $-20°$. Under typical reaction conditions, 1 μg/μl total RNA (20 μg in total) is dissolved in 20 mM HEPES-KOH (pH 7.0) containing 0.025 units/μl of nuclease P1 and 0.002 units/μl of BAP C75 and incubated at 37° for 3 h. The reaction mixture is then ready for LC/MS analysis. It can be also stored at $-20°$ for a short time (<2 weeks). It has been noted that some modified nucleosides tend to produce dinucleotides when this procedure is used (e.g., yWpA for wybutosine [Noma *et al.*, 2006] and mcm^{5}UmpU for 2′-O-methylated nucleosides [Kaneko *et al.*, 2003]).

7. MASS SPECTROMETRIC ANALYSIS OF TOTAL NUCLEOSIDES

An LCQ^{DUO} ion-trap (IT) mass spectrometer (Thermo Fisher Scientific) equipped with an electrospray ionization (ESI) source and HP1100 liquid chromatography system (Agilent Technologies) was used to analyze the nucleosides. The conditions for chromatography have been described previously (Pomerantz and McCloskey, 1990). The nucleosides, prepared as described previously, were loaded onto the LC/MS system. The nucleosides were fractionated with an Inertsil ODS-3 column (2.1 × 250 mm, GL science), with a 3 × 10 mm precolumn cartridge (ODS-3, GL science). The solvent system consisted of 5 mM ammonium acetate (pH 5.3) (solvent A) and 60% acetonitrile (solvent B), and the samples were chromatographed using a flow rate of 150 μl/min with a multistep linear gradient of 1–35% B from 0–35 min, 35–99% B from 35–40 min, 99% B from 40–50 min, 99–1% B from 50–50.1 min, and 1% B from 50.1–60 min. The chromatographic eluent was directly conducted into the ion source without prior splitting. Ions were scanned by use of a positive polarity mode over an m/z range of 103–700 (or 103–900, if detection of dinucleotides is required) throughout the separation. Parameters of the mass spectrometer were tuned with authentic adenosine according to the manufacturer's instructions. The parameters used in this section were sheath gas flow rate,

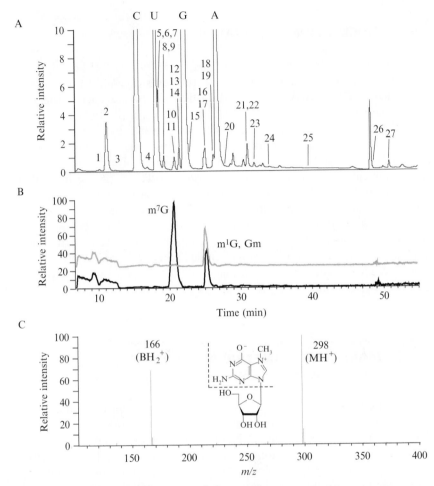

Figure 9.3 LC/MS nucleoside analysis of *E. coli* total RNAs. (A) UV trace chromatogram (at 254 nm) of wild-type *E. coli* A19. Unmodified nucleosides (C, U, G, and A) and modified nucleosides are indicated. The modified nucleosides are numbered: 1, dihydrouridine (D); 2, pseudouridine (Ψ); 3, 5-methylaminomethyluridine (mnm^5U); 4, 3-(3-amino-3-carboxypropyl)uridine (acp^3U); 5, 5-methylaminomethyl-2-thiouridine (mnm^5s^2U); 6, 7-methylguanosine (m^7G); 7, uridine 5-oxyacetic acid (cmo^5U); 8, 2-thiocytidine (s^2C); 9, 5-methylcytidine (m^5C); 10, epoxyqueuosine (oQ); 11, $2'$-O-methylcytidine (Cm); 12, inosine (I); 13, queuosine (Q); 14, lysidine (L); 15, 5-methyluridine (m^5U); 16, 1-methylguanosine (m^1G); 17, $2'$-O-methylguanosine (Gm); 18, N^4-acetylcytidine (ac^4C); 19, 4-thiouridine (s^4U); 20, uridine 5-oxyacetic acid methyl ester ($mcmo^5U$); 21, 2-methyladenosine (m^2A); 22, N^6-threonylcarbamoyladenosine (t^6A); 23, N^6-methyladenosine (m^6A); 24, N^6-methyl-N^6-threonylcarbamoyladenosine (m^6t^6A); 25, N^6,N^6-dimethyladenosine (m_2^6A); 26, N^6-isopentenyladenosine (i^6A); and 27, 2-methylthio-N^6-isopentenyladenosine (ms^2i^6A). (B) Mass chromatogram (black line) at m/z 298 detects m^7G, m^1G, and Gm of wild-type *E. coli*.

95 arb; aux gas flow rate, 0 arb; spray voltage, 4.5 kV; capillary temperature, 200°; capillary voltage, 32 V; tube lens offset, 15 V; multipole 1 offset, −1.5 V; lens voltage, −16 V; and multipole 2 offset, −5 V.

8. LC/MS Profiling of Modified Nucleosides

Mass spectrometer nucleoside analysis of total RNA from the *E. coli* wild-type strain is shown in Fig. 9.3. In the UV (254 nm) and total ion chromatograms (Fig. 9.3A), 27 species of modified nucleosides were clearly detected by their respective retention times. Most of the nucleosides originated from tRNAs, but some (m_2^6A and m^5C) were derived from rRNAs. If a mutant strain lacks a gene responsible for a particular RNA modification, the signal for that modified nucleoside will be absent in this analysis. In the mass chromatogram for m/z 298, peak signals for 7-methylguanosine (m^7G), 1-methylguanosine (m^1G), and 2'-O-methylguanosine (Gm) were observed (Fig. 9.3B). The mass spectrum for m^7G (Fig. 9.3C) reveals the positive ion of the m^7G proton adduct (MH^+, m/z 298) and its base-related fragment ion (BH_2^+, m/z 166). One example of a specific defect in RNA modification is illustrated. When we analyzed the total nucleosides from an *E. coli* strain lacking the *yggH* gene, which is a tRNA (m^7G46)-methyltransferase (De Bie *et al.*, 2003; Okamoto *et al.*, 2004), the peak for m^7G in the mass chromatogram (Fig. 9.3B) was absent. Systematic analyses of a series of mutant strains (ribonucleome analysis) will identify new genes responsible for biogenesis of specific modified nucleosides.

In the case of yeast, nucleoside analysis revealed 23 species of modified nucleosides (Fig. 9.4A). In the mass chromatogram for m/z 336, a signal for N^6-isopentenyladenosine (i^6A) was observed (Fig. 9.4B). In the mass spectrum, the proton adduct of i^6A (MH^+, m/z 336) and its base-related fragment ion (BH_2^+, m/z 204) were both observed (Fig. 9.4C). As an example of ribonucleome analysis, we observed a complete loss of the peak for i^6A in the mass chromatogram (Fig. 9.4B) of total nucleosides from a yeast single-gene disruptant lacking the *MOD5* gene, a tRNA isopentenyltransferase (Dihanich *et al.*, 1987).

The chromatogram (gray line) of an *E. coli* mutant strain (OCL38) lacking *yggH* (*trmB*) shows that m^7G is absent. (C) The mass spectrum of m^7G shows the positively charged proton adduct of m^7G (MH^+, m/z 298) and its base-related ion (BH_2^+, m/z 166), which is produced by spontaneous cleavage of an *N*-glycoside bond (dotted line in the chemical structure) during ionization.

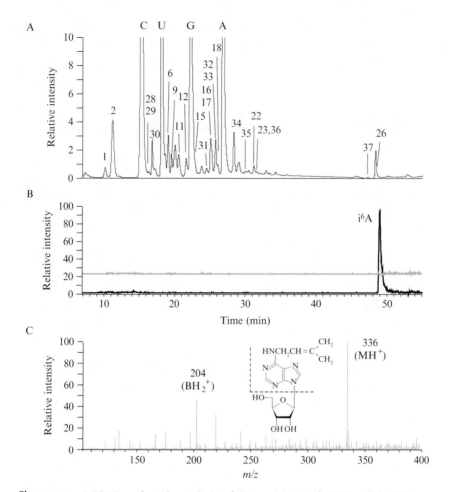

Figure 9.4 LC/MS nucleoside analysis of *S. cerevisiae* total RNAs. (A) UV trace chromatogram (at 254 nm) of wild-type *S. cerevisiae* BY4742. Unmodified nucleosides (C, U, G, and A) and modified nucleosides are indicated. Modified nucleosides in common with *E. coli* are indicated with the same numbers shown in the legend for Fig. 9.3. Additional modified nucleosides appearing in yeast RNAs are numbered as follows: 28, 3-methylcytidine (m^{3}C); 29, 5-carbamoylmethyluridine (ncm^{5}U); 30, 1-methyladenosine (m^{1}A); 31, 1-methylinosine (m^{1}I); 32, 5-methoxycarbonylmethyluridine (mcm^{5}U); 33, N^2-methylguanosine (m^{2}G); 34, N^2,N^2-dimethylguanosine (m$_2^2$G); 35, 2′-*O*-methyladenosine (Am); 36, 5-methoxycarbonylmethyl-2-thiouridine (mcm^{5}s^{2}U); 37, wybutosine (yW). (B) Mass chromatogram (black line) at *m/z* 336 detecting i^{6}A of wild-type *S. cerevisiae*. The chromatogram (gray line) of the *S. cerevisiae* mutant strain Y17332 lacking *YOR274w* (*MOD5*) shows the loss of i^{6}A. (C) The mass spectrum of i^{6}A shows the positively charged proton adduct of i^{6}A (MH$^+$, *m/z* 336) and its base-related ion (BH$_2^+$, *m/z* 204) which is produced by spontaneous cleavage of an *N*-glycoside bond (dotted line in the chemical structure) during ionization.

9. RNase Digestion for RNA Fragment Analysis by LC/MS

RNA fragment analysis by LC/MS is the most powerful technique to determine the positions and species of modified nucleosides in RNA molecules. For fragmentation of RNAs, base-specific ribonucleases (RNase) are used. RNase T_1 is useful for this purpose because of its high specificity (Sato and Egami, 1957). RNase T_1 specifically cuts phosphodiester bonds on the $3'$ side of guanosines in single stranded RNA and yields oligonucleotides with a $3'$-monophosphate through formation of a $2',3'$-cyclic phosphate bond. Therefore, the RNase T_1 digests consist of HnG (H = A, U, or C, n = counting number) oligonucleotides, which possess $5'$-hydroxy and $3'$-phosphate (occasionally $2',3'$-cyclic phosphate) groups, with the exception of the $5'$- and $3'$-terminal fragments. Similarly, the pyrimidine-specific RNase A, which produces RnY fragments (R = A or G; Y = U or C, n = counting number), is also available for RNA fragmentation (Kunitz, 1939). A target RNA (0.1–2 pmol) was digested at 37° for 30 min in 10 μl of a reaction mixture containing 10 mM ammonium acetate (pH 5.3) and 1 unit/μl RNase T_1 (Epicentre), or 10 mM ammonium acetate (pH 7.7) and 1 ng/μl RNase A (Ambion). After digestion, 10 μl of 0.1 M triethylamine-acetate (TEAA) (pH 7.0) was added to the reaction solution and mixed gently by pipetting. The mixtures are immediately applied to the LC/MS system, because long-term storage of these specimens at −20° may result in nonspecific degradation or other damage to the sample. Some modified guanosines, such as N^1-methylguanine, 7-methylguanosine, and $2'$-O-methylguanosine, are resistant to RNase T_1 digestion (Uchida and Egami, 1971).

10. Capillary LC NANO ESI/Mass Spectrometry

RNase T_1 or RNase A digests were analyzed by capillary LC coupled with electrospray ionization (ESI) mass spectrometry. The conditions for chromatography have been described previously (Apffel et al., 1997). We used two types of ESI mass spectrometers, a tandem quadrapole time-of-flight (QqTOF) mass spectrometer (QSTAR® XL, Applied Biosystems) and an ion-trap (IT) mass spectrometer (LCQ Classic, Thermo Fisher Scientific). Both instruments were equipped with nanoelectrospray ionization sources, NANOSPRAY® II (Applied Biosystems) for the QSTAR instrument and a custom-made nanospray ion source for the LCQ instrument. To separate the limited quantity of RNA fragments, we used a splitless nanoflow HPLC system (DiNa, KYA Technologies) equipped with an injection valve (Nanovolume Valve, Valco Instruments). Approximately 500 fmol

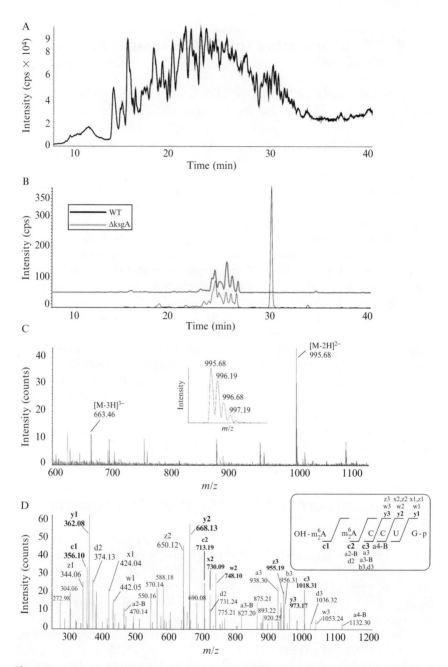

Figure 9.5 RNA fragment analysis of *E. coli* rRNAs. (A) Total ion chromatogram of RNase T₁ digest of *E. coli* rRNAs (cps stands for counts per second). (B) Mass chromatogram at *m/z* 995.68 detecting the modified RNA fragment, m²₆Am²₆ACCUGp in wild-type *E. coli* rRNAs. The chromatogram (gray line) of the *E. coli* mutant strain lacking

of digested RNA mixed with TEAA was loaded onto a nano-LC trap column (C18, ID0.5 × 0.1 mm, KYA Technologies), desalted, and then concentrated with 0.1 M TEAA (pH 7.0). The RNA fragments were eluted from the trap column and directly injected into a HiQ sil C18W-3 capillary column (C18, 3 μm, 120 Å pore size; ID0.15 × 50 mm, KYA Technologies). The solvent system consisted of 0.4 M 1,1,1,3,3,3-hexafluoro-2-propanol (HFIP) (pH 7.0, adjusted with triethylamine) (solvent A) and 0.4 M HFIP (pH 7.0) in 50% methanol (solvent B), and the samples were chromatographed at a flow rate of 500 nl/min with a linear gradient of 0–80% B over 40 min. The chromatographic eluent was sprayed from an energized sprayer tip attached to the capillary column. Ions were scanned with a negative polarity mode over an m/z range of 600–2000 throughout the separation. The parameters for the QSTAR used in this analysis were spray voltage, −2.0 kV; curtain gas, 15; accumulation time, 1 s; mirror voltage, −0.990 kV; plate voltage, −0.330 kV; and grid voltage, 0.422 kV. The parameters for the LCQ Classic used in this analysis were spray voltage, 2.0 kV; heated capillary temperature, 230°; capillary voltage, −4.0 V; tube lens offset, 50.0 V; multipole 1 offset, 6.25 V; lens voltage, 48.0 V; and multipole 2 offset, 8.0 V.

11. RNA FRAGMENT ANALYSIS BY CAPILLARY LC/MS

We describe here an RNA fragment analysis of $E.\ coli$ rRNAs (23S and 16S) as one example of this approach. In the total ion chromatogram (TIC) (Fig. 9.5A), many species of RNA fragments produced by the RNase T_1 digestion were ionized and separated to produce the complex peaks in the chromatogram. A specific ion for an RNA fragment could be identified by mass chromatographic extraction. In the mass chromatogram (Fig. 9.5B), the doubly-charged ion (m/z 995.68) of an RNA fragment containing two N^6,N^6-dimethyladenosine (m_2^6A) was clearly extracted as a distinct peak at a retention time (RT) 30.0 min. In fact, doubly and triply-charged negative ions of this fragment were observed in the mass spectrum (Fig. 9.5C). To determine the sequence position of m_2^6A in the fragment, we chose the doubly charged ion (m/z 995.68) in the mass spectrum as the precursor ion for MS/MS analysis by use of collision-induced dissociation (CID). Many product ions originating from the precursor ion were observed in the CID spectrum (Fig. 9.5D), but a series of c-type and y-type ions were most prominent. According to the nomenclature for product ions of nucleic acids

ksgA lacks the $m_2^6Am_2^6ACCUGp$ oligonucleotide. (C) The mass spectrum shows doubly and triply-charged negative ions of $m_2^6Am_2^6ACCUGp$. (D) The CID spectrum of m_2^6A-$m_2^6ACCUGp$. Product ions, interpreted according to the accepted nomenclature (Mcluckey *et al.*, 1992), are indicated in the spectrum and the fragment sequence.

(Mcluckey *et al.*, 1992), we could determine the sequence of this fragment to be m$_2^6$Am$_2^6$ACCUGp, which corresponds to the 3'-terminal region (positions 1518–1523) of the 16S rRNA. We analyzed rRNA fragments from the *E. coli* single-gene disruptant lacking the *ksgA* gene, which is a methylase for m$_2^6$A1518/1519 (Andresson and Davies, 1980), and observed a complete loss of the peak corresponding to m$_2^6$Am$_2^6$ACCUGp in the mass chromatogram (Fig. 9.5B).

A second example shows the RNA fragment analysis of an isolated tRNA. As shown in Fig. 9.6A, RNase T$_1$ digestion of yeast tRNA^{Phe} produces 14 species of detectable fragments in this analysis. In the mass chromatogram (Fig. 9.6B, left panels), 14 fragments were clearly separated to give distinct peaks. The triply-charged ion (*m/z* 1387.5) of the anticodon-containing fragment (No. 14; ACmUGmAAyWAΨm^{5}CUGp) was extracted as a single peak at RT 33.82 min (Fig. 9.6B, left panels). In the mass spectrum for this fragment, multiply-charged negative ions (−3 to −6) were observed. Assignment of each fragment is listed in Table 9.1. When we analyzed a RNase T$_1$ digest of yeast tRNA^{Phe} isolated from a single-gene disruptant lacking *YOL141w* (BY4742, *YOL141w::kanmx4*), which corresponds to TYW4 (Noma *et al.*, 2006), an enzyme responsible for the last step in yW biogenesis, no signal for the fully modified anticodon-containing fragment (No. 14) was observed, while the other 13 fragments were detected with unchanged molecular masses (Fig. 9.6B, right panels). A new fragment (No. 15; ACmUGmAA[yW-72]AΨm^{5}CUGp) with an *m/z* 1363.5 triply-charged ion appeared (Fig. 9.6B, right panels) at RT 31.69 min. Because a modification intermediate of yW in the *TYW4* disruptant was previously identified as yW-72, this result is entirely consistent with previous observations.

12. DISCUSSION

To search for new genes responsible for RNA modifications, various approaches are available, such as comparative genomics, *in vitro* assays for RNA modifications, and functional genetics. What is most advantageous in ribonucleome analysis is that this approach enables us to identify enzymes mediating RNA modifications, as well as genes that encode proteins without enzymatic activity, such as partner proteins of the enzymes required for RNA recognition or proteins that bind metabolic substrates for RNA modifications. In fact, by use of this approach, a series of Tus proteins responsible for 2-thiouridine formation were identified as mediators of sulfur transfer (Ikeuchi *et al.*, 2006).

Posttranscriptional modifications are important elements that modulate the functions of RNA molecules. Although it is desirable to use fully

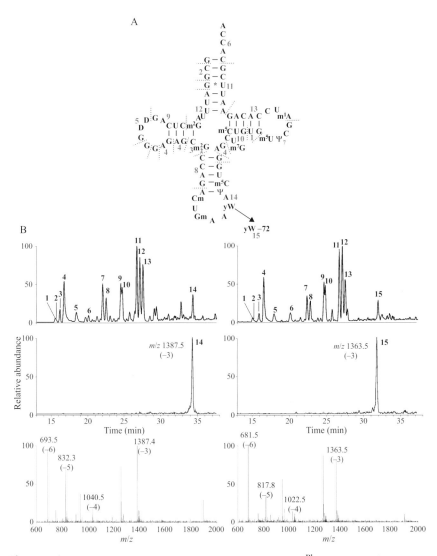

Figure 9.6 RNA fragment analysis of purified yeast tRNA^Phe. (A) Secondary structure of yeast tRNA^Phe with modified nucleosides indicating the RNase T₁ cleavage sites and the numbering of each fragment. Modified nucleosides in this tRNA are: wybutosine (yW), 2′-O-methylguanosine (Gm), 2′-O-methylcytidine (Cm), pseudouridine (Ψ), 5-methylcytidine (m⁵C), 7-methylguanosine (m⁷G), 2-methylguanosine (m²G), N²,N²-dimethylguanosine (m²₂G), dihydrouridine (D), 1-methyladenosine (m¹A), and 5-methyluridine (m⁵U). (B) Mass chromatograms and spectra of RNA fragments of yeast tRNA^Phe digested by RNase T₁. Left and right panels represent the wild-type strain BY4742 and the mutant strain Y16650 lacking *YOL141w* (*TYW4*), respectively. The top panels show the base peak chromatograms and RNA fragments that are numbered as in A. The middle panels show mass chromatograms of the triply-charged ions of the anticodon-containing fragments. The bottom panels show the mass spectra of the anticodon-containing fragments. Multiply-charged ions are indicated.

Table 9.1   List of RNA fragments of yeast tRNA^{Phe} digested by RNase T1

No. RNase T$_1$ fragments	Negative ions (m/z)		Charge state
	Calculated	Observed	
1. UGp	668.1	668.1	-1
2. CGp	667.1	667.1	-1
3. Cm$_2^2$G>p	677.1	677.1	-1
4. AGp	691.1	691.2	-1
5. DDGp	978.1	978.1	-1
6. CACCA$_{-OH}$	1510.3	1510.1	-1
(3'-terminus)	754.6	754.8	-2
7. m^{5}UΨCGp	1293.2	1293.1	-1
	646.1	646.4	-2
8. CCAGp	1301.2	1301.2	-1
	650.1	650.4	-2
9. CUCAGp	1607.2	1607.1	-1
	803.1	803.3	-2
10. m^{7}GUCm^{5}CUGp	1957.3	1957.7	-1
	978.1	978.2	-2
	651.7	651.9	-3
11. AAUUCGp	1937.2	1937.0	-1
	968.1	968.4	-2
	645.1	645.4	-3
12. AUUUAm^{2}Gp	1952.2	1952.1	-1
	975.6	975.8	-2
	650.1	650.4	-3
13. m^{1}AUCCACAGp	1292.2	1291.8	-2
	861.1	861.1	-3
	645.6	645.7	-4
14. ACmUGmAAyWAΨm^{5}CUGp	1387.5	1387.4	-3
	1040.4	1040.5	-4
	832.1	832.3	-5
	693.3	693.5	-6
15. ACmUGmAA(yW-72)	1363.5	1363.5	-3
AΨm^{5}CUGp	1022.4	1022.5	-4
	817.7	817.8	-5
	681.3	681.5	-6

The numbers of the RNA fragments correspond to those in Fig. 9.6A. 3'–terminus of fragment No. 3 (Cm$_2^2$G > p) is a cyclic phosphate.

processed natural RNA molecules obtained from the cell to study the functional aspects of ncRNAs, most RNA research has no choice other than to ignore the precise characterization of RNA modifications. Thus, development of highly efficient and convenient methods for the isolation and analysis of individual RNA species will provide a fundamental strategy to enable characterization of the functional aspects of large numbers of RNA molecules. As described here, mass spectrometry analysis of RNA molecules provides a key technology for the characterization of RNA modifications, as well as the enzymes that are responsible for their biogenesis.

We have previously reported that some human mitochondrial diseases are caused by tRNA modification disorders (Kirino and Suzuki, 2004; Kirino et al., 2004, 2005; Suzuki et al., 2002). In retrospect, it is not surprising that a qualitative disorder of RNA molecules can cause disease, because noncoding RNAs are functional molecules that must mature by undergoing posttranscriptional modifications. Mass spectrometry analysis of RNA molecules will provide a powerful tool to identify novel biomedical markers associated with RNA modifications.

ACKNOWLEDGMENTS

We are grateful to the Suzuki laboratory members, especially S. Kimura, for their experimental support. This work was supported by JSPS Research Fellowships for Young Scientists (to Takeo Suzuki and Yoshiho Ikeuchi), by grants-in-aid for scientific research on priority areas from the Ministry of Education, Science, Sports, and Culture of Japan, and by a grant from the New Energy and Industrial Technology Development Organization (NEDO) (to Tom Suzuki).

REFERENCES

Andresson, O. S., and Davies, J. E. (1980). Some properties of the ribosomal RNA methyltransferase encoded by ksgA and the polarity of ksgA transcription. *Mol. Gen. Genet.* **179**, 217–222.

Apffel, A., Chakel, J. A., Fischer, S., Lichtenwalter, K., and Hancock, W. S. (1997). Analysis of oligonucleotides by HPLC–electrospray ionization mass spectrometry. *Anal. Chem.* **67**, 1320–1325.

Baba, T., Ara, T., Hasegawa, M., Takai, Y., Okumura, Y., Baba, M., Datsenko, K. A., Tomita, M., Wanner, B. L., and Mori, H. (2006). Construction of *Escherichia coli* K-12 in-frame, single-gene knockout mutants: The Keio collection. *Mol. Syst. Biol.* **2**, 2006–2008.

Blow, M. J., Grocock, R. J., van Dongen, S., Enright, A. J., Dicks, E., Futreal, P. A., Wooster, R., and Stratton, M. R. (2006). RNA editing of human microRNAs. *Genome Biol.* **7**, R27.

Chomczynski, P., and Sacchi, N. (1987). Single-step method of RNA isolation by acid guanidinium thiocyanate-phenol-chloroform extraction. *Anal. Biochem.* **162**, 156–159.

De Bie, L. G., Roovers, M., Oudjama, Y., Wattiez, R., Tricot, C., Stalon, V., Droogmans, L., and Bujnicki, J. M. (2003). The yggH gene of *Escherichia coli* encodes a tRNA (m7G46) methyltransferase. *J. Bacteriol.* **185**, 3238–3243.

Dihanich, M. E., Najarian, D., Clark, R., Gillman, E. C., Martin, N. C., and Hopper, A. K. (1987). Isolation and characterization of MOD5, a gene required for isopentenylation of cytoplasmic and mitochondrial tRNAs of *Saccharomyces cerevisiae*. *Mol. Cell Biol.* **7,** 177–184.

Grosjean, H. (2005). "Modification and Editing of RNA: An Overview," pp. 1–23. Springer-Verlag, New York.

Hashimoto, M., Ichimura, T., Mizoguchi, H., Tanaka, K., Fujimitsu, K., Keyamura, K., Ote, T., Yamakawa, T., Yamazaki, Y., Mori, H., Katayama, T., and Kato, J. (2005). Cell size and nucleoid organization of engineered *Escherichia coli* cells with a reduced genome. *Mol. Microbiol.* **55,** 137–149.

Ikeuchi, Y., Shigi, N., Kato, J., Nishimura, A., and Suzuki, T. (2006). Mechanistic insights into sulfur relay by multiple sulfur mediators involved in thiouridine biosynthesis at tRNA wobble positions. *Mol. Cell* **21,** 97–108.

Kaneko, T., Suzuki, T., Kapushoc, S. T., Rubio, M. A., Ghazvini, J., Watanabe, K., Simpson, L., and Suzuki, T. (2003). Wobble modification differences and subcellular localization of tRNAs in Leishmania tarentolae: Implication for tRNA sorting mechanism. *EMBO J.* **22,** 657–667.

Kirino, Y., Goto, Y., Campos, Y., Arenas, J., and Suzuki, T. (2005). Specific correlation between the wobble modification deficiency in mutant tRNAs and the clinical features of a human mitochondrial disease. *Proc. Natl. Acad. Sci. USA* **102,** 7127–7132.

Kirino, Y., and Suzuki, T. (2004). Human mitochondrial diseases associated with tRNA wobble modification deficiency. *RNA Biol.* **1,** 145–148.

Kirino, Y., Yasukawa, T., Ohta, S., Akira, S., Ishihara, K., Watanabe, K., and Suzuki, T. (2004). Codon-specific translational defect caused by a wobble modification deficiency in mutant tRNA from a human mitochondrial disease. *Proc. Natl. Acad. Sci. USA* **101,** 15070–15075.

Kunitz, M. (1939). Isolation from beef pancreas of a crystalline protein possessing ribonuclease activity. *Science* **90,** 112–113.

Li, J., Yang, Z., Yu, B., Liu, J., and Chen, X. (2005). Methylation protects miRNAs and siRNAs from a 3′-end uridylation activity in *Arabidopsis*. *Curr. Biol.* **15,** 1501–1507.

Luciano, D. J., Mirsky, H., Vendetti, N. J., and Maas, S. (2004). RNA editing of a miRNA precursor. *RNA* **10,** 1174–1177.

Mcluckey, S. A., Vanberkel, G. J., and Glish, G. L. (1992). Tandem mass-spectrometry of small, multiply charged oligonucleotides. *J. Am. Soc. Mass Spectrom.* **3,** 60–70.

Miyauchi, K., Ohara, T., and Suzuki, T. (2007). Automated parallel isolation of multiple species of non-coding RNAs by the reciprocal circulating chromatography method. *Nucleic Acids Res.* **35,** e24.

Nakai, Y., Umeda, N., Suzuki, T., Nakai, M., Hayashi, H., Watanabe, K., and Kagamiyama, H. (2004). Yeast Nfs1p is involved in thio-modification of both mitochondrial and cytoplasmic tRNAs. *J. Biol. Chem.* **279,** 12363–12368.

Noma, A., Kirino, Y., Ikeuchi, Y., and Suzuki, T. (2006). Biosynthesis of wybutosine, a hyper-modified nucleoside in eukaryotic phenylalanine tRNA. *EMBO J.* **25,** 2142–2154.

Okamoto, H., Watanabe, K., Ikeuchi, Y., Suzuki, T., Endo, Y., and Hori, H. (2004). Substrate tRNA recognition mechanism of tRNA (m7G46) methyltransferase from *Aquifex aeolicus*. *J. Biol. Chem.* **279,** 49151–49159.

Pomerantz, S. C., and McCloskey, J. A. (1990). Analysis of RNA hydrolyzates by liquid chromatography-mass spectrometry. *Methods Enzymol.* **193,** 796–824.

Rozenski, J., Crain, P. F., and McCloskey, J. A. (1999). The RNA Modification Database: 1999 update. *Nucleic Acids Res.* **27,** 196–197.

Saka, K., Tadenuma, M., Nakade, S., Tanaka, N., Sugawara, H., Nishikawa, K., Ichiyoshi, N., Kitagawa, M., Mori, H., Ogasawara, N., and Nishimura, A. (2005). A complete set of *Escherichia coli* open reading frames in mobile plasmids facilitating genetic studies. *DNA Res.* **12,** 63–68.

Sato, K., and Egami, F. (1957). Studies on ribonucleases in Takadiastase. 1. *J. Biochem.* **44,** 753–767.

Soma, A., Ikeuchi, Y., Kanemasa, S., Kobayashi, K., Ogasawara, N., Ote, T., Kato, J., Watanabe, K., Sekine, Y., and Suzuki, T. (2003). An RNA-modifying enzyme that governs both the codon and amino acid specificities of isoleucine tRNA. *Mol. Cell* **12,** 689–698.

Suzuki, T. (2005). "Biosynthesis and Function of tRNA Wobble Modifications in Fine-tuning of RNA Functions by Modification and Editing," pp. 24–69. Springer-Verlag, New York.

Suzuki, T., and Suzuki, T. Chaplet column chromatography: Isolation of a large set of individual RNAs in a single step. *Methods Enzymol.* In press.

Suzuki, T., Suzuki, T., Wada, T., Saigo, K., and Watanabe, K. (2002). Taurine as a constituent of mitochondrial tRNAs: New insights into the functions of taurine and human mitochondrial diseases. *EMBO J.* **21,** 6581–6589.

Uchida, T., and Egami, F. (1971). Microbial ribonucleases with special reference to RNase T1, T2, N1, and U2. *In* "The Enzymes" (P. D. Boyer, ed.), Vol. 4, pp. 207–250. Academic Press, New York.

Umeda, N., Suzuki, T., Yukawa, M., Ohya, Y., Shindo, H., Watanabe, K., and Suzuki, T. (2005). Mitochondria-specific RNA-modifying enzymes responsible for the biosynthesis of the wobble base in mitochondrial tRNAs. Implications for the molecular pathogenesis of human mitochondrial diseases. *J. Biol. Chem.* **280,** 1613–1624.

Yu, B., Yang, Z., Li, J., Minakhina, S., Yang, M., Padgett, R. W., Steward, R., and Chen, X. (2005). Methylation as a crucial step in plant microRNA biogenesis. *Science* **307,** 932–935.

FURTHER READING

Ohara, T., Sakaguchi, Y., Suzuki, T., Ueda, H., Miyauchi, K., and Suzuki, T. The 3′-termini of Piwi-interacting RNAs are 2′-O-methylated. Submitted.

CHAPTER TEN

CHAPLET COLUMN CHROMATOGRAPHY: ISOLATION OF A LARGE SET OF INDIVIDUAL RNAS IN A SINGLE STEP

Takeo Suzuki *and* Tsutomu Suzuki

Contents

Abstract

RNA molecules mature through various posttranscriptional modifications and editing. To characterize base modifications and terminal chemical structures of fully processed native RNAs, it is necessary to isolate individual RNA species from a limited quantity and complex mixture of cellular RNAs. However, there have been no general and convenient strategies for isolation of individual RNAs. We describe a simple and practical method for effective isolation of multiple individual RNA species named "Chaplet Column Chromatography (CCC)." We successfully isolated all species of mammalian mitochondrial tRNAs with this method.

1. INTRODUCTION

Recent progress in transcriptome analyses of higher organisms has revealed diverse classes of noncoding RNAs (ncRNAs) (Bertone *et al.*, 2004; Carninci *et al.*, 2005; Eddy, 2001; Mattick and Makunin, 2006) that

Department of Chemistry and Biotechnology, Graduate School of Engineering, The University of Tokyo, Bunkyo-ku, Tokyo, Japan

Methods in Enzymology, Volume 425
ISSN 0076-6879, DOI: 10.1016/S0076-6879(07)25010-4

constitute new regulatory elements in various biological processes. In addition to classical ncRNAs, (e.g., tRNA, rRNA, snRNA, and snoRNA), endogenous small ncRNAs, as represented by microRNAs (miRNAs), have been discovered and well characterized (Ambros, 2004; Bartel, 2004). However, many ncRNAs still remain to be characterized.

RNA molecules mature by undergoing posttranscriptional processing, such as capping, splicing, polyadenylation, base modification, editing, and 5′- and/or 3′-end cleavage, all of which are required for their functional expression. Among these, RNA modification (or editing) is a characteristic structural feature of ncRNAs. To obtain quantitative data on cellular RNAs, various hybridization-based techniques, such as Northern hybridization, microarray technology, and RT-PCR, have been routinely used. However, to obtain insights into the functional aspects of ncRNAs, methods that allow characterization of the modifications and terminal chemical structures of fully processed natural RNAs are extremely desirable. Nevertheless, current methods for the isolation of individual RNA species involve complicated and technically difficult procedures. Thus, the development of more efficient and convenient methods for the isolation of individual RNA species is urgently required.

The solid-phase DNA probe method uses affinity resins bound to immobilized DNAs that are complementary to the target RNAs for sequence-specific RNA isolation. A batchwise derivation of this method successfully isolated individual tRNAs from a crude RNA mixture (Tsurui et al., 1994). However, the amount of crude RNA that can be used in this method is limited by the volume of the tube used. In addition, with this method it is difficult to perform sufficient hybridization to entrap the target RNAs. To improve this method, we have developed a "chaplet column chromatography (CCC)" method for the effective isolation of multiple individual RNA species in one step. With this method, we previously isolated several tRNAs from various sources (Kaneko et al., 2003; Kirino et al., 2004; Noma et al., 2006; Soma et al., 2003; Suzuki et al., 2002). Here, we describe the isolation of all species of Bos taurus mitochondrial (mt) tRNAs as a model of how this method can be used to purify minor RNAs.

2. OUTLINE OF CHAPLET COLUMN CHROMATOGRAPHY

The CCC method is a sequence-specific affinity method that uses a DNA complementary to each target RNA (Fig. 10.1A). Each biotinylated DNA specific for a given target tRNA is immobilized on a prepacked streptavidin column, and the columns are connected in tandem. To entrap a target

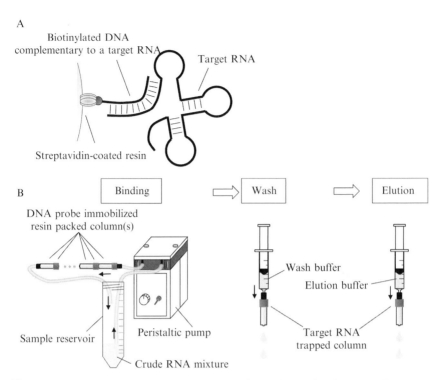

Figure 10.1 (A) Basic principle of the CCC method. A biotinylated DNA probe immobilized on the streptavidin-coated resin is hybridized to a target RNA molecule. (B) Schematic representation of the CCC operation. A crude RNA mixture is continuously passed through the column from a sample reservoir in a closed circuit (binding step). Then, wash or elution buffer is injected into the probe-immobilized column(s) (wash and elution steps). (See color insert.)

RNA, an RNA mixture is continuously circulated through the tandemly connected columns in a closed circuit (Fig. 10.1B) at a high temperature. During this procedure, each target RNA hybridizes to its respective DNA probe in each column. Because 22 columns are joined in series (Fig. 10.2A), all species of mitochondrial tRNAs can be simultaneously isolated from a pool of crude RNA. After washing out nonspecifically bound RNA molecules with high-salt buffer, the trapped RNA is eluted from each column with low-salt buffer at high temperature (Fig. 10.1B). In this procedure, the continuous circulation of the crude RNA mixture is very important for increasing the yields of individual RNAs. Purification scale, size, and number of columns can be changed according to the quantity of target RNA required and the amount of crude RNA available.

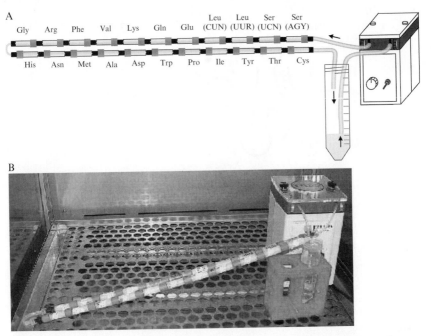

Figure 10.2 Schematic representation (top illustration) and real appearance (bottom photograph) of the CCC column used to isolate 22 species of *Bos taurus* mt tRNAs. The columns are labeled according to their tRNA-binding specificities. Arrows indicate the flow direction. (See color insert.)

3. Materials and Reagents

Peristaltic pump: SJ-1211H (Atto Corporation).

Tubing: silicon tubing with a 2-mm internal diameter, 4-mm outer diameter, and of adequate length to construct the flow path.

Column: HiTrap Streptavidin sepharose HP (1 ml, GE Healthcare).

Fittings: union Luer lock female/M6 male, female/M6 female (GE Healthcare) and 3.2 mm barb to male Luer (Bio-Rad).

DNA probes: as described in *"Immobilization of DNA Probes."*

Immobilization buffer: 400 mM NaCl, 10 mM HEPES-KOH (pH 7.5), and 5 mM EDTA (pH 8.0).

Storage buffer: 1.2 M NaCl, 30 mM HEPES-KOH (pH 7.5), 15 mM EDTA (pH 8.0). and 0.1% NaN$_3$ (Caution: NaN$_3$ is highly toxic. Consult MSDS of this reagent for proper handling instructions.)

Binding buffer: 1.2 M NaCl, 30 mM HEPES-KOH (pH 7.5), and 15 mM EDTA (pH 8.0).

Wash buffer: 100 mM NaCl, 2.5 mM HEPES-KOH (pH 7.5), and 1.25 mM
 EDTA (pH 8.0).
Elution buffer: 20 mM NaCl, 0.5 mM HEPES-KOH (pH 7.5), and 0.25 mM
 EDTA (pH 8.0).

4. PREPARATION OF CRUDE RNA SOLUTION

Crude RNA was obtained from approximately 5 kg of fresh bovine
liver according to an acid phenol extraction method described in the
literature (Suzuki *et al.*, 2002). For successful purification, we recommend
that the RNA be fractionated by anion exchange chromatography to reduce
contamination by nonspecific RNAs. Total RNA was fractionated on an
anion exchange chromatography (DEAE Sepharose Fast Flow, GE Healthcare) column (8 × 73 cm) by applying a linear gradient of NaCl and MgCl$_2$
(200–450 mM and 8–16 mM, respectively) in 20 mM HEPES-KOH
(pH 7.5) at a flow rate of 6.5 ml/min (gravitational flow). The elution positions of the mt tRNAs were monitored by the hybridization assay as described
(Yokogawa *et al.*, 1989). The fractions enriched in mt tRNAs were precipitated with ethanol and adjusted to a concentration of >4 mg/ml with binding
buffer. To ensure smooth circulation, the solution should be at low viscosity,
even if this means that the solution has to be diluted.

5. IMMOBILIZATION OF DNA PROBES

Synthetic 3′-biotinylated DNA probes are widely available commercially from many suppliers. Their sequences were designed to be complementary to the target tRNAs or tRNA genes (Sprinzl *et al.*, 1998) as shown
in Table 10.1. The sequence of each DNA probe should be designed to
complement a thermodynamically unstable region of the target RNA.
Although most of the DNA probes were designed to be complementary
to the 3′-terminal halves of the tRNAs, we recommend that probes be
experimentally selected for their ability to hybridize efficiently to target
tRNAs by use of Northern (or dot) hybridization before use. Each DNA
probe was dissolved in a volume of 1 ml immobilization buffer at a concentration of 0.15 mM and injected with a 1-ml syringe into the streptavidin
column previously equilibrated with 5 ml of immobilization buffer. After
incubation by to-and-fro mixing at room temperature for 1 h, the column
was washed with 10 ml of immobilization buffer. After equilibrating the
column with storage buffer, the column was stored at 4° until use.

Table 10.1 List of probe sequences for *Bos taurus* mitochondrial tRNAs

tRNA species	DNA sequence	Length	Comment
Ala	5′-TAAGGATTGCAAGACTACACCTTACATCAA-3′	30	3′ region
Arg	5′-TTGGTAATTATGAATTAAATCATAATCTAA-3′	30	3′ region
Asn	5′-TAAACACCCTAGCTAACTGGCTTCAATCTA-3′	30	5′ region
Asp	5′-CAAAATTATATAATGTTTTACTAACACCTC-3′	30	5′ region
Cys	5′-AAGCCCCGGCAGAATTGAAGCTGCTTCTCT-3′	30	3′ region
Gln	5′-CTAGAACTATAGGAATCGAACCTACTCCTA-3′	30	3′ region
Glu	5′-AAACCATCGTTGTCATTCAACTACAAGAAC-3′	30	5′ region
Gly	5′-TATTCTTTTTCGGACTAGACCGAAACTAGC-3′	30	3′ region
His	5′-GGTAAATAAGAAGGTAATGAGTTTCTATTG-3′	30	3′ region
Ile	5′-TAGAAATAAGAGGGTTTGAAGCTCTATTAT-3′	30	3′ region
Leu(CUN)	5′-TACTTTTATTTGGAGTTGCACCAATTTTT-3′	30	3′ region
Leu(UUR)	5′-TGTTAAGGAGAGGATTTGAATCTCTGGATA-3′	30	3′ region
Lys	5′-GTTAGTGCTATATAGCTTCTTAGTG-3′	25	5′ region
Met	5′-TAGTACGGGAAGGATATAAACCAACATTTT-3′	30	3′ region
Phe	5′-TGTTTATGGAGTTGGGAGACTCATCTAGGC-3′	30	3′ region
Pro	5′-TCACCATCAACCCCAAAGCTGAAGTTCTA-3′	30	Anticodon arm region
Ser(AGY)	5′-GCAGTTCTTGCATACTTTTTC-3′	21	5′ region
Ser(UCN)	5′-TCAAGCCAACATCATAACCTCTATGTC-3′	27	5′ region
Thr	5′-AGTCTTAGGGAGGTTAGTTGTTCTCCTTCT-3′	30	3′ region
Trp	5′-CAGGAATTAAGTAAATTGTACTTGCTTAGG-3′	30	3′ region
Tyr	5′-TGGTAAAAGAGGAGTCAAACCTCTATCTT-3′	30	3′ region
Val	5′-TCAAGATATTCATAATGAATGAAGTCTTCT-3′	30	3′ region

6. RNA Isolation by the CCC Method

Each DNA-immobilized column was equilibrated with 5 ml of binding buffer and connected to the other columns downstream of a peristaltic pump as illustrated in Fig. 10.2. Approximately 30 ml of crude tRNA solution (20 mg/ml) dissolved in binding buffer was circulated through the columns at 65° in an air-phase incubator (oven) for 30 min at a flow rate of 0.5 ml/min. Under conditions of continuous circulation, the solution was allowed to anneal at room temperature for 80 min (turn off the oven and open the door). The serially-connected columns were then disconnected, and each column was washed with prewarmed (37°) wash buffer with a disposable syringe until the UV absorbance fell below 0.01 A_{260}. After the washing step, a new 5-ml syringe filled with preheated elution buffer (3 ml) at 65° was connected to each column and used to inject 500 μl of elution buffer into the column. Then, the column outlet was plugged with a male Luer plug and the syringe-column-plug assembly was incubated for 5 min at 65°. We found that it was better to use a waterbath to maintain a constant temperature. After unplugging the outlet plug, the elution buffer was immediately flushed from the column by injection with the inlet syringe and fractionated into tubes by every 0.5–1 ml. The elution step should be repeated two or three times. Each mt tRNA was concentrated by ethanol precipitation, and the purity was evaluated by denaturing polyacrylamide gel electrophoresis. As shown in Fig. 10.3, all species of bovine mitochondrial tRNAs yielded major bands, demonstrating the high level of purity that can be obtained with this method, although extra bands were observed in some mt tRNAs preparations. Each tRNA was purified by eluting the major band from the gel and digested with RNase T$_1$. The digests were then subjected to LC/MS by use of ion-trap mass spectrometer, which demonstrated that 22 species of mt tRNAs had been successfully purified by this method. Detailed mass spec data of 22 species of mt tRNAs will be reported in a specialized journal.

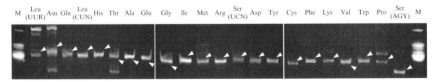

Figure 10.3 Polyacrylamide gel electrophoresis of the isolated 22 mt tRNAs. The gels were stained with ethidium bromide. Lane M, crude RNA mixture before circulation. Each mt tRNA species is indicated by a three-letter abbreviation for the corresponding amino acid. The target RNAs are indicated by arrowheads.

7. DISCUSSION

By use of the CCC method, all 22 species of *Bos taurus* mt tRNAs were successfully and simultaneously isolated in one operation. The difference in the yield and purity for each tRNA probably reflects differences in the hybridization efficiencies and specificities of the DNA probes. We found that temperature during the CCC method was a critical factor for the effective and reproducible isolation of the target RNAs. Because mammalian mt tRNAs have generally low thermostability because of their abnormal secondary structure (Hayashi *et al.*, 1998), a temperature of 65° is considered adequate to denature mt tRNAs. For the isolation of thermostable cytoplasmic tRNAs (e.g., *E. coli*, yeast, and mouse cytoplasmic tRNAs), higher temperatures (e.g., 72°) are required. In addition, the washing temperature should be optimized (e.g., 40–45°) to reduce contamination by nonspecific RNAs.

As described here, the CCC method is a powerful and convenient technique for RNA isolation. However, the high back-pressure developed by connecting the columns in series limits the degree to which this method can be scaled up. Recently, we reported another platform technology named "reciprocal circulating chromatography (RCC)," which enables parallel, automated, and programmable purification (Miyauchi *et al.*, 2007). Further progress in these convenient and high-throughput purification methods will help us to acquire a greater understanding of the functional aspects of ncRNAs.

ACKNOWLEDGMENTS

This work was supported by a JSPS Research Fellowship for Young Scientists (to Takeo Suzuki), and by grants-in-aid for scientific research on priority areas from the Ministry of Education, Science, Sports, and Culture of Japan and by a grant from the New Energy and Industrial Technology Development Organization (NEDO) (to Tom Suzuki).

REFERENCES

Ambros, V. (2004). The functions of animal microRNAs. *Nature* **431,** 350–355.

Bartel, D. P. (2004). MicroRNAs: Genomics, biogenesis, mechanism, and function. *Cell* **116,** 281–297.

Bertone, P., Stolc, V., Royce, T. E., Rozowsky, J. S., Urban, A. E., Zhu, X., Rinn, J. L., Tongprasit, W., Samanta, M., Weissman, S., Gerstein, M., and Snyder, M. (2004). Global identification of human transcribed sequences with genome tiling arrays. *Science* **306,** 2242–2246.

Carninci, P., Kasukawa, T., Katayama, S., Gough, J., Frith, M. C., Maeda, N., Oyama, R., Ravasi, T., Lenhard, B., Wells, C., Kodzius, R., Shimokawa, K., *et al.* (2005). The transcriptional landscape of the mammalian genome. *Science* **309,** 1559–1563.

Eddy, S. R. (2001). Non-coding RNA genes and the modern RNA world. *Nat. Rev. Genet.* **2,** 919–929.

Hayashi, I., Kawai, G., and Watanabe, K. (1998). Higher-order structure and thermal instability of bovine mitochondrial tRNASerUGA investigated by proton NMR spectroscopy. *J. Mol. Biol.* **284,** 57–69.

Kaneko, T., Suzuki, T., Kapushoc, S. T., Rubio, M. A., Ghazvini, J., Watanabe, K., Simpson, L., and Suzuki, T. (2003). Wobble modification differences and subcellular localization of tRNAs in Leishmania tarentolae: Implication for tRNA sorting mechanism. *EMBO J.* **22,** 657–667.

Kirino, Y., Yasukawa, T., Ohta, S., Akira, S., Ishihara, K., Watanabe, K., and Suzuki, T. (2004). Codon-specific translational defect caused by a wobble modification deficiency in mutant tRNA from a human mitochondrial disease. *Proc. Natl. Acad. Sci. USA* **101,** 15070–15075.

Mattick, J. S., and Makunin, I. V. (2006). Non-coding RNA. *Hum. Mol. Genet.* **15**(Suppl. 1), R17–R29.

Miyauchi, K., Ohara, T., and Suzuki, T. (2007). Automated parallel isolation of multiple species of non-coding RNAs by the reciprocal circulating chromatography method. *Nucleic Acids Res.* **35,** e24.

Noma, A., Kirino, Y., Ikeuchi, Y., and Suzuki, T. (2006). Biosynthesis of wybutosine, a hyper-modified nucleoside in eukaryotic phenylalanine tRNA. *EMBO J.* **25,** 2142–2154.

Soma, A., Ikeuchi, Y., Kanemasa, S., Kobayashi, K., Ogasawara, N., Ote, T., Kato, J., Watanabe, K., Sekine, Y., and Suzuki, T. (2003). An RNA-modifying enzyme that governs both the codon and amino acid specificities of isoleucine tRNA. *Mol. Cell* **12,** 689–698.

Sprinzl, M., Horn, C., Brown, M., Ioudovitch, A., and Steinberg, S. (1998). Compilation of tRNA sequences and sequences of tRNA genes. *Nucleic Acids Res.* **26,** 148–153.

Suzuki, T., Suzuki, T., Wada, T., Saigo, K., and Watanabe, K. (2002). Taurine as a constituent of mitochondrial tRNAs: New insights into the functions of taurine and human mitochondrial diseases. *EMBO J.* **21,** 6581–6589.

Tsurui, H., Kumazawa, Y., Sanokawa, R., Watanabe, Y., Kuroda, T., Wada, A., Watanabe, K., and Shirai, T. (1994). Batchwise purification of specific tRNAs by a solid-phase DNA probe. *Anal. Biochem.* **221,** 166–172.

Yokogawa, T., Kumazawa, Y., Miura, K., and Watanabe, K. (1989). Purification and characterization of two serine isoacceptor tRNAs from bovine mitochondria by using a hybridization assay method. *Nucleic Acids Res.* **17,** 2623–2638.

SNO-MEDIATED MODIFICATIONS

CHAPTER ELEVEN

BIOCHEMICAL PURIFICATION OF BOX H/ACA RNPS INVOLVED IN PSEUDOURIDYLATION

John Karijolich, David Stephenson, *and* Yi-Tao Yu

Contents

Abstract

Box H/ACA RNPs, each consisting of four common core proteins and a single unique RNA, are the most complex pseudouridylases yet discovered. The RNA component serves as a guide that directs a target uridine for modification. To study the functions and mechanisms of RNA pseudouridylation, it is desirable to isolate the intact box H/ACA RNP complexes. Purified RNPs will allow further identification and characterization of the RNA component in each RNP complex and permit a systematic analysis of the mechanism by which the enzymes

Department of Biochemistry and Biophysics, University of Rochester Medical Center, Rochester, New York

Methods in Enzymology, Volume 425
ISSN 0076-6879, DOI: 10.1016/S0076-6879(07)25011-6

convert uridines to pseudouridines in a site-specific manner. Over the years, a number of purification techniques have been developed, providing important tools for RNA pseudouridylation research. Here, we describe three of these techniques, including biotin-streptavidin affinity purification by use of biotinylated 5-fluorouridine (5FU)-containing RNA, tandem affinity purification (TAP) by TAP-tagging one of the four core proteins in the complex, and immunoprecipation by use of antibodies against one of the four core proteins.

1. INTRODUCTION

RNA modification is a posttranscriptional process that occurs in all organisms and in a variety of cellular RNAs (Grosjean *et al.*, 1995; Maden, 1990; Ofengand and Fournier, 1998; Reddy and Busch, 1988). One of the most common modifications is pseudouridylation, whereby uridines are converted to pseudouridines through an isomerization reaction (Grosjean *et al.*, 1995; Maden, 1990; Ofengand and Fournier, 1998; Yu *et al.*, 2005). For instance, there are approximately 100 pseudouridines in mammalian rRNAs and 24 in vertebrate spliceosomal snRNAs (Massenet *et al.*, 1998; Ofengand and Fournier, 1998; Reddy and Busch, 1988; Schattner *et al.*, 2004; Yu *et al.*, 2005). Importantly, many pseudouridines are remarkably conserved from species to species and are often clustered in regions that are functionally important (Decatur and Fournier, 2002, 2003; Massenet *et al.*, 1998; Reddy and Busch, 1988; Yu *et al.*, 2005). Together, the abundance, the conservation, and the strategic location of these pseudouridines strongly suggest that they contribute significantly to RNA function. In fact, recent functional analyses have demonstrated that the pseudouridines in rRNAs and spliceosomal snRNAs play important roles in protein translation and pre-mRNA splicing, respectively (Badis *et al.*, 2003; Donmez *et al.*, 2004; King *et al.*, 2003; Yu *et al.*, 1998; Zhao and Yu, 2004).

A large body of evidence indicates that pseudouridylation of eukaryotic and archaeal rRNAs and eukaryotic spliceosomal snRNAs is catalyzed by box H/ACA RNPs (ribonucleoproteins), RNA–protein complexes each consisting of a unique small RNA (box H/ACA RNA) and four core proteins (Cbf5/Nap57/Dyskerin, Nhp2, Gar1, and Nop10) (Balakin *et al.*, 1996; Ganot *et al.*, 1997; Ni *et al.*, 1997; Yu *et al.*, 2005). Because box H/ACA RNPs are by far the most complex pseudouridylases yet discovered (Li and Ye, 2006; Yu, 2006), their mechanism of action—converting a uridine into a pseudouridine—has attracted a great deal of attention. Both computational and experimental analyses have indicated that the RNA component of an eukaryotic box H/ACA RNP forms a unique "hairpin-hinge-hairpin-tail" structure, in which each of the two hairpins contains an internal loop that constitutes a so-called pseudouridylation pocket, directing the conversion of a target uridine to pseudouridine in the substrate (Ganot *et al.*, 1997; Ni *et al.*, 1997; Schattner *et al.*, 2004) (Fig. 11.1). More specifically, the sequence in each pocket serves as a guide that base pairs with its target RNA, precisely positioning the target uridine at the base of the upper stem of the hairpin and leaving it unpaired within the pseudouridylation pocket (Fig. 11.1). When the target uridine is brought to the pocket, Cbf5, the pseudouridylase associated with the box H/ACA guide RNA (Zebarjadian *et al.*, 1999) converts the uridine to pseudouridine (Fig. 11.1).

A large number of box H/ACA RNAs have been identified thus far; however, they still cannot account for the huge number of pseudouridines known to exist in RNAs, especially in spliceosomal snRNAs (Yu *et al.*, 2005). Recently, both computational and experimental screens have identified additional box H/ACA RNAs that have no known targets, suggesting that there might exist many more unidentified pseudouridines in various RNAs (Huttenhofer *et al.*, 2002; Schattner *et al.*, 2004). Taking into account the likely existence of many unidentified pseudouridines, the number of box H/ACA RNAs existing in cells may be far greater than the number we currently know. Likewise, although much has been learned about H/ACA RNP-catalyzed pseudouridylation, the detailed mechanism of this reaction remains largely unknown. To identify a complete set of cellular box H/ACA RNAs and to dissect the molecular mechanism of this fascinating reaction, it is desirable to isolate intact (and more preferably

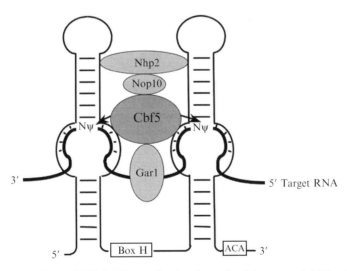

Figure 11.1 A box H/ACA RNP complex is schematized (not to scale). The thin line represents the box H/ACA RNA (the conserved elements, box H, and box ACA, are indicated), and the thick line depicts the target RNA. The arrows indicate two pseudouridines (Ψ) derived from uridines through an isomerization reaction catalyzed by Cbf5. The other three core proteins, Nhp2, Nop10, and Gar1, are also indicated.

functional) box H/ACA RNPs. Here, we focus on three approaches, including biotin-streptavidin affinity purification by use of biotinylated 5-
fluorouridine (5FU)–containing RNA, tandem affinity purification (TAP)
by TAP-tagging one of the four core proteins in the complex, and immunoprecipitation by use of antibodies against one of the four core proteins, to
purifying box H/ACA RNPs.

2. Biotin-Streptavidin Affinity Purification Using Biotinylated 5FU-Containing RNA

2.1. Overview

It has been demonstrated that 5-fluorouracil (5FU), when incorporated into
an RNA, functions as a potent inhibitor that blocks RNA pseudouridylation in a sequence- and site-specific manner (Hoang and Ferre-D'Amare,
2001; Patton *et al.*, 1994; Yu *et al.*, 1998; Zhao and Yu, 2004). For instance,
when injected into *Xenopus* oocytes, U2 RNA containing 5FU in the
branch site recognition region specifically inhibits U2 pseudouridylation
in the branch site recognition region, although pseudouridylation in the
other regions is not affected (Zhao and Yu, 2004). Likewise, *in vivo* incorporation of 5FU into U2 snRNA at the important pseudouridylation sites
within the branch site recognition region effectively inhibits the pseudouridylation of newly synthesized U2 at the respective sites (Zhao and Yu,
2006). The site-specific inhibitory effect of 5FU–containing U2 (or other
RNAs) on RNA pseudouridylation is due to the irreversible interactions
between the 5FU-containing RNA and its site-specific pseudouridylases
(here box H/ACA RNPs) (Hoang and Ferre-D'Amare, 2001; Zhao *et al.*,
2002). It is conceivable that this extremely tight binding sequesters the
enzymes, thus making them unavailable to modify the newly synthesized
RNA (Patton *et al.*, 1994; Yu *et al.*, 1998; Zhao *et al.*, 2002).

It is, therefore, desirable to take advantage of this irreversible binding to
isolate box H/ACA RNPs. By use of a biotinylated 5FU-containing U2,
we, in fact, successfully isolated a number of box H/ACA RNAs responsible for guiding U2 snRNA pseudouridylation (Zhao *et al.*, 2002). In that
work, we first synthesized (by *in vitro* transcription) a fully substituted 5FU
U2 with biotinylated cytidine incorporated at random positions. We then
incubated the 5FU-containing U2 with nuclear extracts prepared from
Xenopus oocytes for a sufficient amount of time, permitting the high-
affinity interaction of the 5FU-containing U2 with its pseudouridylation
enzymes. The complex was then pulled down by streptavidin-coupled
beads and subsequently eluted. The RNA components of the complex
were further purified, $3'-^{32}pCp$ labeled, and resolved on a denaturing
polyacrylamide gel. The RNA profile thus produced revealed a number

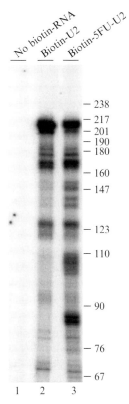

Figure 11.2   Selection of *Xenopus* box H/ACA RNAs by biotinylated 5FU U2-box H/ACA RNP interaction followed by biotin-streptavidin affinity chromatography. Biotinylated U2 (lane 2) or biotinylated fully substituted 5FU U2 (lane 3) was incubated with *Xenopus* oocyte nuclear extracts, and subsequently pulled down by streptavidin beads. Copurified RNAs were eluted, 3′ end-radiolabeled with ^{32}pCp, and resolved on a 6% denaturing gel. Lane 1 is a parallel experiment in the absence of biotinylated U2 or biotinylated 5FU U2. The positions of nucleic acid markers (in nucleotides) are indicated. The band with the slowest mobility is the biotinylated U2 of 189 nucleotides (U2 containing biotin-14-CTP migrates slower than unmodified U2 does).

of novel RNAs compared with those generated with nonsubstituted U2 (Fig. 11.2). Many of them are box H/ACA RNAs, RNA components of box H/ACA RNPs.

Protocols for synthesizing a biotinylated 5FU-containing U2 and isolating U2-specific RNAs from *Xenopus* oocyte nuclear extracts are described in the following. Note that we believe the protocols are not limited to U2-specific H/ACA RNPs or *Xenopus* oocyte nuclear extracts; the protocols are applicable to any other H/ACA RNPs from any other organisms.

2.2. Synthesis of a biotinylated 5FU-containing U2 snRNA

2.2.1. Buffers, reagent, and solutions

$10\times$ transcription buffer: 400 mM Tris-HCl, pH 7.5, 60 mM mgCl$_2$, 20 mM spermidine

T7-U2 Plasmid containing the U2 sequence under the control of the T7 promoter

Nucleotides (NTPs): 10 mM ATP, 10 mM CTP, 10 mM UTP, 10 mM GTP, 10 mM GpppG (Amersham), 10 mM 5-flurouridine-5′-triphosphate (5FUTP, Sierra Bioresearch), 2 mM biotin-14-CTP (Invitrogen)

Radioactive nucleotide: [α^{32}P]GTP (Amersham)

Dithiothreitol (DTT): 0.5 M (Sigma)

RNasin Ribonuclease Inhibitor: 40 units/μl (Promega)

T7 RNA polymerase: 50 units/μl (Epicentre)

RQ1 RNase-free DNase: 1 unit/μl (Promega)

G50 buffer: 20 mM Tris-HCl, pH 7.5, 300 mM sodium acetate, 2 mM EDTA, 0.25% sodium dodecyl sulfate (SDS)

PCA: Tris-HCl (pH 7.5)–buffered phenol/chloroform/isoamyl alcohol (50:49:1)

Ethanol: 100%

Formamide loading buffer: 95% formamide, 10 mM EDTA, 0.1% xylene cyanol FF, 0.1% bromphenol blue

Polyacrylamide gel solution: OmniPur (EMD Chemicals)

Urea (Sigma)

Glycogen: 10 mg/ml (Invitrogen)

Autoclaved distilled water

2.3. Protocol

1. At room temperature, mix together, in a 1.5-ml Microfuge tube, 7 μl of autoclaved distilled water, 10 μl of $10\times$ transcription buffer, 10 μl of a 1 μg/μl linearized T7-U2 plasmid (to synthesize an RNA other than U2, plasmids containing the respective RNA gene may be used), 12 μl each of 10 mM ATP, CTP, 5FUTP (5-flurouridine-5′-triphosphate) and GpppG, 6 μl of 10 mM GTP, 6 μl of 2 mM biotin-14-CTP, 1 μl of 0.1 μCi/μl [α^{32}P]GTP (inclusion of a trace amount of radioactive nucleotide allows calculation of the percentage incorporation and thereby accurate determination of the amount of RNA transcribed), 1 μl of 0.5 M DTT, 1 μl of 40 units/μl RNasin ribonuclease inhibitor, and 10 μl of 50 units/μl of T7 RNA polymerase. As a control, unsubstituted U2 (5FUTP is replaced with UTP) is synthesized in parallel.

2. Remove 1 μl (1% of the reaction) and determine cpm later at step 11 for accurate calculation of the percentage incorporation. Place the

tube containing the rest of the reaction (99 μl) in a 37° waterbath and incubate for 1–2 h.

3. Add 2 μl of 1 unit/μl of RQ1 RNase-free DNase to the tube and continue to incubate at 37° for another 30 min.

4. Add ~150 μl of G50 buffer to the DNase-treated transcription reaction to bring the total volume to ~250 μl. Subsequently, add 2 volumes (~500 μl) of PCA, mix vigorously by vortexing, and then Microfuge for 2 min. Note that because biotin-14-CTP has a long carbon-based linker, an RNA transcript containing a significant number of biotin-14-CTP (depending on the C content in the template and the ratio of biotin-14-CTP/CTP in the transcription mixture) will become highly hydrophobic. Consequently, PCA extraction would result in a significant loss of RNA product. Under this circumstance, the PCA extraction step should be omitted. In cases in which the C content is high in the template, a low ratio of biotin-14-CTP/CTP (lower than that [1:10] used in this protocol, see step 1) can be considered to reduce hydrophobicity.

5. Remove the aqueous phase, mix thoroughly with 2.5 volumes (~630 μl) of 100% ethanol, and place the tube on dry ice for 10 min. Microfuge at 14,000g for 10–15 min. Remove and discard the supernatant promptly.

6. Resuspend the pellet in 10 μl of autoclaved distilled water, mix well with 20 μl of formamide loading buffer, heat at 95° for 3 min, chill on ice immediately, and load the sample on a 6% polyacrylamide–8 M urea gel.

7. Expose the gel marked with ^{32}P-radioisotope on the side to Phosphor-Imager screen for approximately 5 min.

8. Locate and excise the RNA transcript band and place it in a 1.5-ml Microfuge tube.

9. Add 450 μl of G50 buffer to the tube containing the gel slice. Place the tube on dry ice for 5 min and then transfer to room temperature for elution overnight (~16 h).

10. Microfuge the tube containing the gel slice at 14,000g for 15 min. Transfer the supernatant to a new 1.5-ml Microfuge tube. Extract the supernatant with 500 μl of PCA and precipitate the gel-purified RNA with 1000 μl of 100% ethanol (using 1 μl of 10 mg/ml glycogen as carrier), as described in steps 4 and 5.

11. Determine (using a scintillation counter) the total cpm for the precipitated RNA transcript and the sample (1 μl transcription reaction) put aside earlier in step 2. Calculate the percentage of GTP incorporation and the exact amount of RNA transcribed. Resuspend the RNA pellet in autoclaved distilled water at 20 pmol/μl. A 100-μl transcription reaction usually produces ~50 pmol of 5FU-substituted U2, or more than 50 pmol of unsubstituted U2.

2.3.1. Isolation of U2-specific box H/ACA RNAs with affinity chromatography

2.3.1.1. Buffers, reagent, and solutions

2.3.1.1.1. Xenopus *oocyte nuclear extracts*

Buffer A: 100 mM Tris-HCl, pH 8.0, 100 mM ammonium acetate, 5 mM mgCl2, 2 mM DTT, 0.1 mM EDTA, and 0.05% NP-40

Streptavidin agarose beads (Pierce)

Proteinase K: 20 mg/ml (Invitrogen)

2× RNA ligation buffer: 100 HEPES, pH 8.3, 10 μM ATP, 20 mM mgCl$_2$, 6.6 mM DTT, 20% (v/v) dimethyl sulfoxide (DMSO), 30% (v/v) glycerol

[5′-^{32}P]pCp (Amersham)

T4 RNA ligase: 5 units/μl (Amersham)

G50 buffer, autoclaved distilled water, PCA, ethanol, formamide loading buffer, polyacrylamide gel solution, urea (as described earlier)

2.4. Protocol

1. In a 1.5-ml Microfuge tube, mix 2 μl of 20 pmol/μl biotinylated 5FU-containing U2 with ~15 μl of nuclear extracts prepared from 200 isolated *Xenopus* oocyte nuclei. Carry out the binding reaction at room temperature (or at 30°) for 30 min. Here, the protocol also applies to nuclear extracts or cell extracts prepared from other organisms/species. In parallel, set up an identical binding reaction with unsubstituted U2 (as a control).

2. To the binding reaction, add 220 μl of buffer A and 20 μl of buffer A–saturated streptavidin agarose beads (with a bed volume of 10 μl). Gently rotate the tube containing the beads on a rotator at 4° for 2 h.

3. Microfuge the tube briefly (~30 s at ~5000g), and discard the supernatant. Wash the beads four times, each time with 1 ml of buffer A.

4. Add 250 μl of G50 buffer and 10 μl of 20 mg/ml proteinase K to the washed beads, and incubate at 42° for 15–30 min.

5. Transfer the proteinase K–digested sample to 85° for 10 min to allow complete dissociation of box H/ACA RNAs from biotinylated 5FU-containing U2.

6. Microfuge briefly (~30 s at ~5000g), and transfer the supernatant to a new 1.5-ml Microfuge tube.

7. Extract the supernatant twice each with 400 μl PCA, and ethanol precipitate the RNAs with glycogen as carrier, as described earlier.

8. Resuspend the RNA pellet in 2 μl autoclaved distilled water, and place on ice. Add 8 μl of 2× RNA ligation buffer, 5 μl of [5′-^{32}P]pCp, and 1 μl of 5 unit/μl T4 RNA ligase.

9. Transfer the tube of RNA ligation reaction to a 4° water bath, and incubate for ~16 h.

10. Stop the ligation reaction by adding ~250 μl G50 buffer, PCA extract twice and ethanol precipitate the 3'-end radioactively (^{32}pCp) labeled RNAs.

11. Resuspend the 3'^{32}pCp-labeled RNAs in 2 μl of autoclaved distilled water, mix well with 4 μl of formamide loading buffer, heat the sample at 95° for 3 min before loading it on a 7% polyacrylamide–8 M urea gel to resolve RNAs by electrophoresis, as described earlier.

12. Expose the gel marked with ^{32}P-radioisotope on the side to Phosphor-Imager screen for approximately 5 min.

13. Compare the RNA profile generated by 5FU-substituted U2 with that generated by unsubstituted U2, locate and excise the unique RNA bands in the "5FU-substituted U2" lane, and determine the sequences of the RNAs using chemical sequencing (Peattie, 1979) or enzymatic sequencing (Donis-Keller, 1980).

3. TAP Tag Purification of Box H/ACA RNPs

3.1. Overview

The development of the tandem affinity purification (TAP) tag by the Seraphin group was a major advance in the field of protein purification (Rigaut *et al.*, 1999). The original TAP procedure was based on a fusion cassette encoding calmodulin-binding peptide (CBP), a TEV cleavage site, and two IgG binding domains of the *Staphylococcus aureus* Protein A, cloned at either the 5' or 3' end of a particular chromosomal gene (Fig. 11.3). The incorporation of the TAP cassette into the chromosome maintains the expression of the protein at its natural level of expression and, therefore, is ideal for purification and characterization of heteromeric complexes.

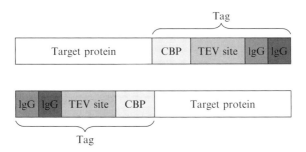

Figure 11.3 Schematic representation of both a C-terminus TAP-tagged fusion protein (top) and an N-terminus TAP-tagged fusion protein (bottom).

The resulting fusion protein and associated components can first be recovered by selection on an IgG matrix. The TEV protease is then used to release the fusion protein from the IgG matrix. The eluant is then incubated with calmodulin-coated beads in the presence of calcium. Last, essentially pure protein can be recovered by the addition of EGTA, a calcium-specific chelator.

Although the original TAP tagging method is still in use today, there also exist several variations of the method. The method used in our laboratory is one such variation (Alexandrov *et al.*, 2005; Ma *et al.*, 2005). Although the two IgG domains are kept unchanged, the TEV protease site is replaced with a 3C protease site, and CBP is substituted with a 6× His tag.

TAP tagging core components of box H/ACA RNPs has enabled a great expansion in our knowledge concerning many aspects of box H/ACA RNP research. TAP tagging of any of the three core proteins, Gar1p, Cbf5p, and Nhp2p, results in the purification of the intact box H/ACA RNPs with a high level of purity (Ma *et al.*, 2005) (Fig. 11.4). Attempts to TAP tag Nop10p have proved to be unsuccessful, possibly because of the small size of Nop10p (Ma and Yu, unpublished data). TAP tag purification of box H/ACA RNPs in conjunction with micrococcal nuclease digestion has led to the development of a highly efficient eukaryotic *in vitro* reconstitution system (Ma *et al.*, 2005). Specifically, we used TAP tagging followed by micrococcal nuclease digestion

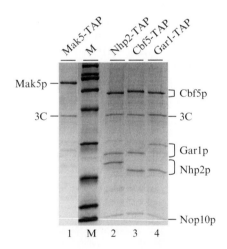

Figure 11.4 The protein compositions of TAP preparations analyzed on an SDS-PAGE gel stained with Coomassie blue. Lanes 2, 3, and 4 contain samples of Nhp2p-TAP, Cbf5p-TAP, and Gar1p-TAP preparations, respectively. Lane 1 is a control containing Mak5p-TAP preparation. The positions of the box H/ACA RNP core proteins, Cbf5p, Gar1p, Nhp2p, and Nop10p (both TAP-tagged and native forms), are indicated on the right. The TAP-tagged proteins migrated more slowly than their native forms. The label 3C indicates protease 3C, which was used during TAP purification. The Mak5p protein band is also indicated. Lane M is a standard protein marker lane.

to show that pseudouridylation of yeast U2 snRNA at position 42 is modified by an RNA–dependent mechanism (Ma *et al.*, 2005). TAP tagged preparations of box H/ACA RNPs are not limited to protein subunit identification and characterization, because they can also lead to the identification of novel box H/ACA RNAs associated with the core proteins. Indeed, TAP tag purification coupled with genomic DNA microarray experiments led to the identification of three new box H/ACA RNAs (Torchet *et al.*, 2005).

The protocols for TAP tagging Cbf5p and the subsequent purification of box H/ACA snoRNPs are described in the following.

3.1.1. Generation of TAP construct by PCR
3.1.1.1. Buffers, reagents, and solutions

Cbf5-TAP-1F primer: 5′ CTA AGA AAT CTA AGA AAA ACC CAG CTT TCT- TGT ACA AAG TGG 3′

Cbf5-TAP-2F primer: 5′ GAC GGT GAT TCT GAG GAA AAG AAA TCT- AAG AAA TCT AAG AAA 3′

Cbf5-TAP-1R primer: ATG AGA TGG AGT GAT GAG AAT CAT ACG ACT- CAC TAT AGG G 3′

Cbf5-TAP-2R primer: AGA AAG CTG TTA AAT ATA TAG GAT GAGT ATG- GAG GA TGA G 3′

Deoxyribonucleotides (10 m*M* dNTPs) (Amersham): 10 m*M* dATP, 10 m*M* dCTP, 10 m*M* dGTP, 10 m*M* dTTP

Taq DNA polymerase (New England Biolabs)

10× Taq DNA polymerase buffer (New England Biolabs)

Plasmid pAVA0258 (Alexandrov *et al.*, 2005)

G50 buffer, PCA, ethanol, autoclaved distilled water (as describer earlier)

3.2. Protocol

1. In a 0.5-ml PCR tube mix 1 μl of 50 ng/μl pAVA0258, 5 μl 10× polymerase buffer, 2 μl 10 m*M* dNTPs, 2 μl of 10 p*M* Cbf5-TAP-1F primer, 2 μl 10 p*M* Cbf5-TAP-1R primer, 37 μl autoclaved distilled water, and 1 μl Taq DNA polymerase.

2. Perform PCR cycles as follows:

   a. Step 1: 95° 2 min (1 cycle)
   b. Step 2: 95° 30 s
   55° 30 s
   72° 30 s (repeat step 2 28 times)
   c. Step 3: 72° 120 s (1 cycle)
   4° (indefinitely)

3. Transfer PCR reaction to a 1.5-ml Microfuge tube. Add 450 μl G50 buffer and 500 μl PCA. Vortex for 30 s. Microfuge the mixture for 3 min at 13,000*g*.

4. Collect top aqueous phase and transfer to a clean 1.5-ml Microfuge tube. Add 1 ml 100% ethanol and vortex 30 s. Leave on dry ice for 10 min.
5. Microfuge sample for 15 min at 14,000*g*. Discard ethanol and allow to air dry for 5 min. Resuspend the pellet in 10 μl autoclaved distilled water.
6. In a 0.5-ml PCR tube, mix 1 μl of the PCR product from the preceding, 5 μl of 10× polymerase buffer, 2 μl of 10 m*M* dNTPs, 2 μl of 10 p*M* Cbf5-TAP-2F primer, 2 μl of 10 p*M* Cbf5-TAP-2R primer, 37 μl of autoclaved distilled water, and 1 μl of Taq DNA polymerase. Perform the PCR reaction and purification of PCR product exactly as described previously in steps 2–5. The purpose of this second PCR reaction is to extend the flanking sequences that are required for the integration of the PCR product into the desired site (3′ end of CBF5 gene) within the yeast chromosomes (transformation and homologous recombination in yeast) (see steps in "Transformation of Cells with TAP Construct").
7. Resuspend the final PCR product in 10 μl autoclaved distilled water. Quantify the DNA by UV/Vis spectroscopy.

3.2.1. Transformation of cells with TAP construct
3.2.1.1. Buffers, reagents, and solutions

YPD media (autoclaved): 10 g yeast extract, 20 g peptone, and 20 g dextrose dissolved in 1 liter of water
Lithium acetate (LiAc): 100 m*M* (autoclaved)
PEG-3350 solution: 15 g PEG-3350 dissolved in 16.5 ml autoclaved distilled water
Lithium acetate–PEG solution: 12 ml PEG-3350 solution, 1.5 ml autoclaved distilled water, and 1.5 ml 1 *M* LiAc
DMSO (Sigma)
SX-Ura dry mixture: 25.1 g yeast nitrogen base without amino acids and ammonium sulfate, 75.4 g ammonium sulfate, 460 mg isoleucine, 2.25 g valine, 300 mg adenine, 300 mg arginine, 300 mg histidine, 450 mg leucine, 450 mg lysine, 300 mg methionine, 750 mg phenylalanine, 300 mg tryptophan, and 450 mg tyrosine. Rotate in a Ball Mill grinder overnight.
SD-Ura plates: 7.5 g of SX-Ura dry mixture, 20 g agar, and 20 g dextrose dissolved in 1 liter of water. After autoclaving, pour 20 ml into each petri dish and allow to solidify overnight.

3.3. Protocol

1. Pick a single colony of yeast (grown on a YPD plate) and inoculate into 5 ml of YPD liquid media and shake at 30° overnight.
2. Next morning dilute yeast culture 1:250 in 25 ml YPD liquid media.
3. Grow cultures to an OD 600 of ~2.0. Pellet cells by centrifugation in a SS-34 rotor for 1 min at 3000*g*. Decant the supernatant.

4. Wash cells two times with 0.5 ml 100 mM LiAc. Resuspend cells in 50 μl of 100 mM LiAc and transfer to a new 1.5-ml Microfuge tube.
5. Add 3–5 μg of the PCR product (see earlier) and mix well by pipetting up and down.
6. Incubate at 30° for 15 min.
7. Add 150 μl lithium acetate-PEG solution and mix well by pipetting up and down.
8. Incubate at 30° for 15 min (can be up to 1.5 h).
9. Add 17 μl DMSO, and heat shock in a 42° waterbath for 15 min.
10. Pellet cells by centrifugation (in Microfuge) for 2 min at 3000g.
11. Remove lithium acetate-PEG solution. Try not to disturb the cell pellet; however, some cell loss is inevitable.
12. Add 600 μl of YPD liquid media, and incubate at 30° for 4 h in a shaker at 200 rpm.
13. Pellet cells by centrifugation (in a Microfuge) for 2 min at 3000g, and remove 500 μl of supernatant. Plate the remaining ~100 μl cells on SD-ura plates and incubate plates at 30° for 2 days. (Colonies should appear on the second day.)

3.3.1. Confirmation of TAP Tag insertion
3.3.1.1. Buffers, reagents, and solutions

Cbf5-internal-5′ primer: 5′ GTC TGA AGA CGG TGA TTC TGA 3′
Cbf5-downstream-3′ primer: 5′ AAA GAA TAC TAC AAG TCG TTG 3′
DNA prep buffer: 2% Triton X100, 1% SDS, 1 mM EDTA, and 0.01 M Tris-HCl, pH 8.0
Glass beads
TE (pH 8.0): 10 mM Tris-HCl, pH 8.0, 1 mM EDTA, pH 8.0
RNase A: 10 mg/ml
Ammonium acetate: 6 M
Yeast genomic DNA
Agarose (Promega)
YPD media, autoclaved distilled water, PCA, ethanol, 10× polymerase buffer, dNTPs, Taq DNA polymerase (as described earlier)

3.4. Protocol

1. Pick a single colony from the SD-ura plate and inoculate into 5 ml of YPD and incubate overnight in a 30° shaker at 200 rpm.
2. Transfer 500 μl of cells to a clean 1.5-ml Microfuge tube and store at −80°. Pellet the remaining cells by centrifugation (in a SS-34 rotor) for 2 min at 3000g.
3. Wash cells with 500 μl autoclaved distilled water.
4. Add 200 μl DNA prep buffer, 200 μl PCA, and 0.3 g glass beads.
5. Vortex for 3 min.

6. Add 200 μl TE (pH 8.0) and Microfuge for 5 min at 13,000g.
7. Transfer supernatant to a clean 1.5-ml Microfuge tube.
8. Add 1 ml 100% ethanol and vortex 30 s.
9. Microfuge for 10 min at 13,000g and discard ethanol.
10. Resuspend pellet in 400 μl TE (pH 8.0).
11. Add 3 μl 10 mg/ml RNase A and incubate for 15 min at 37°.
12. Add 10 μl 6 M ammonium acetate and 1 ml 100% ethanol.
13. Microfuge at 13,000g for 5 min at 4°, and discard ethanol.
14. Air dry for 5 min and resuspend the pellet in 50 μl TE (pH 8.0). Quantify DNA by UV/Vis spectroscopy.
15. In a 0.5-ml PCR tube mix 1 μl of yeast genomic DNA, 5 μl 10× polymerase buffer, 2 μl of 10 mM dNTPs, 1 μl of Taq DNA polymerase, 2 μl of 10 pM Cbf5-internal-5′ primer, 2 μl of 10 pM Cbf5-downstream-3′ primer, and 37 μl of autoclaved distilled water.
16. Perform PCR using the cycle conditions described previously (steps 2–5 in "Generation of TAP Construct by PCR").
17. Run 5 μl of the PCR reaction on a 1% agarose gel. (If the TAP tag is inserted, PCR should generate a specific band with the expected size.)

3.4.1. TAP Tag purification of Cbf5p and associated components
3.4.1.1. Buffers, reagents, and solutions

SD-ura liquid media: 7.14 g of SX-ura mix (as described before) and 20 g of dextrose dissolved in 1 liter of water and autoclaved.

Extraction buffer: 50 mM Tris-HCl, pH 7.5, 1 mM EDTA, 5 mM DTT, 10% glycerol, 1 M NaCl

French press wash buffer: 25 mM Tris-HCl, pH 7.5, 0.5 mM EDTA, 2.5 mM DTT 5% glycerol, 0.5 mM NaCl

IPP-0 buffer: 10 mM Tris-HCl, pH 7.5, 0.1% NP-40

IPP-150 buffer: 10 mM Tris-HCl, pH 7.5, 150 mM NaCl, 0.1% NP-40

IgG beads (Amersham)

3C cleavage buffer: 10 mM Tris-HCl, pH 7.5, 150 mM NaCl, 2.5% glycerol

Talon resin (Invitrogen)

3C protease: 0.7 μg/μl

Imidazol: 300 mM

3.5. Protocol

1. With the 500 μl of cells stored at −80° (see step 2 in "Confirmation of TAP Tag Insertion"), inoculate 5 ml of SD-ura liquid media with a small aliquot of the frozen cells. Incubate overnight in a 30° shaker.
2. Inoculate 1 liter of YPD liquid media with the 5 ml of yeast culture. Incubate overnight in a 30° shaker.
3. Pellet cells by centrifugation in a SLA-3000 rotor at 3000g.

4. Add 1 volume of extraction buffer to 1 volume of cells.
5. Apply sample to the French press cell and bring the cell under the desired pressure (7000–10,000 psi). While maintaining the pressure, adjust the out flow rate to 1 drop every second. Collect the cell lysate in a flask that is kept on ice. Repeat this step two more times (rinsing the French press with 30 ml of French press wash buffer between uses).
6. Centrifuge extract at 20,000g for 40 min.
7. Transfer supernatant to a fresh 50-ml conical tube.
8. Add 5.7 volumes of IPP-0 buffer.
9. Add 200 μl of IgG beads slurry (bed volume 100 μl, prewashed and saturated with IPP-150).
10. Rotate for 2 h at 4°.
11. Briefly centrifuge (in a desktop centrifuge) the mixture at $\sim$3000g to pellet the beads.
12. Transfer the beads to a 1.5-ml Microfuge tube.
13. Wash beads with 1 ml of IPP-150 buffer by centrifugation (in a Microfuge) of mixture at 3000g for 1 min. Repeat three times.
14. Wash beads with 3C cleavage buffer. Repeat three times. After the final wash, remove the supernatant, leaving approximately 500 μl, containing 3C cleavage buffer ($\sim$300 μl) and IgG beads ($\sim$200 μl bed volume), in the Microfuge tube.
15. Transfer the 500 μl beads slurry to a new 0.6-ml tube.
16. Add 12.5 μl of 0.7 μg/μl 3C protease.
17. Rotate overnight at 4°.
18. Pellet the beads in a Microfuge at 3000g for 1min. Transfer the supernatant (approximately 300 μl) into a new 1.5-ml Microfuge tube.
19. Add 20 μl TALON resin (bed volume 10 μl) that has been equilibrated with 3C cleavage buffer.
20. Rotate for 1 h at 4°.
21. Briefly Microfuge ($\sim$30 s at $\sim$5000g) the sample to pellet the resin and remove supernatant.
22. Wash the TALON resin three times with 200 μl 3C cleavage buffer containing 20 mM imidazole (pH 7.0).
23. Elute the protein with 50 μl 3C cleavage buffer containing 200 mM imidazole (pH 7.0) by rotating at RT for 15 min.
24. Microfuge the mixture at 2000g at RT for 2 min. Transfer the supernatant to a new tube, and repeat the elution. (Two elutions will usually elute approximately 80% of the protein.)
25. Pool the two eluants into dialysis tubing (MWCO:12–14000). Dialyze the eluant against 2 liters of 3C cleavage buffer at 4° overnight.
26. After dialysis, add glycerol in a 1:1 ratio and store the sample (purified box H/ACA RNPs) at −20°.

4. Immunoprecipitation of Box H/ACA RNPs

4.1. Overview

Immunoprecipitation (IP) of box H/ACA RNPs is also very useful for structural studies of the RNP and guide RNA identification. This technique uses an antibody specific to an antigen to precipitate the antigen out of solution. After the antibody binds the antigen, the complex is brought out of solution by binding to Protein-A Sepharose beads. This method can be used to precipitate the entire box H/ACA RNP complex or individual subunits. Described here are some examples of how immunoprecipitation has been used to study box H/ACA RNPs.

Immunoprecipitation can be used to identify new box H/ACA guide RNAs. The RNA can be isolated from the complex and identified after precipitating the box H/ACA RNPs from cell extracts. By immunoprecipitating the box H/ACA RNP with an antibody to Gar1, the Kiss group identified 61 novel guide RNAs in HeLa cell extracts (Kiss *et al.*, 2004). The sequence of the pseudouridylation pockets of these RNA were predicted by computer modeling, allowing their targets in rRNA and snRNA to be identified. This study demonstrated that most pseudouridines present in human rRNA could be attributed to box H/ACA RNAs (Kiss *et al.*, 2004).

Immunoprecipitation of box H/ACA RNPs has also been used to identify the components of the RNPs and to study the interactions between the protein subunits and the snoRNA. Nap57, the mammalian Cbf5 homolog, was originally identified by immunoprecipitation by use of a Nopp140 antibody (Meier and Blobel, 1994). In yeast, immunoprecipitation of box H/ACA RNPs by use of antibodies to Gar1p and Cbf5p identified them as components of the RNP complex (Balakin *et al.*, 1996; Watkins *et al.*, 1998). By use of an anti-Gar1 antibody, the Filipowicz group immunoprecipitated box H/ACA RNPs and telomerase from HeLa nuclear and whole cell extracts, demonstrating that Gar1 is a component of the RNP complex and telomerase in human cells (Dragon *et al.*, 2000).

In addition to identifying the individual components of box H/ACA RNPs, immunoprecipitation has been used to identify the minimal number of protein subunits required for box H/ACA RNP activity. Immunoprecipitation with Nap57 antibodies was used to demonstrate that Nopp140, a box H/ACA RNP associated protein, was not required for activity (Wang *et al.*, 2002). In another study that used Nap57 antibodies, it was demonstrated that Naf1, although not a core protein, does associate with box H/ACA RNPs and is likely involved in biogenesis of the complex (Darzacq *et al.*, 2006).

Once the core components had been established, studies began to focus on probing the interactions between the subunits. Immunoprecipitation

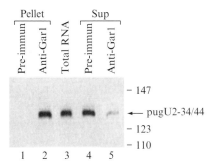

Figure 11.5 Isolation of *Xenopus* box H/ACA RNAs by anti-Gar1 immunoprecipitation. Anti-Gar1 antibodies (lanes 2 and 5) or pre-immune serum (lanes 1 and 4) prebound to protein A-Sepharose beads were incubated with *Xenopus* oocyte nuclear extracts. The beads (pellet, lanes 1 and 2) were then washed and the coprecipitated RNAs eluted. The RNAs were also recovered from the unbound fraction (sup, lanes 4 and 5). The RNAs recovered from both bound (pellet) and unbound (sup) fractions were resolved on a denaturing gel and probed with radiolabeled DNA oligonucleotides complementary to pugU2–34/44, a *Xenopus* box H/ACA RNA. Lane 3 is a control that contained total oocyte nuclear RNA. The arrow indicates the pugU2–34/44 signal.

with Nap57 antibodies was used to show that Gar1 is not necessary for the assembly of the complex (Dragon et al., 2000). The Meier group immunoprecipitated *in vitro* translated box H/ACA RNP subunits in various combinations with the Nap57 antibody (Wang and Meier, 2004). The results demonstrated that Nap57 interacted directly with Nop10. However, Nhp2 could only be precipitated in combination with Nop10, therefore demonstrating no direct interaction between Nap57 and Nhp2.

These studies demonstrate a few of the ways in which immunoprecipitation has been used for box H/ACA RNP research. Our laboratory has used a Gar1 antibody to immunoprecipitate box H/ACA RNPs in *Xenopus* oocyte nuclear extract to isolate the RNA component of the complex (Fig. 11.5).

4.1.1. Prebind antibody to protein A-sepharose
4.1.1.1. Buffers, reagent, and solutions

Net-2: 50 mM Tris-HCl, pH 7.5, 150 mM NaCl, and 0.05% NP-40
Protein A-Sepharose (Amersham)
Anti-Gar1 antibody, 120 μg/μl (Dragon *et al.*, 2000)

4.2. Protocol

1. At room temperature, swell 2.5 mg Protein A-Sepharose beads (PAS) in 500 μl Net-2. Mix Protein A-Sepharose suspension often to prevent settling of the resin.

2. Add 10 μl anti-Gar1 antibody to the swollen Protein A-Sepharose.
3. Rotate 1–2 h at room temperature.
4. Pellet PAS bound anti-Gar1 for approximately 5 s in Microfuge at 5000g.
5. Wash three times with 1 ml of Net-2, resuspending and microfuging each time.
6. Resuspend the antibody-bound beads in 200–300 μl Net-2.

4.2.1. Binding of box H/ACA RNPs in extract to antibody
4.2.1.1. Buffers, reagent, and solutions
4.2.1.1.1. Xenopus oocyte nuclear extracts

IPP: 0.5 M NaCl, 10 mM Tris-HCl, pH 8.0, 0.1% NP-40
Net-2, G50 buffer, proteinase K, PCA, 100% ethanol and RNase-free autoclaved distilled water (as described)

1. Add approximately 20 μl Xenopus oocyte nuclear extract (prepared from ~200–300 oocytes) to the protein A-Sepharose bound anti-Gar1 antibody (in 200–300 μl of Net-2).
2. Rotate at least 1 h at 4°.
3. Microfuge at 5000g for 30 s.
4. Wash the pellet five times with 1 ml of Net-2 or IPP as before. Note that IPP is a more stringent washing buffer (0.5 M NaCl), compared with Net-2 (150 mM NaCl). Washing with IPP sometimes generates better/cleaner results.
5. Resuspend the pellet/beads in 250 μl of G-50 buffer. Vortex occasionally.
6. Add 2 μl of 20 mg/ml Proteinase K.
7. Incubate at 42° for 15–30 min.
8. Extract three times with 500 μl PCA, as described earlier.
9. Add 600 μl of 100% ethanol, and precipitate box H/ACA RNAs, as described earlier.
10. Detect known box H/ACA RNAs of interest by Northern analysis (Zhao et al., 2002) or identify new box H/ACA RNAs by 3'-end ^{32}pCp labeling followed by gel-purification and sequencing (Peattie, 1979), as described earlier.

ACKNOWLEDGMENTS

We thank members of the Yu laboratory for discussion and inspiration. The work described in this article was supported by grants GM62937 and GM078223 (to Y.-T. Yu) from the National Institutes of Health.

REFERENCES

Alexandrov, A., Grayhack, E. J., and Phizicky, E. M. (2005). tRNA m7G methyltransferase Trm8p/Trm82p: Evidence linking activity to a growth phenotype and implicating Trm82p in maintaining levels of active Trm8p. RNA 11, 821–830.

Badis, G., Fromont-Racine, M., and Jacquier, A. (2003). A snoRNA that guides the two most conserved pseudouridine modifications within rRNA confers a growth advantage in yeast. *RNA* **9,** 771–779.

Balakin, A. G., Smith, L., and Fournier, M. J. (1996). The RNA world of the nucleolus: Two major families of small RNAs defined by different box elements with related functions. *Cell* **86,** 823–834.

Darzacq, X., Kittur, N., Roy, S., Shav-Tal, Y., Singer, R. H., and Meier, U. T. (2006). Stepwise RNP assembly at the site of H/ACA RNA transcription in human cells. *J. Cell Biol.* **173,** 207–218.

Decatur, W. A., and Fournier, M. J. (2002). rRNA modifications and ribosome function. *Trends Biochem. Sci.* **27,** 344–351.

Decatur, W. A., and Fournier, M. J. (2003). RNA-guided nucleotide modification of ribosomal and other RNAs. *J. Biol. Chem.* **278,** 695–698.

Donis-Keller, H. (1980). Phy M: An RNase activity specific for U and A residues useful in RNA sequence analysis. *Nucleic Acids Res.* **8,** 3133–3142.

Donmez, G., Hartmuth, K., and Luhrmann, R. (2004). Modified nucleotides at the 5′ end of human U2 snRNA are required for spliceosomal E-complex formation. *RNA* **10,** 1925–1933.

Dragon, F., Pogacic, V., and Filipowicz, W. (2000). *In vitro* assembly of human H/ACA small nucleolar RNPs reveals unique features of U17 and telomerase RNAs. *Mol. Cell. Biol.* **20,** 3037–3048.

Ganot, P., Bortolin, M. L., and Kiss, T. (1997). Site-specific pseudouridine formation in preribosomal RNA is guided by small nucleolar RNAs. *Cell* **89,** 799–809.

Grosjean, H., Sprinzl, M., and Steinberg, S. (1995). Posttranscriptionally modified nucleosides in transfer RNA: Their locations and frequencies. *Biochimie* **77,** 139–141.

Hoang, C., and Ferre-D'Amare, A. R. (2001). Cocrystal structure of a tRNA Psi55 pseudouridine synthase: Nucleotide flipping by an RNA-modifying enzyme. *Cell* **107,** 929–939.

Huttenhofer, A., Brosius, J., and Bachellerie, J. P. (2002). RNomics: Identification and function of small, non-messenger RNAs. *Curr. Opin. Chem. Biol.* **6,** 835–843.

King, T. H., Liu, B., McCully, R. R., and Fournier, M. J. (2003). Ribosome structure and activity are altered in cells lacking snoRNPs that form pseudouridines in the peptidyl transferase center. *Mol. Cell* **11,** 425–435.

Kiss, A. M., Jady, B. E., Bertrand, E., and Kiss, T. (2004). Human box H/ACA pseudouridylation guide RNA machinery. *Mol. Cell Biol.* **24,** 5797–5807.

Li, L., and Ye, K. (2006). Crystal structure of an H/ACA box ribonucleoprotein particle. *Nature* **443,** 302–307.

Ma, X., Yang, C., Alexandrov, A., Grayhack, E. J., Behm-Ansmant, I., and Yu, Y. T. (2005). Pseudouridylation of yeast U2 snRNA is catalyzed by either an RNA-guided or RNA-independent mechanism. *EMBO J.* **24,** 2403–2413.

Maden, B. E. (1990). The numerous modified nucleotides in eukaryotic ribosomal RNA. *Prog. Nucleic Acid Res. Mol. Biol.* **39,** 241–303.

Massenet, S., Mougin, A., and C., B. (1998). Posttranscriptional modifications in the U small nuclear RNAs. *In* "Modification and Editing of RNA" (H. Grosjean, ed.), pp. 201–228. ASM Press, Washington, DC.

Meier, U. T., and Blobel, G. (1994). NAP57, a mammalian nucleolar protein with a putative homolog in yeast and bacteria. *J. Cell Biol.* **127,** 1505–1514.

Ni, J., Tien, A. L., and Fournier, M. J. (1997). Small nucleolar RNAs direct site-specific synthesis of pseudouridine in ribosomal RNA. *Cell* **89,** 565–573.

Ofengand, J., and Fournier, M. (1998). The pseudouridine residues of rRNA: Number, location, biosynthesis, and function. *In* "Modification and Editing of RNA" (H. Grosjean and H. Benne, eds.), pp. 229–253. ASM Press, Washington, DC.

Patton, J. R., Jacobson, M. R., and Pederson, T. (1994). Pseudouridine formation in U2 small nuclear RNA. *Proc. Natl. Acad. Sci. USA* **91,** 3324–3328.

Peattie, D. A. (1979). Direct chemical method for sequencing RNA. *Proc. Natl. Acad. Sci. USA* **76,** 1760–1764.

Reddy, R., and Busch, H. (1988). Small nuclear RNAs: RNA sequences, structure, and modifications. *In* "Structure and Function of Major and Minor Small Nuclear Ribonucleoprotein Particles" (M. L. Birnsteil, ed.), pp. 1–37. Springer-Verlag Press, Heidelberg.

Rigaut, G., Shevchenko, A., Rutz, B., Wilm, M., Mann, M., and Seraphin, B. (1999). A generic protein purification method for protein complex characterization and proteome exploration. *Nat. Biotechnol.* **17,** 1030–1032.

Schattner, P., Decatur, W. A., Davis, C. A., Ares, M., Jr., Fournier, M. J., and Lowe, T. M. (2004). Genome-wide searching for pseudouridylation guide snoRNAs: Analysis of the *Saccharomyces cerevisiae* genome. *Nucleic Acids Res.* **32,** 4281–4296.

Torchet, C., Badis, G., Devaux, F., Costanzo, G., Werner, M., and Jacquier, A. (2005). The complete set of H/ACA snoRNAs that guide rRNA pseudouridylations in *Saccharomyces cerevisiae*. *RNA* **11,** 928–938.

Wang, C., and Meier, U. T. (2004). Architecture and assembly of mammalian H/ACA small nucleolar and telomerase ribonucleoproteins. *EMBO J.* **23,** 1857–1867.

Wang, C., Query, C. C., and Meier, U. T. (2002). Immunopurified small nucleolar ribonucleoprotein particles pseudouridylate rRNA independently of their association with phosphorylated Nopp140. *Mol. Cell Biol.* **22,** 8457–8466.

Watkins, N. J., Gottschalk, A., Neubauer, G., Kastner, B., Fabrizio, P., Mann, M., and Luhrmann, R. (1998). Cbf5p, a potential pseudouridine synthase, and Nhp2p, a putative RNA-binding protein, are present together with Gar1p in all H BOX/ACA-motif snoRNPs and constitute a common bipartite structure. *RNA* **4,** 1549–1568.

Yu, Y. T. (2006). The most complex pseudouridylase. *Structure* **14,** 167–168.

Yu, Y. T., Shu, M. D., and Steitz, J. A. (1998). Modifications of U2 snRNA are required for snRNP assembly and pre-mRNA splicing. *EMBO J.* **17,** 5783–5795.

Yu, Y. T., Terns, R. M., and Terns, M. P. (2005). Mechanisms and functions of RNA-guided RNA modification. *In* "Topics in Current Genetics" (H. Grosjean, ed.), Vol. 12, pp. 223–262. Springer-Verlag, New York.

Zebarjadian, Y., King, T., Fournier, M. J., Clarke, L., and Carbon, J. (1999). Point mutations in yeast CBF5 can abolish *in vivo*. Pseudouridylation of rRNA. *Mol. Cell Biol.* **19,** 7461–7472.

Zhao, X., Li, Z. H., Terns, R. M., Terns, M. P., and Yu, Y. T. (2002). An H/ACA guide RNA directs U2 pseudouridylation at two different sites in the branchpoint recognition region in *Xenopus oocytes*. *RNA* **8,** 1515–1525.

Zhao, X., and Yu, Y. T. (2004). Pseudouridines in and near the branch site recognition region of U2 snRNA are required for snRNP biogenesis and pre-mRNA splicing in *Xenopus oocytes*. *RNA* **10,** 681–690.

Zhao, X., and Yu, Y. T. (2006). Incorporation of 5-fluorouracil into U2 snRNA blocks pseudouridylation and pre-mRNA splicing *in vivo*. *Nucleic Acids Res.* In press.

CHAPTER TWELVE

In Vitro Reconstitution and Affinity Purification of Catalytically Active Archaeal Box C/D sRNP Complexes

Keith Gagnon, Xinxin Zhang, *and* E. Stuart Maxwell

Contents

Department of Molecular and Structural Biochemistry, North Carolina State University, Raleigh, North Carolina

Methods in Enzymology, Volume 425
ISSN 0076-6879, DOI: 10.1016/S0076-6879(07)25012-8

Abstract

Archaeal box C/D RNAs guide the site-specific 2′-O-methylation of target nucleotides in ribosomal RNAs and tRNAs. *In vitro* reconstitution of catalytically active box C/D RNPs by use of *in vitro* transcribed box C/D RNAs and recombinant core proteins provides model complexes for the study of box C/D RNP assembly, structure, and function. Described here are protocols for assembly of the archaeal box C/D RNP and assessment of its nucleotide modification activity. Also presented is a novel affinity purification scheme that uses differentially tagged core proteins and a sequential three-step affinity selection protocol that yields fully assembled and catalytically active box C/D RNPs. This affinity selection protocol can provide highly purified complex in sufficient quantities not only for biochemical analyses but also for biophysical approaches such as cryoelectron microscopy and X-ray crystallography.

1. INTRODUCTION

The box C/D RNAs constitute large populations of small noncoding RNAs found in both eukaryotic and archaeal organisms, where their primary function is to guide the site-specific 2′-O-methylation of nucleotides located in various target RNAs. Guide sequences within each box C/D RNA base pair to complementary sequences in the target RNA, thereby designating specific nucleotides for posttranscriptional modification (Bachellerie *et al.*, 2002; Maxwell and Fournier, 1995; Terns and Terns, 2002; Tollervey, 1996). Archaeal box C/Ds RNAs are defined by highly conserved boxes C and D located near their 5′ and 3′ termini and internally located C′ and D′ boxes (Dennis *et al.*, 2001; Omer *et al.*, 2000). Both the external boxes C and D and the internal C′ and D′ boxes fold to establish kink-turn or K-turn motifs. It is these highly structured K-turn (box C/D) and K-loop (C′/D′) motifs that serve as binding platforms for the box C/D RNP core proteins (Hama and Ferre-D'Amare, 2004; Moore *et al.*, 2004; Omer *et al.*, 2002; Suryadi *et al.*, 2005; Tran *et al.*, 2003). Archaeal core proteins ribosomal protein L7, Nop56/58, and fibrillarin bind both box C/D and C′/D′ motifs to establish individual RNP complexes. It is the core proteins, working in concert with the guide regions located immediately upstream of the D and D′ boxes, that direct the 2′-O-methylation of targeted nucleotides (Omer *et al.*, 2002; Tran *et al.*, 2003).

Investigation of box C/D RNP structure and function has been greatly facilitated in recent years with the establishment of *in vitro* systems that assemble catalytically active archaeal sRNP complexes by use of *in vitro* transcribed sRNAs and recombinant sRNP core proteins purified from bacterial expression systems. Several laboratories, including our own, have used these *in vitro* assembled complexes to investigate the assembly, structure, and methylation function of this RNA-protein enzyme (for a review see Dennis and Omer, 2006). Presented here are detailed protocols for the assembly of a *Methanocaldococcus jannaschii* box C/D sRNP and the assessment of the complex's methylation capabilities. Also described here is a novel sRNP isolation protocol involving three sequential affinity selection steps that use differentially tagged fibrillarin core proteins and an oligonu-cleotide complementary to the sRNA. This purification scheme yields highly purified and fully assembled archaeal box C/D sRNPs in sufficient quantities for not only biochemical analyses but also for biophysical ap-proaches such as cryoelectron microscopy and X-ray crystallography. Although specifically designed for the *M. jannaschii* sR8 box C/D sRNP, this approach can be easily modified for the isolation of other sRNP complexes assembled either *in vitro* or possibly *in vivo* in the cell.

2. Cloning, Expression, and Preparation of *M. jannaschii* Box C/D sRNP Core Proteins

Genes encoding the *M. jannaschii* core proteins L7, Nop56/58, and fibrillarin were PCR-amplified from isolated genomic DNA, inserted into bacterial expression vectors, and recombinant proteins expressed and then purified by use of affinity and cation-exchange chromatography as previ-ously outlined (Tran *et al.*, 2003). The cloning, expression, and purification of each core protein are presented here in greater detail.

2.1. Cloning of *M. jannaschii* L7, Nop56/58, and fibrillarin genes

Ribosomal protein L7, fibrillarin, and Nop56/58 gene coding sequences are PCR amplified from *M. jannaschii* genomic DNA. DNA oligonucleotide primers are synthesized for each core protein gene and used for PCR amplification. The L7 and fibrillarin upstream and downstream primers contain 5' terminal Nde*I* and Bam*HI* restriction sites, respectively, whereas the Nop56/58 primers contain Nco*I* and Bam*HI* restriction sites. After PCR amplification, resulting DNA fragments are digested with the appro-priate restriction endonucleases and ligated into similarly digested pET28a (Novagen) protein expression vectors by standard methods. Plasmid con-structs are transformed into *E. coli* cells and colonies screened for plasmids containing core protein sequences by PCR amplification by use of L7-, Nop56/58-, or fibrillarin-specific primers. Selected colonies are cultured in liquid LB broth with antibiotics and plasmid DNA harvested with Wizard Midiprep Kits (Promega). Expression of recombinant L7 and fibrillarin proteins produces N-terminal and thrombin-cleavable 6×-histidine tags. The expressed Nop56/58 recombinant protein is untagged. To generate FLAG-tagged fibrillarin, the fibrillarin coding sequence is PCR amplified from genomic *M. jannaschii* DNA and ligated into the pET28a vector engineered to encode a FLAG peptide (DYKDDDDK) in place of the His tag. The sequences of all core protein plasmid constructs are verified by DNA sequencing.

2.2. Recombinant core protein expression in bacterial cells

Core proteins are expressed in the Rosetta (DE3) strain of *E. coli* (Novagen), a protein expression host carrying a plasmid encoding rare tRNA codons to help enhance levels of recombinant protein synthesis. Competent cells are transformed with individual plasmids and spread onto LB agar plates containing kanamycin (30 μg/ml) and chloramphenicol (34 μg/ml). A single cell colony is selected and grown in 1 liter of LB broth containing antibiotics with shaking at 37° until the cell culture reaches an optical density at 600 nm (OD_{600}) of 0.8. Expression of His-L7, His-fibrillarin, and FLAG-fibrillarin is initiated by the addition of isopropyl-β-D-thiogalactopyranoside (IPTG) to a final concentration of 1 mM and shaking is continued for 4 h at 37°. For expression of Nop56/58, the transformed cell culture is grown at 37° with shaking to an OD_{600} of 0.5 and then moved to 15° and shaken for an additional 30 min. Nop56/58 expression is induced with the addition of IPTG (0.4 mM) and shaking continued at 15° for an additional 24 h. IPTG induces the expression of a genomically encoded bacteriophage T7 RNA polymerase gene under control of the *lacUV5* promoter. The expressed T7 RNA polymerase transcribes the plasmid-encoded core protein genes under control of the pET28a vector's T7 promoter.

After recombinant core protein expression, cells are pelleted by centrifugation at 10,000g for 10 min at 4° and then resuspended in 5 ml of buffer D (20 mM HEPES, pH 7.0, 100 mM NaCl, 3 mM MgCl$_2$, 20% glycerol [w/v]) per gram of cell paste. A protease inhibitor cocktail (Cocktail Set VII, Calbiochem) at a final concentration of 1% (w/v) and 10 U/ml of Benzonase Nuclease (Novagen) are added to prevent protein degradation and promote nucleic acid degradation, respectively. To lyse cells, the cell suspension is sonicated (Fisher Sonic Dismembrator, Model 150) at maximum power for 30 sec (×3) on ice, with 1-min cooling intervals. The lysate is then mixed by rocking for 1 hr at room temperature to facilitate nucleic acid degradation. Degradation of nucleic acids at this step is crucial for preparation of the Nop56/58 core protein, because contaminating nucleic acids bind this core protein tightly, resulting in protein precipitation and greatly reduced yields. Cell lysates are then separated into soluble and insoluble protein fractions by ultracentrifugation at 38,000g for 30 min at 4°. Most recombinant protein remains in the soluble fraction. Nop56/58 core protein is expressed at lower levels than either L7 or fibrillarin, and only 50% of the total Nop56/58 protein is soluble. However, without low temperature expression, nearly all the Nop56/58 protein is found in the insoluble fraction. These soluble fractions can be stored at $-80°$ for several weeks before continuing with chromatographic enrichment of the individual core proteins.

2.3. Affinity chromatographic isolation of L7 and fibrillarin core proteins

His-tagged L7 (14.7 kDa, 13 kDa without the His tag) and His-tagged fibrillarin (28 kDa) are isolated from soluble protein fractions to >90% purity by use of a single nickel-nitrilotriacetic acid (Ni-NTA) metal-affinity chromatography step (Fig. 12.1A, lanes 1 and 2). The soluble lysate fraction containing recombinant, His-tagged L7 or fibrillarin protein is applied at room temperature to 4 ml of buffer D–equilibrated Ni^{2+}-charged Ni-NTA His-bind resin (Novagen) suspended in a 20 ml (1.5 × 10 cm) chromatography column. This resin has an estimated binding capacity of 8 mg of protein/ml of resin. The column is then washed with 25 bed volumes of buffer D_{300} (buffer D with 300 mM NaCl) containing 40 mM imidazole. Glycerol may be omitted from the wash buffer to increase column flow rates. Bound core protein is eluted with 2 bed volumes of buffer D containing 250 mM imidazole and collected in 1–2 ml fractions. Elution fractions containing the core protein are pooled and dialyzed against 100 volumes of buffer D overnight at $4°$ to remove imidazole. The His-tag may be removed at this point by digesting the isolated core protein with thrombin (~5 U/ml) either before or during dialysis. Protein concentrations are estimated by UV absorbance at 280 nm by use of Beer's Law and extinction coefficients of 5,240 cm^{-1} M^{-1} (experimentally determined) and 29,900 cm^{-1} M^{-1} (calculated) for L7 and fibrillarin, respectively. Use of dye-based assays, such as the Bradford assay to determine protein concentrations, is suitable for fibrillarin. However, L7 is not proportionately stained with respect to protein concentration.

FLAG-tagged fibrillarin (28 kDa) is affinity selected from the soluble cell sonicate with ANTI-FLAG M2 affinity agarose (Sigma). Fibrillarin lysate is applied twice at room temperature to 5 ml of ANTI-FLAG M2 agarose resin previously equilibrated in buffer D and packed in a 20-ml chromatography column (estimated resin binding capacity of 0.5–1.0 mg of fibrillarin per ml of resin). The affinity column with bound fibrillarin is then washed with 25 bed volumes of buffer D_{300} (300 mM NaCl). FLAG-fibrillarin can be eluted with 3 bed volumes of buffer D containing 100 μg/ml of FLAG peptide (Sigma). However, an alternative and more economical approach is to elute bound FLAG-fibrillarin with 0.1 M glycine, pH 3.5, containing 100 mM NaCl. The eluted fractions are collected into 1.5-ml Microfuge tubes containing 1/10 the elution volume of 1 M Tris, pH 8.0 (the Tris pH

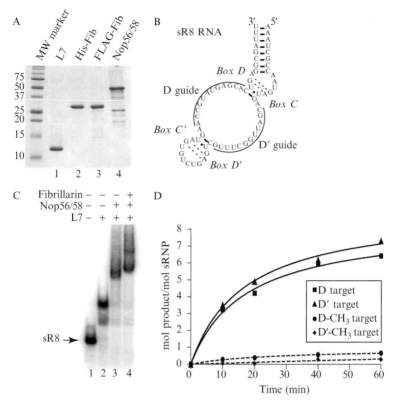

Figure 12.1 Archaeal box C/D recombinant core protein isolation, *in vitro* sRNP assembly, and *in vitro* sRNP-guided nucleotide 2′-O-methylation. (A) Coomassie Brilliant Blue–stained SDS-polyacrylamide gel of isolated *M. jannaschii* box C/D sRNP core proteins used for *in vitro* sRNP assembly. (B) Sequence and folded secondary structure of the *M. jannaschii* sR8 box C/D sRNA used for *in vitro* sRNP assembly. (C) Electrophoretic mobility-shift analysis revealing the hierarchal binding of the sRNP core proteins to 5′-radiolabeled sR8 sRNA in assembly of the box C/D sRNP. (D) *In vitro* methylation of D and D′ target RNAs by the *in vitro* assembled sR8 box C/D sRNP. The lack of methylation at the target nucleotide when this nucleotide is already methylated at the ribose 2′ position (D-CH₃ and D′-CH₃ targets) demonstrates nucleotide-specific modification of each target RNA by the *in vitro* assembled sRNP.

8.0 buffer reestablishes the eluate pH to approximately 7.0). Both elution protocols yield FLAG-tagged fibrillarin that efficiently assembles box C/D sRNPs that are active for sRNP-guided nucleotide methylation. ANTI-FLAG M2 resin has a lower binding capacity than the Ni-NTA resin. Therefore, this chromatography step is typically repeated several times with the same fibrillarin lysate, and the eluted fibrillarin fractions are pooled. Pooled fibrillarin fractions are then concentrated 10–15 fold using sequentially a 50-ml Amicon Centricon Concentrator (10,000 MWCO, Millipore) and a 15-ml spin concentrator (Vivaspin 15R, Vivascience) before dialysis against 100 volumes of buffer D for 16 h at 4°. This single-affinity selection step yields FLAG-tagged fibrillarin at >85% homogeneity (Fig. 12.1A, lane 3). FLAG-tagged fibrillarin concentrations are determined either by a Bradford assay or UV absorbance at 280 nm with a calculated extinction coefficient of 31,400 cm^{-1} M^{-1}. Isolated L7 and fibrillarin protein preparations can be stored at $-80°$ for up to 1 y.

2.4. Isolation of Nop56/58 core protein by cation-exchange chromatography

Purification of recombinant Nop56/58 (48 kDa) by cation-exchange chromatography takes advantage of this core protein's strong positively charged character. Initial chromatographic isolation of Nop56/58 used SP Sepharose Fast Flow cation-exchange resin (Amersham Biosciences) (Tran *et al.*, 2003). We now routinely use heparin agarose, because this resin more efficiently binds Nop56/58, has a higher binding capacity, and yields a more homogeneous protein preparation. Affinity-tagged versions of Nop56/58 previously tested in our laboratory did not bind their affinity resins. Notably, however, an N-terminally His-tagged version of Nop56/58 has been successfully purified with Ni-NTA metal affinity chromatography (developed by the Brown laboratory, Wake Forest University), and this affinity-purified Nop56/58 is comparable to protein isolated by use of cation exchange chromatography (Zhang *et al.*, 2006).

The Nop56/58 lysate is applied twice at room temperature to 10 ml of buffer D–equilibrated heparin agarose (MP Biomedicals) packed in a 40-ml chromatography column (1.5 × 20 cm). The resin with bound Nop56/58 is then washed with 30 bed volumes of buffer D$_{800}$ (800 mM NaCl). Bound Nop56/58 is eluted with 3 bed volumes of high-salt buffer D$_{1300}$ (1.3 M NaCl). The Nop56/58 eluate is diluted with buffer D to a final NaCl concentration of 500 mM and then concentrated to approximately 2 ml by use of the concentration techniques outlined earlier for FLAG-tagged fibrillarin. Concentrated Nop56/58 is then centrifuged at 14,000g at room temperature for 15 min to pellet insoluble protein. Nop56/58 is stored in buffer D$_{500}$ to avoid aggregation and precipitation, although buffers with lower salt concentrations are suitable for lower Nop56/58 concentrations. This single purification step typically yields Nop56/58 protein to approximately 60–70% homogeneity (Fig. 12.1A, lane 4). A Bradford assay estimates the concentration of isolated Nop56/58, and this core protein can be stored at $-80°$ for up to 1 y.

3. CLONING AND *IN VITRO* TRANSCRIPTION OF ARCHAEAL BOX C/D sRNAs

The *M. jannaschi* sR8 box C/D sRNA (Fig. 12.1B) gene was originally PCR amplified from genomic DNA and then cloned into a pUC19 plasmid (Tran *et al.*, 2003). *In vitro* transcription of sR8 sRNA is accomplished by first generating DNA templates from this plasmid by PCR amplification. The upstream PCR primer possesses a T7 promoter sequence (22 nucleotides) at the 5′ terminus followed by the first 22 nucleotides of sR8 coding sequence, whereas the downstream primer is complementary to the last 23 nucleotides of sR8 coding sequence. PCR amplification of the sR8 pUC19 plasmid by use of these primers generates a DNA template for *in vitro* T7 RNA polymerase transcription. A standard 100 μl PCR amplification reaction of 35 cycles produces approximately 8–10 μg of DNA template from 50 ng of plasmid DNA. Ampliscribe Flash T7 Transcription Kits (Epicentre) are used to synthesize sRNA transcripts following the manufacturer's protocol, except that the 37° incubation is extended to 5 h. Transcribed RNA is phenol-chloroform extracted, ethanol precipitated, and then resuspended in Tris-Borate-EDTA (TBE) buffer containing 80% formamide. The RNA is resolved on denaturing 6% polyacrylamide-TBE gels containing 7 *M* urea. RNA bands are visualized by UV shadowing for excision from the gel. sRNA is eluted from the gel slice (×3) at room temperature for 45–60 min with 2 ml of elution buffer (10 m*M* Tris, pH 7.4, 0.3 *M* sodium acetate, 5 m*M* EDTA, 0.1% SDS) per gram of gel by use of the crush-and-soak method. Eluted RNA is ethanol precipitated, resuspended in water, and the RNA concentration determined by absorbance at 260 nm. RNA is then aliquoted, dried, and stored at −80°. This *in vitro* transcription protocol produces approximately 50–80 μg of gel-purified sR8 sRNA per μg of DNA template for a standard 20 μl transcription reaction.

4. *IN VITRO* ASSEMBLY OF THE *M. JANNASCHII* SR8 BOX C/D sRNP COMPLEX

Core protein binding capabilities and sRNP assembly are assessed by use of electrophoretic mobility-shift analysis (EMSA) (Tran *et al.*, 2003). *In vitro* assembly of the *M. jannaschii* sR8 box C/D sRNP is accomplished by incubating 5′-radiolabeled sR8 sRNA (0.2 pmol and 1×10^4 cpm) with 20 pmol of L7, 32 pmol Nop56/58, and 32 pmol fibrillarin in assembly buffer (20 m*M* HEPES, pH 7.0, 150 m*M* NaCl, 1.5 m*M* MgCl$_2$, 10% glycerol) containing tRNA (1.5 mg/ml). Binding capabilities of individual core proteins are determined by sequentially adding L7, Nop56/58, and then fibrillarin as assembly of the complex requires ordered binding of the core proteins. Assembly of the sRNP complex is accomplished by incubating sRNA and core proteins at 75° for 10 min. *In vitro* assembly of the sRNP requires elevated temperatures to facilitate sRNA remodeling required for core protein binding (Gagnon *et al.*, 2006). Partial or completely assembled complexes are resolved on native 4% polyacrylamide gels containing 25 mM potassium phosphate buffer, pH 7.0, and 2% glycerol. After electrophoresis, gels are dried, and assembled RNPs visualized by autoradiography or PhosphorImager analysis. Figure 12.1C shows a representative EMSA analysis of the sequential binding of sRNP core proteins to radiolabeled sR8 sRNA.

5. Assessment of *In Vitro* Assembled *M. jannaschii* Box C/D sRNP Methylation Activity

Methyltransferase activity of the *in vitro* assembled *M. jannaschii* sR8 box C/D sRNP is assessed with an *in vitro* methylation assay. Assembled sRNP complexes are incubated in the presence of the methyl donor S-adenosyl-L-[*methyl*-^{3}H] methionine (SAM) (Amersham Pharmacia) and synthetic target RNA oligonucleotide substrates (Dharmacon) that are complementary to the D or D$'$ guide regions. Methylation activity is assessed by measuring the incorporation of [^{3}H]–CH$_3$ into these target RNAs. Assembly reactions of 80 μl and approximately 0.5 μM assembled sRNP are incubated on ice and mixed with 30 μl of assembly buffer containing 12 μM target RNA substrate(s) and 15 μM SAM (5 μCi of [^{3}H]-SAM at a 1:50 ratio with nonradioactive SAM). This SAM concentration is sufficient for this methylation assay, although recent work has indicated that higher concentrations of SAM can drive the reaction to yield higher levels of target RNA methylation (Hardin and Batey, 2006). We have established that the length of target RNA oligonucleotide substrates affects the level of RNA methylation. Extending the target RNA at both 5$'$ and 3$'$ termini by 4–5 nucleotides beyond that region that base pairs with the sRNA guide sequence significantly increases [^{3}H]–CH$_3$ incorporation (Appel and Maxwell, 2007). Negative controls are target RNAs already possessing a 2$'$-O-CH$_3$ at the target nucleotide or target RNAs with a deoxynucleotide replacing the target ribonucleotide.

Target RNA methylation is initiated by incubating the assembled reactions at 68°. Aliquots of 20 μl are removed at the desired time points and spotted onto 2-cm filter discs (3 M Whatman paper). After drying, the filters are washed in 10% trichloroacetic acid (TCA) and then three times in 5% TCA. Washed and dried filters are suspended in scintillation fluid and counted in a liquid scintillation counter. Results of a typical *in vitro* methylation activity assay are shown in Fig. 12.1D. Methylation activity of the sRNP is reported as moles of methylated target RNA per mole of sRNP. Conversion of cpms to moles of incorporated $[^3H]–CH_3$ is accomplished by spotting 1 μCi of $[^3H]$–SAM onto control filters, determining the cpms/uCi, and then calculating the moles of incorporated CH_3 using the specific activity of the $[^3H]$–SAM (Ci/mole) provided by the manufacturer and the molar ratio of radioactive and nonradioactive SAM in the reaction. This value is then reported with respect to the moles of assembled sRNP in the reaction.

6. Sequential Affinity Chromatographic Purification of *In Vitro* Assembled Box C/D sRNPs

6.1. An overview

Affinity purification with tagged proteins or RNAs has proved to be a powerful approach for isolating multicomponent protein and RNA–protein complexes from isolated cellular extracts (Rigaut *et al.*, 1999; Schimanski *et al.*, 2005; Srisawat and Engelke, 2002; Waugh, 2005). Compared with more traditional fractionation techniques such as gradient sedimentation centrifugation, gel filtration, and ion exchange chromatography, affinity chromatography typically yields highly purified complexes in only one or two isolation steps. Therefore, we have developed a tandem affinity purification protocol for the rapid isolation of *in vitro* assembled archaeal box C/D sRNPs. This protocol uses three affinity selection steps that are designed to isolate fully assembled and catalytically active sRNP.

Archaeal box C/D sRNPs are assembled by use of large preparations of *in vitro* transcribed box C/D sRNA and the three recombinant sRNP core proteins. Assembled sRNPs are first selected by use of an oligonucleotide complementary to the sRNA. Two approaches can be used in this step. An sR8 sRNA engineered with a poly-A tail (14 adenines) at the 3' end is used in sRNP assembly, and selection is carried out with oligo-dT cellulose resin. Alternately, a biotinylated DNA oligonucleotide complementary to the sR8 D guide region is hybridized to the assembled sRNP and then selected with streptavidin resin. For both approaches, the sRNA and bound core proteins are efficiently eluted from the respective resins at elevated temperature. We have noted, however, that sRNP affinity selected with the biotinylated oligonucleotide exhibits a 20–30% reduction in methylation activity guided by the terminal box C/D RNP. Subsequently, assembled complexes are sequentially selected by fibrillarin's FLAG tag and His tag by use of ANTI-FLAG M2 and Ni-NTA affinity resins, respectively. Fibrillarin is the third and final core protein to bind the box C/D and C'/D' motifs. Sequential affinity selection of the *in vitro* assembled sRNP possessing the two tags assures that each isolated complex contains two fibrillarin proteins and is thus a fully assembled sRNP with catalytically active box C/D and C'/D' RNPs. This is particularly important for biophysical analyses such as cryoelectron microscopy or X-ray crystallography, where a homogeneous population of complexes is crucial for analysis.

Selection of the sRNP by use of complementary oligonucleotides as the first selection step effectively eliminates free core proteins not bound to the sRNA. Removal of free proteins is particularly advantageous at this point in the isolation protocol. The highly charged character of free Nop56/58 can cause aggregation problems. Protein aggregation may be minimized by use of higher salt buffers and working in the presence of very low concentrations of SDS ($\sim$0.007%), nonionic detergents (0.1% Triton X-100), or nonspecific RNA (1–2 mg/ml tRNA). These added components could increase sRNP yields, although small amounts of detergent may be carried through the purification process and be present in the final purified sRNP fraction. Nop56/58 and fibrillarin also efficiently dimerize in the absence of sRNA and sequential isolation of the sRNP complexes solely by means of the tagged fibrillarin core proteins results in isolation not only of the sRNP but also of free Nop56/58-fibrillarin dimers. ANTI-FLAG M2 affinity chromatography is the second step in sRNP purification, because this resin exhibits low binding capacity and low elution efficiency. By placing this affinity step second in the purification protocol, we are able to use larger amounts of resin and elute the sRNP in larger elution volumes for more efficient recovery. As the final selection step, the Ni-NTA resin binds the His-tagged fibrillarin with high affinity. The large capacity of this resin facilitates elution of more concentrated sRNP in smaller final volumes. We have found that this order of affinity selection steps is most efficient for purification of fully assembled and methylation-competent sRNP.

A flowchart for the tandem affinity selection and purification of *in vitro* assembled archaeal box C/D sRNP complexes is presented in Fig. 12.2. The starting amounts of assembled sRNP may be reduced or scaled up, depending on the quantity of purified sRNP desired. For the particular sRNP purification experiment shown here, additional components such as detergents were omitted. sRNP obtained from the final Ni-NTA affinity column can be exchanged with any buffer of choice during concentration, depending on the requirements of the planned experiments, although buffers of higher ionic strength help to reduce sRNP aggregation. sRNP fractions applied to each resin, collected flow through fractions, and subsequently eluted sRNP fractions are analyzed on an SDS-polyacrylamide gel and shown in Fig. 12.3A.

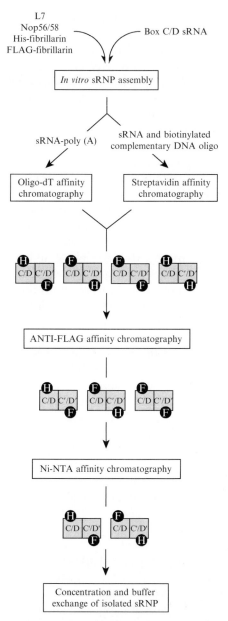

Figure 12.2 Sequential affinity purification of *in vitro* assembled archaeal box C/D sRNP.

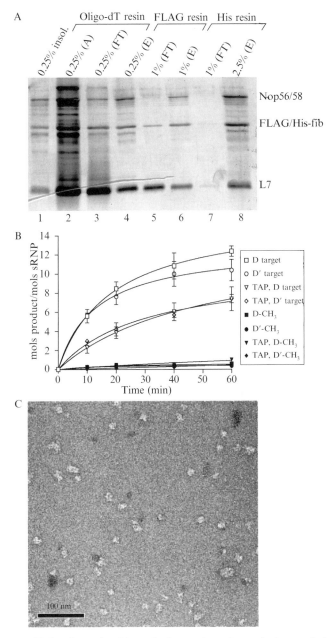

Figure 12.3 SDS-polyacrylamide gel electrophoretic analysis, methyltransferase activity, and electron microscopy of *in vitro* assembled *M. jannaschii* sR8 box C/D sRNPs. (A) SDS-polyacrylamide gel electrophoretic analysis of sRNP fractions. Silver-stained sRNP fractions analyzed from the individual affinity columns include the sRNP sample applied (A); the flow-through fraction of unbound material (FT);

6.2. Affinity chromatography buffers

10× Binding buffer (BB): 100 mM HEPES, pH 7.0, 1.0 M NaCl
Buffer R: 20 mM HEPES, pH 7.0, 0.1 M NaCl, 1 mM EDTA
Buffer D: 20 mM HEPES, pH 7.0, 3 mM MgCl$_2$, 100 mM NaCl, 20% glycerol
Buffer E: 20 mM HEPES, pH 7.0, 1.5 mM MgCl$_2$, 500 mM NaCl
Buffer G: 0.1 M glycine, pH 3.5, 100 mM NaCl

6.3. Preparation of the affinity chromatography resins

1. Oligo-dT cellulose (Ambion): Equilibrate 4 ml of resin packed in a 20 ml (1.5 × 10 cm) chromatography column with buffer E.
1A. Streptavidin agarose (Novagen): Equilibrate 3 ml of resin packed in a 20 ml (1.5 × 10 cm) chromatography column with buffer D. Bind 3 mg of biotinylated DNA oligonucleotide (Integrated DNA Technologies) complementary to the D guide sequence (under-lined) by incubating the oligonucleotide (5′-(biotin)-ACAGTCAT CGCTTGCTCATACGGTCTC-3′) and resin for 10 min at room temperature. Then equilibrate the resin with buffer E.
2. ANTI-FLAG M2 agarose (Sigma): Equilibrate 3.0 ml of resin packed in a 20 ml (1.5 × 10 cm) chromatography column with buffer E.
3. Ni-NTA His-Bind Resin (Novagen): Charge 1.5 ml of resin packed in a 20-ml chromatography column (1.5 × 10 cm) with 5 ml of 50 mM NiSO$_4$. Then equilibrate the resin with buffer E containing 25 mM imidazole.

6.4. Assembly of the sR8 sRNP complex

sRNP are assembled *in vitro* by use of *in vitro* transcribed sRNA and recombi-nant core proteins, with the amount of assembled sRNP dependent on the desired amount of purified complex. Table 12.1 lists the components re-quired for assembling 12 nmol (∼2.5 mg) of sRNP complex. Approximately

---

affinity column–bound and eluted sRNP fraction (E). Percentage of the total sample volume used for each electrophoretic analysis is indicated above the lane. (B) Assessment of the methyltransferase activity of tandem affinity-purified box C/D sRNP (TAP) compared with *in vitro* assembled but not affinity-selected complexes. Both affinity-selected and nonpurified sR8 box C/D sRNP methylate D and D′ targets, indicating fully assembled and catalytically active box C/D and C′/D′ RNPs. D and D′ targets possessing a 2′-O-CH$_3$ at the target nucleotide (negative controls) are not methylated, demonstrating nucleotide-specific modification for both affinity-purified and nonpurified complexes. (C) Electron micrograph of *in vitro* assembled and tandem affinity-purified *M. jannaschii* sR8 box C/D sRNP. Electron micrograph courtesy of Franziska Bleichert, Vinzenz Unger, and Susan Baserga (Yale University).

Table 12.1 *In vitro* sRNP assembly

Reaction component	Concentration	Volume (μl)	Approx. mass (μg)
L7	100 μM	250	325
Nop56/58	100 μM	250	1200
His-fibrillarin	100 μM	125	350
FLAG-fibrillarin	100 μM	125	350
Buffer D^a	1$\times$	500	—
poly (A) sRNA	38 μM	320	320
NaCl	5 M	155	—
Binding buffer^b	10$\times$	250	—
H$_2$O	—	525	—
Final		2500	2545

^{a} 20 mM HEPES, pH 7.0, 0.1 M NaCl, 3 mM MgCl$_2$, 20%(w/v) glycerol.
^{b} 100 mM HEPES, pH 7.0, 1.0 M NaCl.

12 nmol (320 μg) of sR8 sRNA, 24 nmol of L7 (325 μg), 24 nmol of Nop56/58 (1200 μg), and 12 nmol (350 μg) each of FLAG-tagged and His-tagged fibrillarin are incubated in a final assembly volume of 2.5 ml. Core proteins are stored in buffer D except for Nop56/58, which is stored in buffer D containing 0.5 M NaCl to maintain protein solubility. The final assembly buffer (20 mM HEPES, pH 7.0, 0.5 M NaCl, 1.5 mM MgCl$_2$, and 10% glycerol [w/v]) contains 0.5 M NaCl to minimize protein and sRNP aggregation. After mixing components, the assembly reaction is heated to 75° for 5 min and then cooled to room temperature. Centrifugation of the assembly reaction at 14,000g for 5 min at room temperature removes insoluble materials from the supernatant fraction containing assembled sRNP. Shown in Fig. 12.3A are the sRNP fractions obtained from the sequential affinity selection steps.

6.5. Tandem affinity purification of *in vitro* assembled box C/D sRNP

6.5.1. Affinity selection step 1: oligo dT cellulose chromatography

1. Mix the assembled sRNP ($\sim$12 nmol in 2.5 ml; Fig. 12.3A, lane 2) with 4 ml of oligo dT cellulose resin equilibrated with buffer E in a 20-ml chromatography column.
2. Bind the sRNP to the oligo dT resin by rocking the column for 5 min at room temperature and then continue rocking for 30 min at 4°.
3. Begin chromatography by collecting the flow through fraction (Fig. 12.3A, lane 3) and then washing the oligo dT resin with 20 bed volumes of cold (4°) buffer E. Maintain the column at 4° during washing to stabilize hydrogen bonding of the poly (A) tail to the oligo dT cellulose.

4. Transfer the column with bound sRNP to a 50° incubator and heat the column and resin for 20 min.
5. Elute the sRNP with 4 bed volumes of buffer E heated to 50° (Fig. 12.3A, lane 4).
6. Cool the sRNP eluate to room temperature for subsequent affinity selection on ANTI-FLAG M2 agarose.
7. Regenerate the oligo dT cellulose resin by washing sequentially with 5 bed volumes each of buffer R, 0.1 M NaOH, and H_2O.
8. Equilibrate the resin with buffer R containing 0.05% sodium azide and store at 4°. (For long-term storage, after the water wash, rinse the resin with ethanol, dry the resin, and store at −20°.)

6.5.2. Alternative affinity selection step 1: streptavidin affinity chromatography

1. Mix the assembled sRNP with 3 ml of buffer E–equilibrated streptavidin resin bound with the biotinylated DNA oligonucleotide and suspend in a 20-ml chromatography column.
2. Bind the sRNP to the streptavidin resin by rocking the column for 5 min at room temperature and then continue rocking for 30 min at 4°.
3. Begin chromatography by collecting the flow through fraction and then washing the resin with 20 bed volumes of cold buffer E. Maintain the column with resin at 4° during washing to stabilize hybridization of the biotinylated DNA oligonucleotide to the sRNA.
4. Transfer the affinity column with resin-bound sRNP to a 60° incubator and equilibrate the column for 20 min at this elevated temperature.
5. Elute bound sRNP with 4 bed volumes of buffer E heated to 60°.
6. Cool the eluted sRNP fraction to room temperature for subsequent affinity selection on ANTI-FLAG M2 agarose.
7. Regenerate the DNA oligonucleotide-streptavidin resin by washing with 6 bed volumes of buffer R heated to 60°.
8. Equilibrate the resin in buffer R containing 0.05% sodium azide and store at 4°.

6.5.3. Affinity selection step 2: ANTI-FLAG M2 affinity chromatography

1. Mix the cooled sRNP eluate (12–16 ml) with 3 ml of buffer E–equilibrated ANTI-FLAG M2 agarose suspended in a 20-ml chromatography column.
2. Bind the sRNP to the agarose resin by incubating for 15 min at room temperature.

3. Collect the flow through (Fig. 12.3A, lane 5) and reapply this eluate to the affinity column twice more, each time incubating the sRNP eluate with the resin for 15 min at room temperature.

4. Wash the agarose resin with 20 bed volumes of buffer E at room temperature.

5. Release the bound sRNP by resuspending the agarose resin in 1 bed volume of buffer E containing 110 μg/ml of FLAG peptide (Sigma) and incubate for 10 min at room temperature.

6. Collect the column eluate containing the sRNP.

7. Resuspend the agarose resin twice more in 1 bed volume of buffer E with FLAG peptide, each time incubating the resin for 10 min at room temperature before collecting the eluate.

8. Pool all three sRNP eluate fractions (Fig. 12.3A, lane 6) for subsequent Ni-NTA affinity chromatography.

9. Regenerate the agarose resin by washing twice with 4 bed volumes of buffer G. Do not allow this resin to be suspended in buffer G for more than 20 min. Wash the resin with buffer R.

10. Equilibrate the agarose resin in buffer R containing 0.05% sodium azide and store at 4°.

6.5.4. Affinity selection step 3: Ni-NTA His-Bind resin affinity chromatography

1. Mix the pooled eluates from ANTI-FLAG M2 affinity chromatography with 1.5 ml of buffer E–equilibrated Ni-NTA His-Bind resin suspended in a 20-ml chromatography column and incubate for 10 min at room temperature.

2. Collect the column flow through fraction (Fig. 12.3A, lane 7) and reapply the eluate to the resin, incubating another 10 min at room temperature.

3. Wash the affinity resin with 20 bed volumes of buffer E containing 25 mM imidazole.

4. Release the bound sRNP by resuspending the resin in 2 bed volumes of buffer E containing 200 mM imidazole and incubating for 5 min at room temperature.

5. Collect the eluate fraction.

6. Resuspend the resin in an additional 2 bed volumes of buffer E containing 200 mM imidazole and incubate for an additional 5 min at room temperature.

7. Collect the second eluate fraction.

8. Pool the two eluates (this pooled eluate may be stored overnight at 4°).

9. Concentrate the pooled eluates to between 200 and 500 μl by use of a 15-ml spin concentrator (Vivaspin 15R, Vivascience) and a microspin concentrator (Microcon YM-3, Millipore). During concentration, an

exchange of buffers to the desired final buffer may be accomplished. Dialysis is not recommended at this step.

10. Regenerate the Ni–NTA His–Bind resin by washing with 5 bed volumes of 100 mM EDTA, pH 7.0, followed by washing with water.
11. Equilibrate the Ni–NTA His Bind resin in 20% ethanol and store at 4°.

The yield of purified sRNP is estimated by assessing the amount of RNA and/or protein contained in a small aliquot of complex resolved on polyacrylamide gels. Alternately, the amount of RNA in a phenol-extracted aliquot of sRNP is determined by absorbance at 260 nm. Typically 200–300 pmol (40–60 μg) of purified sRNP is obtained from approximately 2.5 mg of assembled complex. Typical losses are observed for each affinity step, with major losses being at the second and third selection step for the FLAG-tagged and His-tagged fibrillarin proteins. At these steps, significant amounts of sRNP are lost as unselected complexes, because these sRNP are assembled with similarly tagged but unselected fibrillarin proteins at both box C/D and C'/D' RNPs. Assessment of box C/D RNP-guided nucleotide methylation activity of the purified complexes reveals methylation of target RNAs for both box C/D and C'/D' RNPs comparable to that of unpurified sRNPs (Fig. 12.3B). Determination of sRNP methylation capabilities for complexes suspended in buffer E with elevated NaCl concentration has no discernible effect on the complex's enzymatic activities. Electron microscopy of isolated sRNP reveals sRNP of homogeneous size with some larger complexes (Fig. 12.3C). The larger complexes are aggregated sRNP, and their presence can be diminished with the addition of ionic (SDS, heparin) compounds, although elevated concentrations can destabilize the sRNP complex.

7. Concluding Remarks

In vitro assembled archaeal sRNPs provide a model complex for the investigation of box C/D RNP assembly, structure, and function. Study of these minimal, yet catalytically active, RNA–protein enzymes will help define the fundamental principles behind RNA-guided nucleotide modification. The archaeal sRNP also serves as a prototype box C/D RNA-guided nucleotide modification enzyme for understanding the more structurally and functionally complex eukaryotic snoRNPs. Affinity purification of completely assembled and catalytically active sRNPs can now provide not only complexes for more detailed biochemical and functional studies but also for various biophysical approaches requiring larger amounts of material. Of particular advantage is the fact that the isolated complexes are homogeneous in composition, a prerequisite for approaches such as cryoelectron microscopy and X-ray crystallography.

The sequential affinity approach described here may well have broader applications in the study of RNA-guided nucleotide modification complexes. This same isolation protocol should be easily adapted to isolate other box C/D complexes and, in principle, *in vitro* assembled H/ACA sRNPs. Perhaps more exciting is the possibility to use this affinity protocol for the isolation of *in vivo* assembled RNP complexes. Expression of tagged core proteins in various cell lines should make affinity selection of the corresponding RNP complexes from cell lysates or various cellular fractions possible. This could ultimately lead to the identification of additional or "accessory" proteins associated with a family of complexes, thus defining novel proteins important for RNP biogenesis and/or function. Also possible may be the selection of specific RNP complexes from a homogenous RNP population by use of oligonucleotides complementary to a given sRNA or snoRNA species. Again, such an approach could lead to the identification of accessory proteins unique to a specific RNP complex. Although these suggested approaches have yet to be tested, they have the potential to greatly facilitate more detailed examinations of the diverse and highly conserved populations of RNA-guided nucleotide modification complexes.

ACKNOWLEDGMENTS

We thank James Brown for providing *Methanocaldococcus jannaschii* genomic DNA and Franziska Bleichert for critical reading of the manuscript. We are grateful to Franziska Bleichert, Vinzenz Unger, and Susan Baserga (Yale University) for providing electron micrographs of affinity-purified *M. jannaschii* sR8 box C/D sRNPs. This work was supported by NSF grant MCB 0543741 (E.S.M.).

REFERENCES

Appel, C. D., and Maxwell, E. S. (2007). Structural features of the guide: Target RNA duplex required for archaeal box C/D sRNA-guided nucleotide 2′-O-methylation. *RNA* e-pub. In press.

Bachellerie, J. P., Cavaille, J., and Huttenhofer, A. (2002). The expanding snoRNA world. *Biochimie* **84,** 775–790.

Dennis, P. P., Omer, A., and Lowe, T. (2001). A guided tour: Small RNA function in Archaea. *Mol. Microbiol.* **40,** 509–519.

Dennis, P. P., and Omer, A. (2006). Small non-coding RNAs in Archaea. *Curr. Opin. Microbiol.* **8,** 685–694.

Gagnon, K. T., Zhang, X., Agris, P. F., and Maxwell, E. S. (2006). Assembly of the archaeal box C/D sRNP can occur via alternative pathways and requires temperature-facilitated sRNA remodeling. *J. Mol. Biol.* **362,** 1025–1042.

Hama, T., and Ferre-D'Amare, A. R. (2004). Structure of protein L7Ae bound to a k-turn derived from an archaeal box H/ACA sRNA at 1.8 Å resolution. *Structure* **12,** 893–903.

Hardin, J. W., and Batey, R. T. (2006). The bipartite architecture of the sRNA in an archaeal box C/D complex is a primary determinant of specificity. *Nucleic Acids Res.* **34,** 5039–5051.

Maxwell, E. S., and Fournier, M. J. (1995). The small nucleolar RNAs. *Annu. Rev. Biochem.* **64,** 897–934.

Moore, T., Zhang, Y., Fenley, M. O., and Li, H. (2004). Molecular basis of box C/D RNA-protein interactions: Cocrystal structure of archaeal L7Ae and a box C/D RNA. *Structure* **12,** 807–818.

Omer, A. D., Lowe, T. M., Russell, A. G., Ebhardt, H., Eddy, S. R., and Dennis, P. P. (2000). Homologs of small nucleolar RNAs in Archaea. *Science* **288,** 517–522.

Omer, A., Ziesche, S., Ebhardt, H., and Dennis, P. (2002). *In vitro* reconstitution and activity of a C/D box methylation guide ribonucleoprotein complex. *Proc. Natl. Acad. Sci. USA* **99,** 5289–5294.

Rigaut, G., Shevchenko, A., Rutz, B., Wilm, M., Mann, M., and Seraphin, B. (1999). A generic protein purification method for protein complex characterization and proteome exploration. *Nat. Biotech.* **17,** 1030–1032.

Schimanski, B., Nguyen, T. N., and Gunzl, A. (2005). Highly efficient tandem affinity purification of trypanosome protein complexes based on a novel epitope combination. *Euk. Cell* **4,** 1942–1950.

Srisawat, C., and Engelke, D. R. (2002). RNA affinity tags for purification of RNAs and ribonucleoprotein complexes. *Methods* **26,** 156–161.

Suryadi, J., Tran, E. J., Maxwell, E. S., and Brown, II, B. A. (2005). The crystal structure of *Methanocaldococcus jannaschii* multifunctional L7Ae RNA-binding protein reveals an induced-fit interaction with the box C/D RNAs. *Biochemistry* **44,** 9657–9672.

Terns, M., and Terns, R. (2002). Small nucleolar RNAs: Versatile *trans*-acting molecules of ancient evolutionary origin. *Gene Expr.* **10,** 17–39.

Tollervey, D. (1996). Small nucleolar RNAs guide ribosomal RNA methylation. *Science* **273,** 1056–1057.

Tran, E. J., Zhang, X., and Maxwell, E. S. (2003). Efficient RNA 2′-O-methylation requires juxtaposed and symmetrically assembled archaeal box C/D and C′/D′ RNPs. *EMBO J.* **22,** 3930–3940.

Waugh, D. S. (2005). Making the most of affinity tags. *Trends Biotech.* **23,** 316–320.

Zhang, X., Champion, E. A., Tran, E., Brown, B. A., II, Baserga, S. J., and Maxwell, E. S. (2006). The coiled-coil domain of the Nop56/58 core protein is dispensable for sRNP assembly but is critical for archaeal box C/D sRNP-guided nucleotide methylation. *RNA* **12,** 1092–1103.

CHAPTER THIRTEEN

Identifying Effects of snoRNA-Guided Modifications on the Synthesis and Function of the Yeast Ribosome

Wayne A. Decatur, Xue-hai Liang, Dorota Piekna-Przybylska, *and* Maurille J. Fournier

Contents

Department of Biochemistry and Molecular Biology, University of Massachusetts, Amherst, Massachusetts

Methods in Enzymology, Volume 425
ISSN 0076-6879, DOI: 10.1016/S0076-6879(07)25013-X

Abstract

The small nucleolar RNAs (snoRNAs) are associated with proteins in ribonucleoprotein complexes called snoRNPs ("snorps"). These complexes create modified nucleotides in preribosomal RNA and other RNAs and participate in nucleolytic cleavages of pre-rRNA. The various reactions occur in site-specific fashion, and the mature rRNAs are ultimately incorporated into cytoplasmic ribosomes. Most snoRNAs exist in two structural classes, and most members in each class are involved in nucleotide modification reactions. Guide snoRNAs in the "box C/D" class target methylation of the 2′-hydroxyl moiety, to form 2′-O-methylated nucleotides (Nm), whereas guide snoRNAs in the "box H/ACA" class target specific uridines for conversion to pseudouridine (Ψ). The rRNA nucleotides modified in this manner are numerous, totaling approximately 100 in yeast and twice that number in humans. Although the chemistry of the modifications and the factors involved in their formation are largely explained, very little is known about the influence of the copious snoRNA-guided nucleotide modifications on rRNA activity and ribosome function. Among eukaryotic organisms the sites of rRNA modification and the corresponding guide snoRNAs have been best characterized in *S. cerevisiae*, making this a model organism for analyzing the consequences of modification. This chapter presents approaches to characterizing rRNA modification effects in yeast and includes strategies for evaluating a variety of specific rRNA functions. To aid in planning, a package of bioinformatics tools is described that enables investigators to correlate guide function with targeted ribosomal sites in several contexts. Genetic procedures are presented for depleting modifications at one or more rRNA sites, including ablation of all Nm or Ψ modifications made by snoRNPs, and for introducing modifications at novel sites. Methods are also included for characterizing modification effects on cell growth, antibiotic sensitivity, rRNA processing, formation of various rRNP complexes, translation activity, and rRNA structure within the ribosome.

1. INTRODUCTION

The rRNA modifications created by snoRNPs and the corresponding guide snoRNAs have been best explained and experimentally validated in budding yeast. Approximately 95% of the *S. cerevisiae* snoRNAs known at this writing are involved in the modification of ribosomal RNA. Forty-three box C/D snoRNAs guide 2′-O-methylation (Nm) at 53 specific sites (Davis and Ares, 2006; Lowe and Eddy, 1999), and 28 box H/ACA snoRNAs guide pseudouridine (Ψ) formation at 45 specific sites (Schattner *et al.*, 2004;

Torchet *et al.*, 2005; Decatur and Schnare, unpublished results; Piekna-Przybylska, Decatur, and Fournier, RNA in press). In addition to the Nm and Ψ modifications created by snoRNPs, yeast cytoplasmic rRNA contains a ribose methylated nucleotide and a pseudouridine that are introduced by RNA-independent enzymes (Lapeyre and Purushothaman, 2004; Decatur, Schnare, Gray and Fournier, unpublished results), as well as 10 nucleotides with modified bases (Decatur and Fournier, 2002).

The snoRNPs in each class consist of one snoRNA and four class-specific core proteins. For the C/D snoRNPs, the core proteins are Nop1p, Nop58p, Nop56p, and Snu13p, and for the H/ACA snoRNPs, the proteins include Cbf5p, Nop10p, Gar1p, and Nhp2p (Bertrand and Fournier, 2004; Fatica and Tollervey, 2002, 2003; Meier, 2005, 2006; Yu *et al.*, 2005). In each type of modifying-snoRNP, the catalytic component is an associated core protein: Nop1p (fibrillarin) in the box C/D snoRNPs and Cbf5p (dyskerin) in the H/ACA snoRNPs. Despite the great number of snoRNA-guided rRNA modifications and presence in conserved functional regions of the ribosome, their role(s) remains elusive.

Because of its powerful genetic system and the fact that apparently all of the guide snoRNAs that target rRNA have been identified, *S. cerevisiae* is an ideal model organism for examining the function of the snoRNA-guided modifications. Although yeast has ~100 snoRNA-guided modifications, the total is approximately half the number estimated for humans (Maden, 1990; Ofengand and Bakin, 1997; Ofengand and Fournier, 1998; Yang *et al.*, 2006). Importantly, the snoRNP machinery and overall pattern of modification is highly conserved in eukaryotic cells, including humans. Thus, results obtained from the budding yeast are relevant to a wide range of organisms. Strikingly, the *Archaea* also possess similar RNA-dependent modifying machines (Dennis and Omer, 2005; Omer *et al.*, 2000; Rozhdestvensky *et al.*, 2003; Zago *et al.*, 2005), indicating that the snoRNPs are of ancient origin.

Here, we present approaches and methods for investigating the effects of snoRNA-guided modification on the synthesis and function of the yeast ribosome, including several genetic strategies for blocking or altering the functions of the snoRNAs. In addition to providing methods for analyzing rRNA and ribosome functions, we also present details for the construction and analysis of strains where subsets of modifications in specific functional regions of the ribosome are genetically ablated.

2. EXPERIMENTAL STRATEGIES

Traditionally, disrupting gene function and altering activity through genetics has been a powerful means to explain fundamental principles and mechanisms, and the snoRNP field is no exception. During the course of studying yeast snoRNAs and the associated proteins, our laboratory has used several strategies to deplete varying modifications, ranging from just a few to all of a certain type, and to guide modifications to novel rRNA sites as well. The genetic manipulations involved can generally be divided into three types: (1) targeting modifications to new sites in rRNA, (2) blocking modification at a global level by mutating the catalytic proteins, and (3) disrupting modification at selected sites by depleting snoRNAs in particular combinations. The first two strategies are covered briefly, in part because they have been described previously and in part because most investigators will be interested in the roles of specific natural modifications. Thus, most of the chapter is devoted to strategies for depleting one or more preselected modifications from particular regions of the ribosome and evaluating the consequences in a variety of contexts. In the course of our own studies, we have seen a broad range of *in vivo* defects for cells lacking various subsets of modifications. Impairment has been observed, variously, in all of the activities described in the following, including cell growth rate, temperature, and drug sensitivity; the processing rate of pre-rRNA; accumulation of mature rRNA and ribosomes; amino acid incorporation rate; translation fidelity; and rRNA folding properties with the ribosome (King *et al.*, 2003; Decatur, Liang, Piekna-Przybylska, and Fournier, unpublished results).

2.1. The use of engineered snoRNAs to guide novel ribose methylations and pseudouridines in rRNA

Our laboratory developed a general method to easily provide Nm guide snoRNAs with new site specificity to allow methylation of rRNA sites that are not normally methylated. The method involves altering the guide element of an engineered host snoRNA gene that is expressed under control of a galactose-inducible promoter. On induction, the novel snoRNA is produced, assembled into a snoRNP, and directs modification to a nucleotide specified by the new (13nt) guide element. Because this method has been described in a "Methods" chapter (Liu *et al.*, 2001), we refer the reader to that source and to a study in which this strategy was used to mutagenize the backbone of rRNA and analyze the resulting effects (Liu and Fournier, 2004).

Engineered H/ACA box snoRNAs have also been used in yeast to target pseudouridylation to novel rRNA sites (Bortolin *et al.*, 1999; King, McCully, and Fournier, unpublished results). Our laboratory has been successful in such work; however, the approach is not as straightforward as simply changing the guide elements, most likely because of deficiencies in our knowledge. Whereas that manipulation is sufficient for retargeting C/D guide snoRNAs, altering the two short guides in the box H/ACA snoRNAs inexplicably was not sufficient to achieve universal success for several test sites and guide snoRNAs examined. Clearly, more systematic analysis is required to resolve this important issue.

2.2. Interfering with the activity of the catalytic snoRNP proteins

The *NOP1* gene is essential in yeast and was determined to be the catalytic component of the C/D box snoRNPs on the basis of: (1) point mutations, (2) crystallographic results showing remarkable similarity between an archaeal ortholog and known methylases, and (3) methylation *in vitro* for a reconstituted archaeal snoRNP-like RNP (Omer *et al.*, 2002; Schimmang *et al.*, 1989; Wang *et al.*, 2000). In the yeast genetic characterization, several temperature-sensitive, lethal point mutations were generated (Tollervey *et al.*, 1993). These included two mutations that severely inhibit 2′-O-methylation of pre-rRNA at a global level (*nop1*P219S and *nop1*A175V). Residue 219 corresponds to a highly conserved proline, and altering alanine 175 may interfere with binding of the cofactor S-adenosyl-L-methionine. Screens identifying mutations that cause synthetic lethality with the temperature-sensitive alleles have been useful for identifying several interacting proteins (Berges *et al.*, 1994; Gautier *et al.*, 1997; Lafontaine and Tollervey, 1999).

CBF5 is also an essential gene in yeast and point mutations in regions with homology to conserved motifs in known pseudouridine synthases established the Cbf5p protein (dyskerin) as the pseudouridine synthase in the box H/ACA snoRNPs (Lafontaine *et al.*, 1998; Zebarjadian *et al.*, 1999). Pertinent to this chapter, two such mutant strains supported growth, albeit poorly, yet lacked detectable pseudouridine in the rRNA. One of these strains (*cbf5*D95A; Zebarjadian *et al.*, 1999) has been used in further studies by us to determine whether particular pseudouridines in the small rRNAs are guided by snoRNAs (Decatur, Schnare, Gray, and Fournier, unpublished results; Piekna-Przybylska and Fournier, unpublished results). In this determination, the state of pseudouridylation at a site is compared for both the Cbf5p-point mutant strain and control parental strain. This approach and strain has also been used to demonstrate that a guide snoRNA (snR81) targets pseudouridylation of one of the three pseudouridines in yeast U2 splicing snRNA; the other two are targeted by typical protein enzymes (Ma *et al.*, 2005).

3. ANALYZING snoRNA-GUIDED MODIFICATIONS IN SELECTED REGIONS OF THE RIBOSOME

In this section, we focus on subtracting (or adding) modifications selectively and in various combinations from particular regions of the ribosome. Regions known or suspected of being functionally important are of obvious interest, but all modification sites need to be examined. Our approach to designing a study is described first, on the basis of knowledge of the yeast guide snoRNAs and the structure of the ribosome. A section on genetic manipulations altering modifications is next, and this is followed by details of methods for assessing the modification effects on cell growth, rRNA and ribosome synthesis, and ribosome function.

3.1. Selecting modification sites for genetic depletion analysis

Several converging factors make it now both feasible and desirable to systematically design collections of strains that lack single and multiple modifications in particular regions of the ribosome. This is achieved by deletion or otherwise blocking the guide function of the corresponding snoRNA(s). Part of the motivation for blocking multiple modifications is that disruption of most individual guide snoRNAs does not cause a notable change in growth (e.g., Balakin *et al.*, 1993; King *et al.*, 2003; Lowe and Eddy, 1999; Schattner *et al.*, 2004; Thompson *et al.*, 1988; Zagorski *et al.*, 1988). This situation suggests that most individual modifications may have only subtle effects, although more in-depth examination may reveal substantial effects on particular processes. The first study of this type, carried out for Ψs in the region of the peptidyl transferase reaction center of the large subunit, revealed a translation defect for a particular A-loop Ψ (King *et al.*, 2003). In any case, examining and dissecting combinational effects of modification is important.

Such an approach is now possible because of the recent advances in identifying the yeast snoRNAs (Davis and Ares, 2006; Lowe and Eddy, 1999; Schattner *et al.*, 2004; Torchet *et al.*, 2005) and determining the structure of the ribosome. In regard to the latter, monumental progress has been made in describing ribosome structure both by crystallographic means and by cryoelectron microscopy (Ban *et al.*, 2000; Berisio *et al.*, 2003; Klein *et al.*, 2001; Ogle *et al.*, 2002; Schuwirth *et al.*, 2005; Selmer *et al.*, 2006; Yusupova *et al.*, 2001). With this information, the positions of the modified nucleotides targeted by snoRNAs (and vice versa) can be correlated with specific structural and functional regions of the ribosome (Fig. 13.1). Until high-resolution structures are available for the yeast ribosome, we favor transposing sites of yeast modifications onto 3D models of other ribosomes. This is achieved by determining equivalent nucleotides and highlighting these sites in 3D models by use of molecular visualization software (e.g., Decatur and Fournier [2002]; Piekna-Przybylska, Decatur, and Fournier, RNA, in press). Models for prokaryotic ribosome structures have been solved to atomic resolution, and the modifications in yeast rRNA are almost exclusively in the highly conserved regions, arguing that this method of analysis is, indeed, valid. In fact, much of the RNA and proteins of the high-resolution crystallographically determined prokaryotic structures have been fit into low-resolution models of the yeast ribosome on the basis of cryoelectron microscopy (Spahn *et al.*, 2001, 2004).

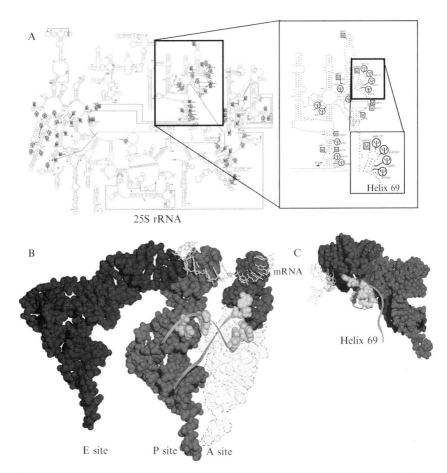

Figure 13.1 Correlating yeast snoRNA-guided modifications of rRNA with functional regions in the ribosome. (A) The secondary structure of yeast 25S rRNA with nucleotide modifications labeled. "Ψ" indicates pseudouridine; "M" marks Nm methylation. The insets show increasingly greater detail of the multifunctional helix 69, which is part of an intersubunit bridge. (B) The deduced locations of the snoRNA-guided modified nucleotides in helix 69 are shown as lightly shaded van der Waals radii spheres. The mRNA and the A-, P-, E- site tRNAs found in the 2.8-Å model of the ribosome (Selmer *et al.*, 2006) are labeled. Only the anticodon stem loop of the A-site tRNA is visible in the 2.8-Å model. Thus, the locations of the remaining portions of the A-site tRNA—from the 5.5 Å model of the ribosome (Yusupov *et al.*, 2001)—is outlined in dashed lines and "muted." (C) 90° rotation of panel (B) around the *y*-axis.

We have designed tools that facilitate analysis of yeast modifications relative to regions of the ribosome and integrated them into an upgraded version of the yeast snoRNA database (Piekna-Przybylska, Decatur, and Fournier, RNA, in press). The database is available at http://people. biochem.umass.edu/sfournier/fournierlab/snornadb/. The positions of the

Ψ and Nm modifications are featured in the sequence and secondary structure of the rRNA, and most importantly, in the deduced 3D structure of the ribosome. In each case, the corresponding guide snoRNA is also identified. The interactive, deduced 3D maps use structures with tRNAs bound so that functional regions are conspicuous, and owing to the Jmol applet (Herráez, 2006), only a modern JAVA-enabled browser is necessary for full functionality. Other parts of the snoRNA database provide details needed for genetic manipulation of the snoRNA genes. This is a significant advance and valuable resource, because it combines genetic and structural data from several sources in a form that can be accessed easily and used to help design and interpret studies such as those outlined in this chapter.

3.2. Design and construction of test strains

The first experimental step in evaluating the significance of a snoRNA-guided modification(s) in the ribosome is to construct a yeast strain lacking the modification(s) of interest (see Figs. 13.2 and 13.3 for an overview of a typical study). Two main approaches can be considered for creating test strains. One option is to block expression of a snoRNA gene by means of gene deletion/disruption. The second is to inactivate the snoRNA guide sequence that targets the snoRNP to the rRNA segment to be modified; disrupting this interaction is sufficient to block the desired modification. Blocking expression is most appropriate when the snoRNA is known to carry only a single guide element that only targets the modification of interest. However, if a snoRNA uses two guide elements to direct modifications to two (or more) positions in rRNA, mutating the corresponding guide sequence in the snoRNA gene is most desirable. Alternately, if deletion of an entire snoRNA leads to elimination of an additional modification, the desired modification can be maintained by providing an artificial guide snoRNA (Fig. 13.4A). The technology is well established for the C/D guide snoRNAs and involves insertion of a customized guide sequence into a designed host snoRNA (Liu and Fournier, 2004; Liu *et al.*, 2001). In principle, such manipulations can be done for the box H/ACA snoRNAs; however, as noted previously, we have not found engineering guides for Ψ to be straightforward as of yet (unpublished results and see Bortolin *et al.*, 1999).

When multiple deletions of snoRNA genes are to be performed, it is most convenient to use a parental yeast strain devoid of several functioning marker genes, such as *URA3, HIS3, ADE2, LEU2,* and *TRP1*. Genetic manipulations that rely on such marker genes are well established and particularly effective for deleting genes (Burke *et al.*, 2000). With such an approach, it is important to proceed with a well-thought-out plan for sequentially inactivating several snoRNA genes, because only *URA3* can be used multiple times. It is recommended to save one or two marker genes if

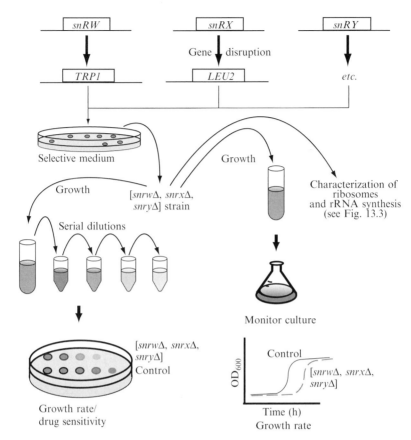

Figure 13.2 Schematic of the genetic and microbiological aspects of a typical study in which selected modification guide snoRNAs are disrupted (see text for details). The flow diagram depicts types of analyses that would be performed to characterize a test strain depleted of one or more modifications. Analyses include screening of growth rate at different temperatures with solid and liquid media and with different ribosome-based antibiotics. (See color insert.)

mutant snoRNA genes are to be provided or, for example, if a plasmid-encoded reporter gene will be used for functional analysis of the mutant ribosomes (see "Analysis of Translation Fidelity"). Alternate genetic approaches without this constraint have been described for disrupting multiple genes in yeast (Gueldener *et al.*, 2002; Johnston *et al.*, 2002; Storici *et al.*, 1999).

Deletion of a snoRNA gene can be performed in one step by use of homologous recombination (Burke *et al.*, 2000). In this approach, a selectable marker gene replaces the gene or coding sequence targeted for disruption. Cells are transformed with PCR amplicons containing the marker gene flanked by sequences homologous to those flanking the snoRNA gene.

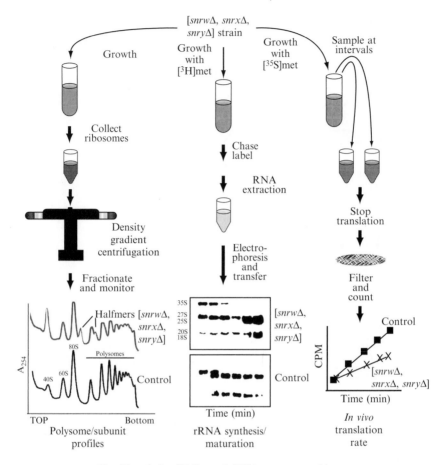

Figure 13.3   Schematic of the biochemical characterization of a typical modification-deficient test strain. Analyses include: (1) rRNA processing and accumulation, (2) ribosome assembly and stability, and (3) rate of amino acid incorporation *in vivo*. Other possible analyses are not depicted, such as translation fidelity and chemical probing of rRNA structure in the ribosome. (See color insert.)

Yeast transformation is carried out with a standard protocol that uses lithium acetate and polyethylene glycol (Gietz and Woods, 2002; Gietz *et al.*, 1992). Cassettes of selectable gene markers can be amplified by PCR from, for example, the now publicly available plasmids in the pRS series (ATCC) (Sikorski and Hieter, 1989). Importantly, the sequence of oligonucleotides for amplification should match at least 18–22 nts of the template at the 3′ end and extend for 45–60 nts adjacent to the target snoRNA gene at the 5′ end (Johnston *et al.*, 2002). The flanking homologous sequences direct integration of the marker gene at the locus targeted, and in most colonies

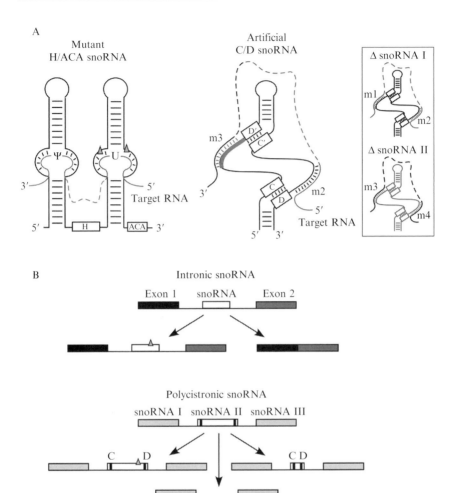

Figure 13.4 Genetic approaches to altering guide snoRNA function in yeast. Manipulations include depleting natural modifications and targeting modifications to new rRNA sites. (A) Mutations of guide elements (grey triangles) disrupt base pairing, with target RNA leading to loss of modification, as shown for an H/ACA RNA (left panel). New guide elements (grey thick line) can be introduced into an engineered snoRNA to maintain modification lost by deletion of a dual-guide snoRNA(s), as shown for a C/D snoRNA (right panel). m1–m4 are methylations guided by normal snoRNAs I and II; an artificial guide snoRNA is shown that restores the m2 and m3 modifications. (B) Schematic representation for inactivating intronic or polycistronic snoRNAs. The protein encoded by the exons can be expressed from a gene with a mutation(s) in the snoRNA sequence (grey triangles) or from a gene that lacks the intron. To inactivate a particular snoRNA in a polycistronic snoRNA coding unit, the corresponding guide element(s) can be mutated (grey triangles) or the coding sequence can be deleted.

selected, the marker gene will replace the snoRNA gene. If additional gene markers are required in a multiple snoRNA deletion project, excision of *URA3* can be performed by plating on medium containing 5-fluoroorotic acid plates (Burke *et al.*, 2000). Alternately, the neomycin–resistance gene can be used as an additional marker with G418 (Geneticin; from Sigma) as a selective agent (Jimenez and Davies, 1980).

Approximately one third of the yeast snoRNAs (25) are encoded in introns of protein genes or in polycistronic gene units with other snoRNAs. In these cases, extra steps and consideration are needed, as outlined in Fig. 13.4B, to ensure expression of either the protein or the other snoRNAs in the transcription unit. Leaving C and D boxes with very little to no sequence between them can ensure proper processing of adjacent snoRNAs in the same transcript (Caffarelli *et al.*, 1996; Cavaille and Bachellerie, 1996; Watkins *et al.*, 1996).

3.3. Analysis of growth and drug sensitivity

Impaired growth has been observed for many yeast strains with mutations in rRNA, ribosomal proteins, and transacting proteins involved in rRNA maturation and ribosome assembly (Buchhaupt *et al.*, 2006; Cochella and Green, 2004; Dresios *et al.*, 2000; Ferreira-Cerca *et al.*, 2005; Tabb-Massey *et al.*, 2003; van Beekvelt *et al.*, 2000; Venema and Tollervey, 1999). Loss of rRNA modifications is also known to influence rRNA production and stability and function of the completed ribosome. As noted previously, certain point mutations in the catalytic snoRNP proteins effectively block formation of an entire class of modification, and these cause severe impairment of growth (Tollervey *et al.*, 1993; Zebarjadian *et al.*, 1999). Examples in which growth is impaired by loss of one or a few modifications are still sparse. In one case, lack of a few pseudouridines in the peptidyl transferase center had synergistic negative effects beyond that caused by loss of a snoRNP that also participates in rRNA processing (King *et al.*, 2003). In another study, growth was impaired with loss of two neighboring $2'$-O-methylations in the A-loop of the large ribosomal subunit (Lapeyre and Purushothaman, 2004).

Initial screening should consider possible cell morphological changes for mutant strains. The size of yeast cells and their behavior during the cell division cycle can be monitored by microscopy (e.g., Guo *et al.*, 2000; Halme *et al.*, 2004; King *et al.*, 2003; Mullen *et al.*, 2001; Rethinaswamy *et al.*, 1998). We first analyze the growth properties of mutant cells on solid medium and at different temperatures, where differences in colony size relative to wild-type cells streaked on the same plate can reveal an altered growth rate. More sensitive is a dilution-spot assay. Here, test and control strains are incubated in liquid medium at $30°$ until they reach an OD_{600} of 3.0 (5.5×10^7 cells/ml) and are then diluted 1:5 serially in water. The highest cell concentration in each series is 5×10^6 cells/ml (OD_{600} of ~0.4).

Cells are spotted on a plate of appropriate medium with either a multipronged transfer device or by pipetting 5 μl of each dilution; plates are incubated at 15°, 30°, and 37°. In these and the other analyses described later, it is important to examine multiple isolates and to use suitable control strains to avoid potential artifacts related to differences in nutritional marker genes. This means keeping the genetic background of cells being compared as isogenic as possible, with the exception of the snoRNA genes under study. Complementation by reintroducing the snoRNA genes is an important control.

More comprehensive studies of growth rate at different temperatures are performed in liquid medium. The increase in the number of cells over time can be determined by optical density (OD_{600}) measurements, counting colonies of diluted samples plated on solid medium, or by direct counting of diluted cells by light microscopy with a cell counting chamber (i.e., a hemocytometer) (Burke *et al.*, 2000). For test strains with similar growth curves, a direct growth-competition experiment is the best approach for detecting small differences (e.g., Badis *et al.* [2003]; King *et al.* [2003]). The only requirement in this assay is that the test strain(s) harbors a unique selective gene marker, for example, *LEU2*, so that the two types of cells can be readily distinguished when plated on selective medium. In this example, two strains to be compared are grown separately in the same medium supplemented with 60 mg/liter of leucine to an OD_{600} of 1.0 (1.85×10^7 cells/ml). Equal numbers of cells from each strain are mixed and diluted into 3 ml of the same medium to an OD_{600} of 0.05 (6.5×10^5 cells/ml), and culturing is continued with periodic sampling through multiple growth cycles, as needed. To calculate the number of viable cells through these cycles, triplicate aliquots are taken and plated separately onto selective (−Leu) and nonselective (+Leu) medium. The plates are incubated at 30°. Calculating the number of colonies through a full cycle and over the course of several generations will reveal the fraction of mutant cells in the coculture and whether a strain has a growth disadvantage. Trials at various temperatures (e.g., 15°, 25°, or 36°) may also reveal differences in viability for the competing strains.

Ribosomes containing modification-deficient rRNA may not cause impaired growth at optimal culturing conditions in the laboratory. However, culturing under more extreme conditions, such as the higher or lower temperatures identified previously, or in the presence of particular antibiotics could enhance differences and cause an aberrant growth phenotype. Several antibiotics bind directly to ribosomal subunits and cause structural alterations that influence ribosome function (Hermann, 2005; Moore and Steitz, 2003, 2005; Ogle and Ramakrishnan, 2005; Tenson and Mankin, 2006). Screening growth of test strains in the presence of drugs at a concentration near the threshold level of sensitivity can reveal ribosome differences that are otherwise not apparent (e.g., King *et al.* [2003]; Vos *et al.* [2004]).

Table 13.1 Ribosome-based antibiotics for analyzing drug sensitivity of mutant eukaryotic ribosomes

Antibiotic	Specificity	Threshold[a]	Diffusion assay[a]
Paromomycin	SSU	600 μg/ml[b]	250 mg/ml[c]
Neomycin	SSU	>500 μg/ml[d]	250 mg/ml
Hygromycin B	SSU	100 μg/ml[b,e]	20 mg/ml[c]
Geneticin (G418)	SSU	60–100 μg/ml[e,f]	~40 mg/ml[g]
Streptomycin	SSU	>100 μg/ml[d]	10 mM^h
Cycloheximide	LSU	2 μg/ml[d]	2–5 m$M^{h,i}$
Anisomycin	LSU	20–50 μg/ml[e,j]	2–10 mg/ml[c,i]
Sparsomycin	LSU	20 μg/ml[f]	2 mg/ml[i]

[a] Threshold sensitivity concentration in solid medium and working concentrations in disk diffusion assays (10 μl/disk) are only approximate and can differ for different parental yeast strains.
[b] Benard *et al.* (1998).
[c] Sobel and Wolin (1999).
[d] Shockman and Lampen (1962).
[e] Spence *et al.* (2000).
[f] Meskauskas and Dinman (2001).
[g] Hendrick *et al.* (2001).
[h] Kinzy and Woolford (1995).
[i] Dinman and Kinzy. (1997).
[j] Fleischer *et al.* (2006).

   Drug sensitivity screening can be performed effectively on plates with solid medium supplemented with antibiotic, which binds to either the small or large subunit (Table 13.1). The concentration of drug used should be determined empirically, because different parental yeast strains can differ in sensitivity. Test and control strains are grown in liquid medium at 30° to an OD_{600} of 3.0 (5.5×10^7 cells/ml). The concentration is adjusted to 5×10^6 cells/ml (OD_{600} of ~0.4) and then diluted 1:5 serially. Each dilution is spotted as for the dilution-spot assays earlier, and the plate is incubated at the desired temperature for 1 or 2 days. An alternate method for testing sensitivity is to dispense a solution of the drug onto a sterile 0.25-inch paper filter disk on solid medium that has been seeded with the strain to be analyzed (0.3 ml of cells at OD_{600} of 0.4 per plate, well spread). The zone of growth inhibition around the disk reflects the degree of drug sensitivity.
   Interpreting altered drug sensitivity is not straightforward, because the effects may not stem solely from ribosome changes in the drug-binding region. Remote effects can reasonably be expected because of the complex nature of the interplay between and within the subunits; these relationships are only beginning to be understood from the crystallographic and cryo-em structural studies. Although changes in ribosome-based drug effects can be expected to stem from altered ribosome structure, secondary effects must be considered as well, including anomalous subunit ratios or a substantial difference in ribosome levels because of changes in ribosome synthesis or turnover. Methods to address these issues are covered in the next section.

4. DETERMINING EFFECTS ON rRNA AND RIBOSOME BIOGENESIS

In yeast, the 35S pre-rRNA is first cleaved at sites A_0 and A_1 in the $5'$ ETS to generate the 32S pre-rRNA. Subsequent cleavage at site A_2 in the ITS1 separates 20S and 27S precursors of the small and large subunit rRNAs, respectively. The 20S pre-rRNA is cleaved to mature 18S rRNA in the cytoplasm, whereas the 27S species is processed into the mature 5.8S and 25S rRNAs in the nucleus through two alternative pathways (for review, see Fatica and Tollervey, 2002; Nazar, 2004; Venema and Tollervey, 1999). A brief schematic representation of the processing pathways is shown in Fig. 13.5A. *In vivo* rRNA processing is conveniently monitored by

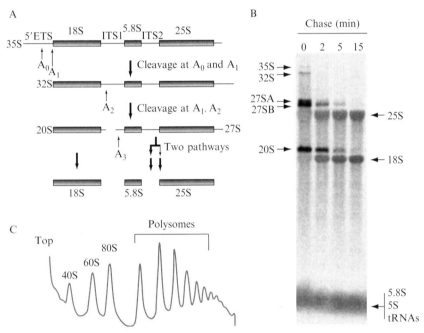

Figure 13.5 Determining effects of modification on rRNA processing and formation of ribosomal complexes. (A) Schematic of major pre-rRNA processing cleavages in yeast rRNA. (B) Pulse-chase analysis of pre-rRNA processing. Test cells are incubated with [^{3}H]-methionine (or [^{3}H]-uracil) to label pre-rRNA. An excess of unlabeled methionine is then added, and cells are sampled at various times. Total RNA is separated on a 1.2% agarose gel, the RNA is transferred to a membrane, and the pattern is visualized by fluorography. The bands representing the major precursor species and mature 18S and 25S rRNAs are indicated. (C) Analysis of a polysome profile. Extracts prepared from test cells are fractionated through a 7–47% sucrose gradient, and ribosomal complexes are determined by measuring UV absorbance with a flow cell. The identities of the major ribosomal complexes are indicated. (See color insert.)

examining the precursors and products through pulse-chase labeling with [methyl-^{3}H]methionine, which donates the labeled methyl group to pre-rRNA during methylation. Alternately, pre-rRNAs can be labeled with [5,6-^{3}H]uracil, which is incorporated into rRNA during transcription, minimizing variations that could occur if methylation is defective. The commonly detected intermediates are 35S, 32S, 27SA, and 27SB for the large subunit rRNAs and the 20S precursor for 18S rRNA. The approach described in the following is based on Li *et al.* (1990) and Tollervey (1987). In addition, pre-rRNA processing can also be examined by Northern and primer extension assays, the latter for more precise information about cleavage accuracy (e.g., Henry *et al.*, 1994; King *et al.*, 2001, 2003; Kos and Tollervey, 2005).

4.1. Monitoring rRNA processing by *in vivo* pulse-chase labeling

4.1.1. Cell growth and *in vivo* labeling

Test strains are grown in selective medium lacking methionine for 12–14 h, diluted into 10 ml of the same, prewarmed medium to an OD_{600} of 0.2–0.3 (2.5–3.8 × 10^6 cells/ml), and then growth continued until the OD_{600} reaches 0.8–1.0 (early mid-log phase; 1.3–1.8 × 10^7 cells/ml), approximately 3–6 h generally. If a test strain has a different growth rate, the initial concentration should be adjusted accordingly to obtain similar final concentrations (we recommend that the differences in OD_{600} should be <20%). Next, 2.4 OD_{600} units of cells (~5 × 10^7 cells/ml) are collected for each strain and the volume adjusted to 3 ml with prewarmed medium. To begin labeling, 100 μl (1 μCi/μl, ~1.4 mmol) [methyl-^{3}H]methionine (70–85 Ci/mmol; Perkin-Elmer) is then added and mixed immediately. After 3 min of labeling, the radioactivity is chased with 200 μl 60 mM unlabeled D,L-methionine. At time points of 0, 2, 5, and 15 min after addition of the unlabeled methionine, 800 μl of cell culture are taken and frozen in liquid nitrogen immediately. The sampling interval for each strain should be the same. Labeling with 100 μl of [5,6-^{3}H]uracil (30–50 Ci/mmol; Perkin-Elmer) is performed in a similar way, except that the medium lacks uracil, and the chase is performed with 1 mg/ml unlabeled uracil.

4.1.2. RNA extraction

RNA is prepared from the labeled cells by hot phenol extraction (e.g., Köhrer and Domdey [1991]; Wise [1991]), or another preferred method. Some laboratory members isolate RNA with a procedure that features TRI Reagent (Sigma). The latter procedure is carried out according to manufacturer's instructions with accommodations for yeast lysis. In the TRI Reagent procedure, cells are thawed on ice, collected by centrifugation, and washed once with ice-cold, sterile water. The cell pellet is resuspended in 100 μl RNase-free lysis buffer (50 mM Tris-Cl, pH 7.4; 100 mM NaCl; 10 mM EDTA, in water containing 0.1% diethyl pyrocarbonate [DEPC; from Sigma]), followed by addition of 10 μl of 10% SDS and 100 μl acid-washed glass beads (425–600 microns; Sigma). Cells are broken by vigorous vortex mixing at high speed for 3 min, and 900 μl TRI Reagent is added and mixed by vigorous shaking. The samples are incubated at room temperature for 7 min, and 200 μl chloroform is added and mixed by vigorous shaking. After incubating 5 min at room temperature, samples are centrifuged at 4°, 11,000g for 15 min. The upper phase (550 μl) for each sample is transferred to a clean tube and mixed with 500 μl isopropanol. After a 5-min incubation, RNA is precipitated by centrifugation at 12,000g, 4°, for 10 min. The RNA pellet is washed once with 300 μl 75% ethanol, air-dried, and dissolved in 10 μl DEPC (0.1%) water.

4.1.3. Denaturing gel electrophoresis

Three microliters total RNA from each sample, or if preferred, equivalent counts per minute (CPMs) of RNA, is diluted in 5 μl water containing 0.1% DEPC. Next, 10 μl formamide, 3 μl formaldehyde, and 2 μl 10× MOPS buffer (0.4 M MOPS, 0.1 M NaOAC, and 10 mM EDTA, in DEPC water, adjusted to pH 7.0) are added to each sample and mixed. After 5 min at 65° and 2 min cooling on ice, the following are added: 2 μl loading buffer (50% Ficoll-400 [Sigma] and 1% bromophenol blue) and 0.5 μl 10 mg/ml ethidium bromide (EB). Samples are loaded on a 1.2% denaturing agarose gel prepared with 17% (v/v) formaldehyde and 1× MOPS. Electrophoresis is performed in a chemical hood with 1× MOPS buffer at 55–70 volts. After photographing under ultraviolet (UV) light, the gel is soaked in 50 mM NaOH for 20 min at room temperature, washed twice with water, and soaked in 2× SSC buffer (0.3 M NaCl, 30 mM sodium citrate, pH 7.0) for 30 min. RNA is transferred to a membrane (GeneScreen Plus® [Perkin-Elmer]) in a chemical hood for 14–16 h by passive downward transfer with blotting paper saturated with 10× SSC buffer on the top.

The membrane is air-dried and cross-linked under UV at 0.125 joules, or other low energy setting, for 2 × 30 sec, once on each side of the membrane. Although GeneScreen Plus does not require crosslinking, in our experience it is best to always include this step if rehybridization is anticipated. Results from pulse-chase analysis are visualized by autoradiography by use of a storage-phosphor imaging system or X-ray film. For film autoradiography, the membrane is sprayed three times with Enhance Spray® (Perkin-Elmer), and air-dried for 20 min each time. The RNA side of the membrane should be next to the film for exposure.

4.1.4. Semi-denaturing gel electrophoresis

The denaturing electrophoresis procedure described previously is used to disrupt secondary and tertiary RNA structure; however, it involves the use of formamide and formaldehyde. To minimize the use of these toxic reagents, gel electrophoresis can alternately be carried out with a semidenaturing gel, because the rRNA species in yeast are well characterized, and there is no need to determine the exact size in all cases. Three microliters RNA ($\sim$3 μg) or RNA with equivalent radioactivity is mixed with 3 μl RNA loading buffer (80% [v/v] formamide; 1 mM EDTA; 0.1% bromophenol blue; 0.1% xylene cyanol), heated at 65° for 5 min, and cooled on ice for 2 min. Next, RNA samples are loaded on a 1.2% agarose gel prepared with TBE buffer (90 mM Tris base; 90 mM boric acid; 20 mM EDTA) and separated at 80 V constant with TBE as the buffer. The gel is then soaked in 50 mM NaOH with 2 μg/ml EB for 20 min. The gel is washed briefly by immersion in water, the RNA pattern is photographed under UV light, and the RNA is transferred to a membrane as described previously. Results from pulse-chase analysis are visualized in the same manner as for denaturing gel electrophoresis. This approach yields excellent resolution with sharp rRNA bands; one such example is shown in Fig. 13.5B.

Pre-rRNA processing defects can also be examined by conventional Northern analysis to detect different processing intermediates by use of specific oligonucleotides (e.g., Fatica *et al.*, 2002; King *et al.*, 2001). Exact cleavage sites are detected by primer extension analysis, to evaluate the processing accuracy, with oligonucleotides downstream of the site to be examined (e.g., Fatica *et al.*, 2003; Henry *et al.*, 1994; King *et al.*, 2001).

4.2. Determining the steady-state level of rRNA

Yeast strains are grown as previously to an OD_{600} of $\sim$1.0 (with differences between strains less than 20%). One OD_{600} unit of cells is collected for each strain in triplicate, and RNA is prepared and analyzed independently to minimize experimental error. If cells from strains differ in size, it is important to know the cell concentration rather than to rely on optical density to collect equal number of cells. Cell concentration can be determined by viewing samples under a light microscope with a cell counting chamber.

RNA is extracted and dissolved in 10 μl water with 0.1% DPEC. Three microliters RNA is mixed with 3 μl RNA loading buffer, heated at 65° for 5 min, cooled on ice, and separated on a 1.2% denaturing or semidenaturing agarose gel. After electrophoresis, the gel is soaked in 50 mM NaOH with 2 μg/ml EB for 20 min, washed twice with water, and rRNA is visualized under UV light. The staining intensity can be used to evaluate the relative levels of rRNA if the staining signal is not saturated. For hybridization, RNA is transferred to a membrane (GeneScreen Plus® [Perkin-Elmer]) by use of passive elution as described previously. Prehybridization is performed at 42° for 15 min with 5–10 ml Rapid-hyb™ hybridization buffer (Amersham). Next, 10 μl of labeled oligonucleotide (see later) is added, and hybridization is performed at 42° for 30 min. The membrane is washed twice with wash buffer (2× SSC, pH 7.0; 0.1% SDS) at 42° for 10 min, and then washed once with water for 15 min at room temperature. The result is visualized by storage-phosphor or X-ray film autoradiography. With the Rapid-hyb hybridization buffer and this approach, hybridization and washing for the Northern RNA blotting can be completed in 90 min.

Labeling of primers at the 5′ end is performed with T4 polynucleotide kinase (PNK). Typically, 20 pmol of oligonucleotide are labeled in a 20-μl reaction with 0.5 μl (gamma-^{32}P) ATP ($\sim$0.05 mCi), 2 μl PNK buffer, and 1 μl (10 unit) PNK (Biolabs). The reaction is performed at 37° for 1 h. The kinase is inactivated by heating at 95° for 1 min. Ten microliters of primer labeled in this way is used directly for hybridization without further purification, and a low background of nonspecific hybridization is generally obtained. For detecting RNAs of very low abundance, such as pre-rRNAs, the labeled probe from the kinase reaction should be purified by polyacrylamide gel electrophoresis or by passage through Sephadex G-25 to remove the shorter ($<$12 nts) oligonucleotide products and unincorporated radiolabeled nucleotide. U2 snRNA ($\sim$1.2 kb in *S. cerevisiae*) can be used as an internal control to normalize sample loading. A semidenaturing gel is suitable for detecting rRNAs and U2 snRNA by hybridization. The membrane can be reused for hybridization after boiling for 8 min with wash buffer.

4.3. Evaluating rRNA stability with *in vivo* labeling

To determine effects on rRNA stability, the optimal approach would be to use a mutant strain in which transcription by RNA polymerase I can be inhibited conditionally (e.g., Nogi *et al.*, 1991; Oakes *et al.*, 1993). We have not yet used such strains in our modification depletion analyses. Rather, we use transient labeling of the rRNA to examine stability. This approach cannot be used to determine the exact half-life of an rRNA; however, it is very satisfactory for use in evaluating rRNA stability between strains.

Cells are grown in medium without methionine to an OD_{600} of 0.8–1.0 (1.3–1.8 × 10^7 cells/ml). Four OD_{600} units of early- to mid-log phase cells in 5 ml are taken for each sample. To minimize error, each test strain should be examined with duplicate (or more) samples. To each sample 100 μl (methyl-^{3}H)methionine (1 μCi/μl, $\sim$1.4 mmol; Perkin-Elmer) is added, and pulse-labeling is performed at room temperature for 10 min. Subsequently, 200 μl of 60 mM unlabeled D,L-methionine is added to each culture, and a chase is performed at room temperature for 20 min to ensure the maturation of the labeled rRNA. Next, cells are pelleted and washed twice with fresh medium to remove the unincorporated [methyl-^{3}H]methionine. The pellet is resuspended in 5 ml medium supplemented with 300 μl of 60 mM unlabeled methionine. Cells are incubated at 30°, and 1 ml is taken at time points of 0, 20, 40, and 60 h. Immediately after collection, cells are pelleted, washed once with water, and stored at $-80°$.

The cell pellets are thawed on ice, and RNA is prepared with the TRI Reagent (or other method) and dissolved in 20 μl DEPC water. The RNA concentration is determined for each sample by measuring A_{260}. Next, an equal volume (25%, 5 μl) is taken for each sample, and the amounts of RNA adjusted to be equivalent with unlabeled total RNA, to a final volume of 7 μl; this manipulation minimizes signal variation. Next, 7 μl RNA loading buffer (80% [v/v] formamide; 1 mM EDTA; 0.1% bromophenol blue; 0.1% xylene cyanol) is added and mixed. After heating at 65° for 5 min, the samples are cooled on ice and loaded on a 1.2% agarose gel prepared with 1× TBE buffer. Electrophoresis, transfer, and visualization are performed as described previously. The amount of signal is quantified with ImageJ (NIH) or similar software, and the relative level of labeled rRNA at different time points is calculated.

5. INVESTIGATING RIBOSOMAL COMPLEXES

Differences in nucleotide modification could affect the patterns of various precursor ribosomal rRNP complexes as well as mature subunits, ribosomes, and polysomes. Ribosomal complexes in living cells are conveniently examined by sedimentation through sucrose gradients followed by measuring the absorbance of fractions at 254 nm. The major complexes detected are: 40S and 60S free subunits, 80S ribosomes, and polysome complexes involved in translation elongation that contain more than one 80S ribosome (Warner and Knopf, 2002). Substantial defects in subunit assembly or stability may be detected in this way (e.g., Benard *et al.*, 1998; Charollais *et al.*, 2003; Deshmukh *et al.*, 1995; Emery *et al.*, 2004; King *et al.*, 2003; West *et al.*, 2005), although the resolving power is only modest. An imbalance in the ratio of subunits, specifically a deficiency in large subunits, or a defect in association of subunits can lead to formation of "halfmer" polysomes. These correspond to mRNA with one or more 80S ribosomes and a small subunit bound at the beginning of the mRNA, which appear as "shoulder" peaks to the typical 80S and polysome peaks (Helser *et al.*, 1981). The ratio and abundance of subunits in the cell is evaluated by dissociating the mono- and polyribosomes (Foiani *et al.*, 1991).

5.1. Polysome profiling

Test strains are grown in 20 ml of appropriate medium for 14–16 h, and then diluted to 150–200 ml in the same prewarmed medium to an OD_{600} of 0.1–0.2 (1.3–2.5 × 10^6 cells/ml). Growth is continued to OD_{600} of 0.8–1.0 (1.3–1.8 × 10^7 cells/ml; with the difference in OD_{600} between strains <20%). Cycloheximide (Sigma #C7698) is added with shaking to a final concentration of 100 μg/ml, and cultures are incubated on ice for 10 min to arrest translation. Others use 5 min on ice or centrifuge samples immediately after addition of cycloheximide (Baim *et al.*, 1985). We assume it is important to be consistent in these manipulations and to chill the mixture rapidly; commonly, we add an equal volume of ice-chilled water supplemented with the final concentration of cycloheximide directly to the culture mixture.

Cells are pelleted at 4° by centrifugation and washed once with 20 ml ice-cold lysis buffer (10 mM Tris-Cl, pH 7.4; 100 mM NaCl; 30 mM MgCl$_2$; 100 μg/ml cycloheximide; 200 μg/ml heparin in 0.1% DEPC water). The pellet is then resuspended in 1 ml lysis buffer, transferred to 15-ml polypropylene conical tubes, and a quarter volume of acid-washed glass beads (425–600 microns; Sigma) is added. The cells are then broken by 12 cycles of vigorous vortexing for 30 sec, followed by 30 sec on ice. Debris is removed by centrifugation at 5000g at 4° for 5 min (SLA-1500 rotor). The supernatant is transferred to a prechilled Microfuge tube, and further cleared by centrifugation at 12,000g (11,000 rpm with 9.2 cm radius rotor), 4°, for 8 min. The supernatant (extract) is transferred to a clean tube, and the A$_{260}$ of a 1000-fold dilution is measured.

An amount of extract corresponding to 10 A$_{260}$ units is loaded on to a ~12 ml 7%–47% sucrose gradient buffered with 50 mM Tris-Cl, pH 7.0; 50 mM NH$_4$Cl; 12 mM MgCl$_2$; and 1 mM DTT (added freshly). A larger volume gradient allows better resolution of more polysomes. Centrifugation is performed at 39,000 rpm (261,000g) for 2 h and 18 min or 41,000 rpm (288,000g), 4°, for 2 h, using a SW41 TI rotor (Beckman). The peaks corresponding to the ribosomal complexes are detected by monitoring with absorbance at 254 nm with a flow cell, with the following parameters in our system: baseline 6, sensitivity 0.5, and lamp current 340 mA (ISCO UA-5 monitor). The absorbance is measured every ~0.1 sec with Logger Pro (Ver. 2.2, Tufts University and Vernier software) and plotted with Excel (Microsoft) software. A sample polysome profile is shown in Fig. 13.5C.

5.2. Distinguishing the elongating "80S" particle

The 80S ribosomal complexes in the gradient profiles include intact ribosomes bound to mRNA, ribosome couples not bound to mRNA, and 80S preinitiation complexes. The 80S ribosome engaged in elongation (bound to mRNA) is resistant to high salt, whereas the other 80S complexes can be dissociated into 40S and 60S subunits at high salt concentration (Chuang et al., 1997; Zhong and Arndt, 1993).

Ten A$_{260}$ units of extract prepared as earlier for polysome profiles is mixed with an equal volume of high salt buffer (10 mM Tris-Cl, pH 7.0; 1.9 M NaCl; 30 mM MgCl$_2$; 50 μg/ml cycloheximide; 200 μg/ml heparin), to a final concentration of 1 M NaCl. The extract is incubated on ice for 1 h to dissociate the unbound 80S couples. The extract is then loaded onto an 11 ml 7%–47% sucrose gradient buffered with 50 mM Tris-Cl, pH 7.0, 50 mM NH$_4$Cl, 12 mM MgCl$_2$, 1M NaCl, and 1 mM DTT. Centrifugation is performed at 39,000 rpm (261,000g), 4°, for 3 h with a

SW41 TI rotor, and the absorbance is measured and plotted as described previously for polysome profiling.

5.3. Determining the 40S/60S ratio

To examine the relative ratio of 40S and 60S subunits, 80S and polysome complexes need to be dissociated completely into subunits. To measure the level of the two subunits, test strains are grown in 100 ml medium to OD_{600} of 0.8–1.0 (1.2–1.8×10^7 cells/ml), pelleted, washed twice with 10 ml buffer C (50 mM Tris-Cl, pH 7.4; 50 mM NaCl; and 1 mM DTT), and resuspended in 500 μl buffer C. After lysis by glass beads, 500 μl buffer C is added, and extract is prepared subsequently as described previously for polysome profiling. Two A_{260} units of extract are loaded on a 10 ml 7%–47% sucrose gradient prepared with buffer C. Centrifugation is performed at 39,000 rpm (261,000g), 4°, for 3 h with a SW41 TI rotor (Beckman). Absorbance is measured and plotted as described previously.

6. CHARACTERIZING RIBOSOME FUNCTION

Loss of nucleotide modifications could influence ribosome activity directly or indirectly. Modifications that are concentrated in regions known to be important for ribosome function are obvious candidates, but modifications in other regions could be important too. At present, our screening of ribosome activity is limited to *in vivo* analyses, in particular measuring the rate and capacity of protein synthesis and examining translational fidelity. Protein synthesis assays can also be carried out *in vitro* (e.g., Algire *et al.*, 2002; Iizuka and Sarnow, 1997; Iizuka *et al.*, 1994; Tarun *et al.*, 1997).

6.1. Analysis of the rate of protein synthesis

The translation rate of ribosomes *in vivo* is assessed by measuring [^{35}S] methionine incorporation. Test and control strains are incubated at 30° in minimal medium (Burke *et al.*, 2000) deficient in methionine until an OD_{600} value of approximately 0.8 ($\sim 1.3 \times 10^7$ cells/ml) is reached. Ten μCi of [^{35}S]methionine (1175 Ci/mmol; Perkin-Elmer) is added to 10 ml of culture. At this point (zero time point), 0.5 ml of labeled cells is taken and mixed immediately with 0.2 ml ice-cold stop solution (70 μl of unlabeled 60 mM D,L-methionine, 1 μl of 10 mg/ml cycloheximide, and 129 μl of 50% TCA) (Liu and Fournier, 2004). Next, four or five samples of labeled cells are taken at intervals of one to a few minutes, depending on the severity of the effects with modification loss. Potential complications related to limiting methionine can be avoided by labeling cells with [^{35}S]methionine mixed with unlabeled methionine (50 μM final concentration in culture).

For ease in handling multiple samples or with long sampling periods, cells taken at each time point can be moved to $-80°$ instead of being left on ice.

After incubation on ice for 10 min, samples are heated to $70°$ for 20 min and next filtered through 25-mm Whatman GF/C glass microfiber filters. With spray bottles, filters are washed three times with approximately 5 ml of ice-cold 5% TCA and then three times with approximately 5 ml of cold 95% ethanol. The filters are dried and then analyzed by scintillation counting immersed in 5 ml of Scintisafe™ fluid (Fisher Scientific). Background activity is subtracted, and incorporation values are normalized to the approximate number of cells. The rate of amino acid incorporation is plotted against time, generally listed as counts per minute or (better) pmol or nmol per minute. Where differences are observed, potential secondary effects on amino acid uptake should also be considered (e.g., Tanudjojo et al. [2002]; Liang and Fournier, unpublished).

6.2. Analyzing translation fidelity

Studies on translation fidelity are typically performed by use of a plasmid-borne reporter gene system that can express an easily assayable protein, usually an enzyme, such as β-galactosidase or luciferase (e.g., Chauvin et al., 2005; Hirabayashi et al., 2006; Jacobs and Dinman, 2004; Robert and Brakier-Gingras, 2003). Very sensitive bicistronic reporter systems are available for analysis of recoding events in yeast (e.g., Bidou et al., 2000; Jacobs and Dinman, 2004; Meskauskas et al., 2005; Namy et al., 2004; Stahl et al., 1995, 2004; Vimaladithan and Farabaugh, 1998). One paradigm for such systems features a set of test sequences between the open reading frames of β-galactosidase and luciferase; test sequences include stop codons with various termination signals or programmed frameshift sites (e.g., Bidou et al., 2000; Namy et al., 2004, 2005; Stahl et al., 1995, 2004; Vimaladithan and Farabaugh, 1998). Fidelity of translation is calculated from the degree of stop codon read-through or frameshifting expressed as activity of luciferase, normalized to the level of β-galactosidase activity. β-Galactosidase activity can be measured with a standard assay system. Reagents for lysing yeast cells and assaying activity are commercially available (Yeast β-Galactosidase Assay Kit, Pierce, No 75768). Light production with a luciferase reporter gene is measured with a luminometer.

In yeast, the sequences immediately flanking stop codons influence the efficiency of translation termination; thus, termination efficiency in test strains can be analyzed for stop codons with weak, average, or strong termination signals (Bertram et al., 2001; Bonetti et al., 1995; Namy et al., 2001; Williams et al., 2004). In ade2 test strains, the red pigmentation that accumulates during growth as the adenine concentration decreases (e.g., Ugolini and Bruschi [1996]) can be eliminated by including an ADE2 marker gene on the expression-assay plasmid.

7. Chemical Probing of rRNA Structure

Differences in rRNA structure can be examined with a variety of chemical modifying agents both *in vivo* and *in vitro*. We prefer *in vivo* probing, so that results are obtained for natural rRNP complexes, although this is a mixed population. Our reagent of choice for *in vivo* probing is dimethylsulfate (DMS), which methylates accessible A (N1) and C (N3) sites of RNA. Another reagent, a-keto-β-ethoxybutyraldehyde (kethoxal) has also been used for *in vivo* structural probing of other rRNAs (Balzer and Wagner, 1998); however, permeabilized cells are needed, and we do not know whether such treatment is appropriate for yeast. DMS is small and readily taken up by cells, and probing can be done in standard culture medium. A method for probing with DMS (Mereau *et al.*, 1997) is described, followed by brief comments about chemical probes available for screening rRNA structure *in vitro*.

7.1. DMS modification

Strains are grown in minimal, selective medium to OD_{600} of 1.5–2.0 (2.8–3.7×10^7 cells/ml), with the differences in OD_{600} between test strains $<20\%$. Approximately 25–30 ml of culture is diluted with 45 ml of pre-warmed medium to an OD_{600} of 1 (1.8×10^7 cells/ml), and four samples of 10 ml are transferred to 50-ml polypropylene tubes. Next, fresh DMS (Sigma) is added to each tube to final concentrations of 0, 40, 80, 160 mM, respectively, to identify a suitable condition for modification probing (the molarity of pure DMS is 10.6 M). Reaction mixtures are incubated for 2 min at room temperature with rocking in a chemical hood. Next, ice-cold 2-mercaptoethanol is added to a final concentration of 0.7 M (the molarity of pure mercaptoethanol is 14.3 M). After mixing by vigorous shaking, 5 ml water-saturated, ice-cold isoamyl alcohol is added and mixed. The cells are collected by centrifugation at 3000g for 10 min, and the supernatant is carefully removed and transferred to a container marked for DMS waste. The cell pellet is washed once with 0.7 ml wash buffer (10 mM MgCl$_2$, 3 mM CaCl$_2$, 10 mM Tris-HCl pH 7.5, 250 mM sucrose, and 0.7 M mercaptoethanol) and resuspended in 150 μl RNase-free lysis buffer (50 mM Tris-Cl, pH 7.4; 100 mM NaCl; 10 mM EDTA, in water containing 0.1% DEPC). Fifteen microliters 10% SDS and 150 μl glass beads are added, and RNA is prepared with TRI Reagent as described previously in the section on RNA extraction. The RNA pellet is dissolved in 15 μl DEPC water and the concentration determined from the measurement of A_{260}.

It should be noted that DMS is toxic, and care must be taken as per the manufacturer's recommendations (http://www.dupont.com/dms/safety/ppe.html); inhalation is hazardous, and it is readily absorbed through the skin. Thus, DMS should be handled only in a chemical hood with double nonlatex gloves. The isoamyl alcohol and 2-mercaptoethanol used in the procedure are also volatile and noxious. Waste from this experiment should be stored in closed, marked containers, and good hood ventilation is also important.

7.2. *In vitro* probing

Both chemical reagents and enzymes are used for probing RNA *in vitro* as reviewed elsewhere (Brunel and Romby, 2000; Ehresmann *et al.*, 1987; Stern *et al.*, 1988). Procedures for chemical probing of rRNA in ribosomal complexes have been described for several reagents (e.g., DMS, N-cyclohexyl-N′-beta-[4-methylmorpholinium] ethylcarbodiimide [CMCT], and a-keto-β-ethoxybutyraldehyde [kethoxal]). These particular agents have been used in yeast ribosome studies (e.g., Alkemar and Nygard, 2006; Briones and Ballesta, 2000; Briones *et al.*, 1998; Hogan *et al.*, 1984a,b; Lempereur *et al.*, 1985; Velichutina *et al.*, 2000), and with bacterial ribosomes as well (as reviewed in Brunel and Romby, 2000; Ehresmann *et al.*, 1987; Moazed *et al.*, 1986; Stern *et al.*, 1988). Other chemical probes featured with bacterial ribosomes are Fe^{2+}-EDTA, which attacks accessible single and double-stranded RNA in a sequence-nonspecific manner (Brenowitz *et al.*, 2002; Heilek *et al.*, 1995), and Pb^{2+}, which cleaves single-stranded regions (Lindell *et al.*, 2002). Enzymatic probing features nucleases such as S1, T1, and V1, and are described elsewhere (e.g., Balakin *et al.*, 1996; Ehresmann *et al.*, 1987; Knapp, 1989; Normand *et al.*, 2006; Nygard *et al.*, 2006; Stern *et al.*, 1988).

7.3. Primer extension analysis

The DMS modification levels at particular sites can be detected by primer extension analysis, because A and C nucleotides methylated by DMS cause extension to pause or stop one nucleotide before the modified site. Oligonucleotides are labeled with T4 PNK as described previously, except that 80 pmol oligonucleotide is labeled in a 20 μl reaction. A primer labeled in this way usually works well without further purification. Equal amounts of RNA (5 μg) prepared from DMS-modified cells are adjusted to 8 μl with DEPC water, mixed with 2 μl of labeled primer (1:6 dilution of the labeling reaction), and annealed at 65° for 10 min. Although such an abridged denaturing/annealing step is fine in most cases, for highly structured regions a more standard denaturing/annealing sequence (80° for 3 min and then cool to room temperature gradually for 10 min) may be necessary. After chilling on ice for 2 min, 2.5 μl 100 mM DTT, 2 μl 25 mM $MgCl_2$, 2 μl 5 mM dNTPs, 5 μl 5× AMV RT buffer (USB), 3 units of AMV reverse transcriptase (USB), and 3 units of RNasin (Promega) in 1 μl are added to each reaction. The extension reaction is performed at 37° for 1 h, and the products are precipitated with ethanol (3 vol of 100% ethanol and 0.3 M NaOAC, pH 5.2), washed once with 75% ethanol, and analyzed on an 8% polyacrylamide-7 M urea sequencing gel (Sequagel-8; National Diagnostics). The gel is dried, and results are visualized by autoradiography by use of a storage-phosphor imaging system or X-ray film.

7.4. Primer extension sequencing

To determine the position of the DMS-modified nucleotides, the gel should include a set of sequencing reactions performed with the same radiolabeled oligonucleotide. Primer extension sequencing is conducted as follows (Lowe and Eddy, 1999; McPheeters *et al.*, 1986): 3 μl total RNA (6–10 μg) is mixed with 1 μl 10× RT(-)Mg buffer (0.5 M Tris-Cl pH 8.6; 0.6 M NaCl; 0.1 M DTT), 2 μl of 1:6 diluted primer, and 4 μl DEPC water to a total volume of 10 μl. Samples are heated at 65° for 5 min, cooled on ice for 2 min. Two μl portions are transferred to four tubes, each contains 1 μl 5× dNTPs (1.7 mM each of the four nucleotides in 1× RT[+]Mg buffer [50 mM Tris-Cl, pH 8.6; 60 mM NaCl; 6 mM MgCl$_2$; 10 mM DTT]), 1 μl RVT mix (2 unit/μl reverse transcriptase [USB], and 2 unit/μl of RNasin [Promega] in 1× RT[+]Mg buffer), and 1 μl of the appropriate 5× ddNTP (1 mM of either ddT, ddC, ddG, or ddA in 1× RT[+]Mg buffer). The extension reactions are performed at 37° for 30 min and stopped by addition of 5 μl of RNA loading buffer as described previously for fractionation with a semidenaturing gel. Samples are analyzed on an 8% polyacrylamide-7 M urea gel as described previously, alongside the probing reactions.

ACKNOWLEDGMENTS

The procedures described were adapted from the literature, with alterations made by former laboratory mates in many cases. We gratefully acknowledge these colleagues and others who have provided good advice over the years. This work is supported by a grant to M.J.F. from the US Public Health Service (GM19351).

REFERENCES

Algire, M. A., Maag, D., Savio, P., Acker, M. G., Tarun, S. Z., Jr., Sachs, A. B., Asano, K., Nielsen, K. H., Olsen, D. S., Phan, L., Hinnebusch, A. G., and Lorsch, J. R. (2002). Development and characterization of a reconstituted yeast translation initiation system. *RNA* **8**, 382–397.

Alkemar, G., and Nygård, O. (2006). Probing the secondary structure of expansion segment ES6 in 18S ribosomal RNA. *Biochemistry (Mosc.)* **45**, 8067–8078.

Badis, G., Fromont-Racine, M., and Jacquier, A. (2003). A snoRNA that guides the two most conserved pseudouridine modifications within rRNA confers a growth advantage in yeast. *RNA* **9**, 771–779.

Baim, S. B., Pietras, D. F., Eustice, D. C., and Sherman, F. (1985). A mutation allowing an mRNA secondary structure diminishes translation of *Saccharomyces cerevisiae* iso-1-cytochrome c. *Mol. Cell. Biol.* **5**, 1839–1846.

Balakin, A. G., Schneider, G. S., Corbett, M. S., Ni, J., and Fournier, M. J. (1993). snR31, snR32, and snR33: Three novel, non-essential snRNAs from *Saccharomyces cerevisiae*. *Nucleic Acids Res.* **21**, 5391–5397.

Balakin, A. G., Smith, L., and Fournier, M. J. (1996). The RNA world of the nucleolus: Two major families of small RNAs defined by different box elements with related functions. *Cell* **86**, 823–834.

Balzer, M., and Wagner, R. (1998). A chemical modification method for the structural analysis of RNA and RNA-protein complexes within living cells. *Anal. Biochem.* **256**, 240–242.

Ban, N., Nissen, P., Hansen, J., Moore, P. B., and Steitz, T. A. (2000). The complete atomic structure of the large ribosomal subunit at 2.4 Å resolution. *Science* **289**, 905–920.

Benard, L., Carroll, K., Valle, R. C., and Wickner, R. B. (1998). Ski6p is a homolog of RNA-processing enzymes that affects translation of non-poly(A) mRNAs and 60S ribosomal subunit biogenesis. *Mol. Cell. Biol.* **18**, 2688–2696.

Berges, T., Petfalski, E., Tollervey, D., and Hurt, E. C. (1994). Synthetic lethality with fibrillarin identifies Nop77p, a nucleolar protein required for pre-rRNA processing and modification. *EMBO J.* **13**, 3136–3148.

Berisio, R., Schluenzen, F., Harms, J., Bashan, A., Auerbach, T., Baram, D., and Yonath, A. (2003). Structural insight into the role of the ribosomal tunnel in cellular regulation. *Nat. Struct. Biol.* **10**, 366–370.

Bertram, G., Innes, S., Minella, O., Richardson, J., and Stansfield, I. (2001). Endless possibilities: Translation termination and stop codon recognition. *Microbiology* **147**, 255–269.

Bertrand, E., and Fournier, M. J. (2004). The snoRNPs and Related Machines: Ancient Devices That Mediate Maturation of rRNA and Other RNAs. *In* "The Nucleolus" (M. O. J. Olson, ed.), pp. 223–257. Eurekah.com and Kluwer Academic/Plenum Publishers, Georgetown, TX; New York, NY.

Bidou, L., Stahl, G., Hatin, I., Namy, O., Rousset, J. P., and Farabaugh, P. J. (2000). Nonsense-mediated decay mutants do not affect programmed -1 frameshifting. *RNA* **6**, 952–961.

Bonetti, B., Fu, L., Moon, J., and Bedwell, D. M. (1995). The efficiency of translation termination is determined by a synergistic interplay between upstream and downstream sequences in *Saccharomyces cerevisiae*. *J. Mol. Biol.* **251**, 334–345.

Bortolin, M. L., Ganot, P., and Kiss, T. (1999). Elements essential for accumulation and function of small nucleolar RNAs directing site-specific pseudouridylation of ribosomal RNAs. *EMBO J.* **18**, 457–469.

Brenowitz, M., Chance, M. R., Dhavan, G., and Takamoto, K. (2002). Probing the structural dynamics of nucleic acids by quantitative time-resolved and equilibrium hydroxyl radical "footprinting." *Curr. Opin. Struct. Biol.* **12**, 648–653.

Briones, C., and Ballesta, J. P. (2000). Conformational changes induced in the *Saccharomyces cerevisiae* GTPase-associated rRNA by ribosomal stalk components and a translocation inhibitor. *Nucleic Acids Res.* **28**, 4497–4505.

Briones, E., Briones, C., Remacha, M., and Ballesta, J. P. (1998). The GTPase center protein L12 is required for correct ribosomal stalk assembly but not for *Saccharomyces cerevisiae* viability. *J. Biol. Chem.* **273**, 31956–31961.

Brunel, C., and Romby, P. (2000). Probing RNA structure and RNA-ligand complexes with chemical probes. *Methods Enzymol.* **318**, 3–21.

Buchhaupt, M., Meyer, B., Kotter, P., and Entian, K. D. (2006). Genetic evidence for 18S rRNA binding and an Rps19p assembly function of yeast nucleolar protein Nep1p. *Mol. Genet. Genomics* **276**, 273–284.

Burke, D., Dawson, D., and Stearns, T. (2000). "Methods in Yeast Genetics: A Cold Spring Harbor Laboratory Course Manual 2000 Edition." Cold Spring Harbor Laboratory Press, Cold Spring Harbor, New York.

Caffarelli, E., Fatica, A., Prislei, S., De Gregorio, E., Fragapane, P., and Bozzoni, I. (1996). Processing of the intron-encoded U16 and U18 snoRNAs: The conserved C and D boxes control both the processing reaction and the stability of the mature snoRNA. *EMBO J.* **15**, 1121–1131.

Cavaille, J., and Bachellerie, J. P. (1996). Processing of fibrillarin-associated snoRNAs from pre-mRNA introns: An exonucleolytic process exclusively directed by the common stem-box terminal structure. *Biochimie* **78**, 443–456.

Charollais, J., Pflieger, D., Vinh, J., Dreyfus, M., and Iost, I. (2003). The DEAD-box RNA helicase SrmB is involved in the assembly of 50S ribosomal subunits in *Escherichia coli*. *Mol. Microbiol.* **48**, 1253–1265.

Chauvin, C., Salhi, S., Le Goff, C., Viranaicken, W., Diop, D., and Jean-Jean, O. (2005). Involvement of human release factors eRF3a and eRF3b in translation termination and regulation of the termination complex formation. *Mol. Cell Biol.* **25**, 5801–5811.

Chuang, R. Y., Weaver, P. L., Liu, Z., and Chang, T. H. (1997). Requirement of the DEAD-box protein ded1p for messenger RNA translation. *Science* **275**, 1468–1471.

Cochella, L., and Green, R. (2004). Isolation of antibiotic resistance mutations in the rRNA by using an *in vitro* selection system. *Proc. Natl. Acad. Sci. USA* **101**, 3786–3791.

Davis, C. A., and Ares, M., Jr. (2006). Accumulation of unstable promoter-associated transcripts upon loss of the nuclear exosome subunit Rrp6p in *Saccharomyces cerevisiae*. *Proc. Natl. Acad. Sci. USA* **103**, 3262–3267.

Decatur, W. A., and Fournier, M. J. (2002). rRNA modifications and ribosome function. *Trends Biochem. Sci.* **27**, 344–351.

Dennis, P. P., and Omer, A. (2005). Small non-coding RNAs in Archaea. *Curr. Opin. Microbiol.* **8**, 685–694.

Deshmukh, M., Stark, J., Yeh, L. C., Lee, J. C., and Woolford, J. L., Jr. (1995). Multiple regions of yeast ribosomal protein L1 are important for its interaction with 5 S rRNA and assembly into ribosomes. *J. Biol. Chem.* **270**, 30148–30156.

Dinman, J. D., and Kinzy, T. G. (1997). Translational misreading: Mutations in translation elongation factor 1alpha differentially affect programmed ribosomal frameshifting and drug sensitivity. *RNA* **3**, 870–881.

Dresios, J., Derkatch, I. L., Liebman, S. W., and Synetos, D. (2000). Yeast ribosomal protein L24 affects the kinetics of protein synthesis and ribosomal protein L39 improves translational accuracy, while mutants lacking both remain viable. *Biochemistry (Mosc.)* **39**, 7236–7244.

Ehresmann, C., Baudin, F., Mougel, M., Romby, P., Ebel, J. P., and Ehresmann, B. (1987). Probing the structure of RNAs in solution. *Nucleic Acids Res.* **15**, 9109–9128.

Emery, B., de la Cruz, J., Rocak, S., Deloche, O., and Linder, P. (2004). Has1p, a member of the DEAD-box family, is required for 40S ribosomal subunit biogenesis in *Saccharomyces cerevisiae*. *Mol. Microbiol.* **52**, 141–158.

Fatica, A., Cronshaw, A. D., Dlakic, M., and Tollervey, D. (2002). Ssf1p prevents premature processing of an early pre-60S ribosomal particle. *Mol. Cell* **9**, 341–351.

Fatica, A., Oeffinger, M., Tollervey, D., and Bozzoni, I. (2003). Cic1p/Nsa3p is required for synthesis and nuclear export of 60S ribosomal subunits. *RNA* **9**, 1431–1436.

Fatica, A., and Tollervey, D. (2002). Making ribosomes. *Curr. Opin. Cell Biol.* **14**, 313–318.

Fatica, A., and Tollervey, D. (2003). Insights into the structure and function of a guide RNP. *Nat. Struct. Biol.* **10**, 237–239.

Ferreira-Cerca, S., Poll, G., Gleizes, P. E., Tschochner, H., and Milkereit, P. (2005). Roles of eukaryotic ribosomal proteins in maturation and transport of pre-18S rRNA and ribosome function. *Mol. Cell* **20**, 263–275.

Fleischer, T. C., Weaver, C. M., McAfee, K. J., Jennings and Link, A. J. (2006). Systematic identification and functional screens of uncharacterized proteins associated with eukaryotic ribosomal complexes. *Genes Dev.* **20,** 1294–1307.

Foiani, M., Cigan, A. M., Paddon, C. J., Harashima, S., and Hinnebusch, A. G. (1991). GCD2, a translational repressor of the GCN4 gene, has a general function in the initiation of protein synthesis in *Saccharomyces cerevisiae. Mol. Cell. Biol.* **11,** 3203–3216.

Gautier, T., Berges, T., Tollervey, D., and Hurt, E. (1997). Nucleolar KKE/D repeat proteins Nop56p and Nop58p interact with Nop1p and are required for ribosome biogenesis. *Mol. Cell. Biol.* **17,** 7088–7098.

Gietz, D., St Jean, A., Woods, R. A., and Schiestl, R. H. (1992). Improved method for high efficiency transformation of intact yeast cells. *Nucleic Acids Res.* **20,** 1425.

Gietz, R. D., and Woods, R. A. (2002). Transformation of yeast by lithium acetate/single-stranded carrier DNA/polyethylene glycol method. *Methods Enzymol.* **350,** 87–96.

Gueldener, U., Heinisch, J., Koehler, G. J., Voss, D., and Hegemann, J. H. (2002). A second set of loxP marker cassettes for Cre-mediated multiple gene knockouts in budding yeast. *Nucleic Acids Res.* **30,** e23.

Guo, B., Styles, C. A., Feng, Q., and Fink, G. R. (2000). A *Saccharomyces* gene family involved in invasive growth, cell-cell adhesion, and mating. *Proc. Natl. Acad. Sci. USA* **97,** 12158–12163.

Halme, A., Bumgarner, S., Styles, C., and Fink, G. R. (2004). Genetic and epigenetic regulation of the FLO gene family generates cell-surface variation in yeast. *Cell* **116,** 405–415.

Heilek, G. M., Marusak, R., Meares, C. F., and Noller, H. F. (1995). Directed hydroxyl radical probing of 16S rRNA using Fe(II) tethered to ribosomal protein S4. *Proc. Natl. Acad. Sci. USA* **92,** 1113–1116.

Helser, T. L., Baan, R. A., and Dahlberg, A. E. (1981). Characterization of a 40S ribosomal subunit complex in polyribosomes of *Saccharomyces cerevisiae* treated with cycloheximide. *Mol. Cell. Biol.* **1,** 51–57.

Hendrick, J. L., Wilson, P. G., Edelman, I. I., Sandbaken, M. G., Ursic, D., and Culbertson, M. R. (2001). Yeast frameshift suppressor mutations in the genes coding for transcription factor Mbf1p and ribosomal protein S3: Evidence for autoregulation of S3 synthesis. *Genetics* **157,** 1141–1158.

Henry, Y., Wood, H., Morrissey, J. P., Petfalski, E., Kearsey, S., and Tollervey, D. (1994). The 5′ end of yeast 5.8S rRNA is generated by exonucleases from an upstream cleavage site. *EMBO J.* **13,** 24–52–24-63.

Hermann, T. (2005). Drugs targeting the ribosome. *Curr. Opin. Struct. Biol.* **15,** 355–366.

Herráez, A. (2006). Biomolecules in the Computer. Jmol to the Rescue. *Biochem. Mol. Biol. Educ.* **34,** 255–261.

Hirabayashi, N., Sato, N. S., and Suzuki, T. (2006). Conserved loop sequence of helix 69 in *Escherichia coli* 23 S rRNA is involved in A-site tRNA binding and translational fidelity. *J. Biol. Chem.* **281,** 17203–17211.

Hogan, J. J., Gutell, R. R., and Noller, H. F. (1984a). Probing the conformation of 18S rRNA in yeast 40S ribosomal subunits with kethoxal. *Biochemistry (Mosc.)* **23,** 3322–3330.

Hogan, J. J., Gutell, R. R., and Noller, H. F. (1984b). Probing the conformation of 26S rRNA in yeast 60S ribosomal subunits with kethoxal. *Biochemistry (Mosc.)* **23,** 3330–3335.

Iizuka, N., Najita, L., Franzusoff, A., and Sarnow, P. (1994). Cap-dependent and cap-independent translation by internal initiation of mRNAs in cell extracts prepared from *Saccharomyces cerevisiae. Mol. Cell Biol.* **14,** 7322–7330.

Iizuka, N., and Sarnow, P. (1997). Translation-competent extracts from *Saccharomyces cerevisiae*: Effects of L-A RNA, 5′ cap, and 3′ poly(A) tail on translational efficiency of mRNAs. *Methods* **11,** 353–360.

Jacobs, J. L., and Dinman, J. D. (2004). Systematic analysis of bicistronic reporter assay data. *Nucleic Acids Res.* **32,** e160.

Jimenez, A., and Davies, J. (1980). Expression of a transposable antibiotic resistance element in *Saccharomyces. Nature* **287,** 869–871.

Johnston, M., Riles, L., and Hegemann, J. H. (2002). Gene disruption. *Methods Enzymol.* **350,** 290–315.

King, T. H., Decatur, W. A., Bertrand, E., Maxwell, E. S., and Fournier, M. J. (2001). A well-connected and conserved nucleoplasmic helicase is required for production of box C/D and H/ACA snoRNAs and localization of snoRNP proteins. *Mol. Cell Biol.* **21,** 7731–7746.

King, T. H., Liu, B., McCully, R. R., and Fournier, M. J. (2003). Ribosome structure and activity are altered in cells lacking snoRNPs that form pseudouridines in the peptidyl transferase center. *Mol. Cell* **11,** 425–435.

Kinzy and Woolford. (1995). Increased expression of *Saccharomyces cerevisiae* translation elongation factor 1 alpha bypasses the lethality of a TEF5 null allele encoding elongation factor 1 beta. *Genetics* **141,** 481–489.

Klein, D. J., Schmeing, T. M., Moore, P. B., and Steitz, T. A. (2001). The kink-turn: A new RNA secondary structure motif. *EMBO J.* **20,** 4214–4221.

Knapp, G. (1989). Enzymatic approaches to probing of RNA secondary and tertiary structure. *Methods Enzymol.* **180,** 192–212.

Köhrer, K., and Domdey, H. (1991). Preparation of high molecular weight RNA. *Methods Enzymol.* **194,** 398–405.

Kos, M., and Tollervey, D. (2005). The putative RNA helicase Dbp4p is required for release of the U14 snoRNA from preribosomes in *Saccharomyces cerevisiae. Mol. Cell* **20,** 53–64.

Lafontaine, D. L., Bousquet-Antonelli, C., Henry, Y., Caizergues-Ferrer, M., and Tollervey, D. (1998). The box H + ACA snoRNAs carry Cbf5p, the putative rRNA pseudouridine synthase. *Genes Dev.* **12,** 527–537.

Lafontaine, D. L., and Tollervey, D. (1999). Nop58p is a common component of the box C + D snoRNPs that is required for snoRNA stability. *RNA* **5,** 455–467.

Lapeyre, B., and Purushothaman, S. K. (2004). Spb1p-directed formation of Gm2922 in the ribosome catalytic center occurs at a late processing stage. *Mol. Cell* **16,** 663–669.

Lempereur, L., Nicoloso, M., Riehl, N., Ehresmann, C., Ehresmann, B., and Bachellerie, J. P. (1985). Conformation of yeast 18S rRNA. Direct chemical probing of the 5′ domain in ribosomal subunits and in deproteinized RNA by reverse transcriptase mapping of dimethyl sulfate-accessible. *Nucleic Acids Res.* **13,** 8339–8357.

Li, H. D., Zagorski, J., and Fournier, M. J. (1990). Depletion of U14 small nuclear RNA (snR128) disrupts production of 18S rRNA in *Saccharomyces cerevisiae. Mol. Cell Biol.* **10,** 1145–1152.

Lindell, M., Romby, P., and Wagner, E. G. (2002). Lead(II) as a probe for investigating RNA structure *in vivo. RNA* **8,** 534–541.

Liu, B., and Fournier, M. J. (2004). Interference probing of rRNA with snoRNPs: A novel approach for functional mapping of RNA *in vivo. RNA* **10,** 1130–1141.

Liu, B., Ni, J., and Fournier, M. J. (2001). Probing RNA *in vivo* with methylation guide small nucleolar RNAs. *Methods* **23,** 276–286.

Lowe, T. M., and Eddy, S. R. (1999). A computational screen for methylation guide snoRNAs in yeast. *Science* **283,** 1168–1171.

Ma, X., Yang, C., Alexandrov, A., Grayhack, E. J., Behm-Ansmant, I., and Yu, Y. T. (2005). Pseudouridylation of yeast U2 snRNA is catalyzed by either an RNA-guided or RNA-independent mechanism. *EMBO J.* **24,** 2403–2413.

Maden, B. E. (1990). The numerous modified nucleotides in eukaryotic ribosomal RNA. *Prog. Nucleic Acid Res. Mol. Biol.* **39,** 241–303.

McPheeters, D. S., Christensen, A., Young, E. T., Stormo, G., and Gold, L. (1986). Translational regulation of expression of the bacteriophage T4 lysozyme gene. *Nucleic Acids Res.* **14,** 5813–5826.

Meier, U. T. (2005). The many facets of H/ACA ribonucleoproteins. *Chromosoma* **114,** 1–14.

Meier, U. T. (2006). How a single protein complex accommodates many different H/ACA RNAs. *Trends Biochem. Sci.* **31,** 311–315.

Mereau, A., Fournier, R., Gregoire, A., Mougin, A., Fabrizio, P., Luhrmann, R., and Branlant, C. (1997). An *in vivo* and *in vitro* structure-function analysis of the *Saccharomyces cerevisiae* U3A snoRNP: Protein-RNA contacts and base-pair interaction with the pre-ribosomal RNA. *J. Mol. Biol.* **273,** 552–571.

Meskauskas, A., and Dinman, J. D. (2001). Ribosomal protein L5 helps anchor peptidyl-tRNA to the P-site in *Saccharomyces cerevisiae*. *RNA* **7,** 1084–1096.

Meskauskas, A., Petrov, A. N., and Dinman, J. D. (2005). Identification of functionally important amino acids of ribosomal protein L3 by saturation mutagenesis. *Mol. Cell Biol.* **25,** 10863–10874.

Moazed, D., Stern, S., and Noller, H. F. (1986). Rapid chemical probing of conformation in 16 S ribosomal RNA and 30 S ribosomal subunits using primer extension. *J. Mol. Biol.* **187,** 399–416.

Moore, P. B., and Steitz, T. A. (2003). The structural basis of large ribosomal subunit function. *Annu. Rev. Biochem.* **72,** 813–850.

Moore, P. B., and Steitz, T. A. (2005). The ribosome revealed. *Trends Biochem. Sci.* **30,** 281–283.

Mullen, J. R., Kaliraman, V., Ibrahim, S. S., and Brill, S. J. (2001). Requirement for three novel protein complexes in the absence of the Sgs1 DNA helicase in *Saccharomyces cerevisiae*. *Genetics* **157,** 103–118.

Namy, O., Hatin, I., and Rousset, J. P. (2001). Impact of the six nucleotides downstream of the stop codon on translation termination. *EMBO Rep.* **2,** 787–793.

Namy, O., Lecointe, F., Grosjean, H., and Rousset, J.-P. (2005). Translational recoding and RNA modifications. *In* "Fine-Tuning of RNA Modifications by Modification and Editing" (H. Grosjean, ed.), Vol. 12, pp. 309–340. Springer-Verlag, New York.

Namy, O., Rousset, J. P., Napthine, S., and Brierley, I. (2004). Reprogrammed genetic decoding in cellular gene expression. *Mol. Cell* **13,** 157–168.

Nazar, R. N. (2004). Ribosomal RNA processing and ribosome biogenesis in eukaryotes. *IUBMB Life* **56,** 457–465.

Nogi, Y., Yano, R., and Nomura, M. (1991). Synthesis of large rRNAs by RNA polymerase II in mutants of *Saccharomyces cerevisiae* defective in RNA polymerase I. *Proc. Natl. Acad. Sci. USA* **88,** 3962–3966.

Normand, C., Capeyrou, R., Quevillon-Cheruel, S., Mougin, A., Henry, Y., and Caizergues-Ferrer, M. (2006). Analysis of the binding of the N-terminal conserved domain of yeast Cbf5p to a box H/ACA snoRNA. *RNA* **12,** 1868–1882.

Nygard, O., Alkemar, G., and Larsson, S. L. (2006). Analysis of the secondary structure of expansion segment 39 in ribosomes from fungi, plants and mammals. *J. Mol. Biol.* **357,** 904–916.

Oakes, M., Nogi, Y., Clark, M. W., and Nomura, M. (1993). Structural alterations of the nucleolus in mutants of *Saccharomyces cerevisiae* defective in RNA polymerase I. *Mol. Cell Biol.* **13,** 2441–2455.

Ofengand, J., and Bakin, A. (1997). Mapping to nucleotide resolution of pseudouridine residues in large subunit ribosomal RNAs from representative eukaryotes, prokaryotes, archaebacteria, mitochondria and chloroplasts. *J. Mol. Biol.* **266,** 246–268.

Ofengand, J., and Fournier, M. J. (1998). The pseudouridine residues of ribosomal RNA: Number, location, biosynthesis, and function. *In* "Modification and Editing of RNA: The Alteration of RNA Structure and Function" (H. Grosjean and R. Benne, eds.). ASM Press, Washington, DC.

Ogle, J. M., Murphy, F. V., Tarry, M. J., and Ramakrishnan, V. (2002). Selection of tRNA by the ribosome requires a transition from an open to a closed form. *Cell* **111,** 721–732.

Ogle, J. M., and Ramakrishnan, V. (2005). Structural insights into translational fidelity. *Annu. Rev. Biochem.* **74,** 129–177.

Omer, A. D., Lowe, T. M., Russell, A. G., Ebhardt, H., Eddy, S. R., and Dennis, P. P. (2000). Homologs of small nucleolar RNAs in Archaea. *Science* **288,** 517–522.

Omer, A. D., Ziesche, S., Ebhardt, H., and Dennis, P. P. (2002). In vitro reconstitution and activity of a C/D box methylation guide ribonucleoprotein complex. *Proc. Natl. Acad. Sci. USA* **99,** 5289–5294.

Rethinaswamy, A., Birnbaum, M. J., and Glover, C. V. (1998). Temperature-sensitive mutations of the CKA1 gene reveal a role for casein kinase II in maintenance of cell polarity in *Saccharomyces cerevisiae. J. Biol. Chem.* **273,** 5869–5877.

Robert, F., and Brakier-Gingras, L. (2003). A functional interaction between ribosomal proteins S7 and S11 within the bacterial ribosome. *J. Biol. Chem.* **278,** 44913–44920.

Rozhdestvensky, T., Tang, T., Tchirkova, I., Brosius, J., Bachellerie, J.-P., and Hüttenhofer, A. (2003). Binding of L7Ae protein to the K-turn of archaeal snoRNAs: A shared RNA binding motif for C/D and H/ACA box snoRNAs in Archaea. *Nucleic Acids Res.* **31,** 869–877.

Schattner, P., Decatur, W. A., Davis, C. A., Ares, M., Jr., Fournier, M. J., and Lowe, T. M. (2004). Genome-wide searching for pseudouridylation guide snoRNAs: Analysis of the *Saccharomyces cerevisiae* genome. *Nucleic Acids Res.* **32,** 4281–4296.

Schimmang, T., Tollervey, D., Kern, H., Frank, R., and Hurt, E. C. (1989). A yeast nucleolar protein related to mammalian fibrillarin is associated with small nucleolar RNA and is essential for viability. *EMBO J.* **8,** 4015–4024.

Schuwirth, B. S., Borovinskaya, M. A., Hau, C. W., Zhang, W., Vila-Sanjurjo, A., Holton, J. M., and Cate, J. H. (2005). Structures of the bacterial ribosome at 3.5 Å resolution. *Science* **310,** 827–834.

Selmer, M., Dunham, C. M., Murphy, F. V., 4th, Weixlbaumer, A., Petry, S., Kelley, A. C., Weir, J. R., and Ramakrishnan, V. (2006). Structure of the 70S ribosome complexed with mRNA and tRNA. *Science* **313,** 1935–1942.

Shockman, G. D., and Lampen, J. O. (1962). Inhibition by antibiotics of the growth of bacterial and yeast protoplasts. *J. Bacteriol.* **84,** 508–512.

Sikorski, R. S., and Hieter, P. (1989). A system of shuttle vectors and yeast host strains designed for efficient manipulation of DNA in *Saccharomyces cerevisiae. Genetics* **122,** 19–27.

Sobel, S. G., and Wolin, S. L. (1999). Two yeast La motif-containing proteins are RNA-binding proteins that associate with polyribosomes. *Mol. Biol. Cell* **10,** 3849–3862.

Spahn, C. M., Beckmann, R., Eswar, N., Penczek, P. A., Sali, A., Blobel, G., and Frank, J. (2001). Structure of the 80S ribosome from *Saccharomyces cerevisiae*-tRNA-ribosome and subunit-subunit interactions. *Cell* **107,** 373–386.

Spahn, C. M., Gomez-Lorenzo, M. G., Grassucci, R. A., Jorgensen, R., Andersen, G. R., Beckmann, R., Penczek, P. A., Ballesta, J. P., and Frank, J. (2004). Domain movements of elongation factor eEF2 and the eukaryotic 80S ribosome facilitate tRNA translocation. *EMBO J.* **23,** 1008–1019.

Spence, J., Gali, R. R., Dittmar, G., Sherman, F., Karin, M., and Finley, D. (2000). Cell cycle-regulated modification of the ribosome by a variant multiubiquitin chain. *Cell* **102,** 67–76.

Stahl, G., Bidou, L., Rousset, J. P., and Cassan, M. (1995). Versatile vectors to study recoding: Conservation of rules between yeast and mammalian cells. *Nucleic Acids Res.* **23,** 1557–1560.

Stahl, G., Salem, S. N., Chen, L., Zhao, B., and Farabaugh, P. J. (2004). Translational accuracy during exponential, postdiauxic, and stationary growth phases in *Saccharomyces cerevisiae. Eukaryot. Cell* **3,** 331–338.

Stern, S., Moazed, D., and Noller, H. F. (1988). Structural analysis of RNA using chemical and enzymatic probing monitored by primer extension. *Methods Enzymol.* **164,** 481–489.

Storici, F., Coglievina, M., and Bruschi, C. V. (1999). A 2–mm DNA-based marker recycling system for multiple gene disruption in the yeast *Saccharomyces cerevisiae. Yeast* **15,** 271–283.

Tabb-Massey, A., Caffrey, J. M., Logsden, P., Taylor, S., Trent, J. O., and Ellis, S. R. (2003). Ribosomal proteins Rps0 and Rps21 of *Saccharomyces cerevisiae* have overlapping functions in the maturation of the 3′ end of 18S rRNA. *Nucleic Acids Res.* **31,** 6798–6805.

Tanudjojo, N., Soedigdo, P., and Soedigdo, S. (2002). Uptake rate measurement of some amino acids on normal and treated yeast cells to xenobiotics using 14C labelled amino acid. *Food Nutr. Bull.* **23,** 61–65.

Tarun, S. Z., Jr., Wells, S. E., Deardorff, J. A., and Sachs, A. B. (1997). Translation initiation factor eIF4G mediates *in vitro* poly(A) tail-dependent translation. *Proc. Natl. Acad. Sci. USA* **94,** 9046–9051.

Tenson, T., and Mankin, A. (2006). Antibiotics and the ribosome. *Mol. Microbiol.* **59,** 1664–1677.

Thompson, J. R., Zagorski, J., Woolford, J. L., and Fournier, M. J. (1988). Sequence and genetic analysis of a dispensable 189 nucleotide snRNA from *Saccharomyces cerevisiae. Nucleic Acids Res.* **16,** 5587–5601.

Tollervey, D. (1987). A yeast small nuclear RNA is required for normal processing of pre-ribosomal RNA. *EMBO J.* **6,** 4169–4175.

Tollervey, D., Lehtonen, H., Jansen, R., Kern, H., and Hurt, E. C. (1993). Temperature-sensitive mutations demonstrate roles for yeast fibrillarin in pre-rRNA processing, pre-rRNA methylation, and ribosome assembly. *Cell* **72,** 443–457.

Torchet, C., Badis, G., Devaux, F., Costanzo, G., Werner, M., and Jacquier, A. (2005). The complete set of H/ACA snoRNAs that guide rRNA pseudouridylations in *Saccharomyces cerevisiae. RNA* **11,** 928–938.

Ugolini, S., and Bruschi, C. V. (1996). The red/white colony color assay in the yeast *Saccharomyces cerevisiae*: Epistatic growth advantage of white ade8–18, ade2 cells over red ade2 cells. *Curr. Genet.* **30,** 485–492.

van Beekvelt, C. A., Kooi, E. A., de Graaff-Vincent, M., Riet, J., Venema, J., and Raue, H. A. (2000). Domain III of *Saccharomyces cerevisiae* 25 S ribosomal RNA: Its role in binding of ribosomal protein L25 and 60 S subunit formation. *J. Mol. Biol.* **296,** 7–17.

Velichutina, I. V., Dresios, J., Hong, J. Y., Li, C., Mankin, A., Synetos, D., and Liebman, S. W. (2000). Mutations in helix 27 of the yeast *Saccharomyces cerevisiae* 18S rRNA affect the function of the decoding center of the ribosome. *RNA* **6,** 1174–1184.

Venema, J., and Tollervey, D. (1999). Ribosome synthesis in *Saccharomyces cerevisiae. Annu. Rev. Genet.* **33,** 261–311.

Vimaladithan, A., and Farabaugh, P. J. (1998). Identification and analysis of frameshift sites. *Methods Mol. Biol.* **77,** 399–411.

Vos, H. R., Faber, A. W., de Gier, M. D., Vos, J. C., and Raue, H. A. (2004). Deletion of the three distal S1 motifs of *Saccharomyces cerevisiae* Rrp5p abolishes pre-rRNA processing at site A(2) without reducing the production of functional 40S subunits. *Eukaryot. Cell* **3,** 1504–1512.

Wang, H., Boisvert, D., Kim, K. K., Kim, R., and Kim, S. H. (2000). Crystal structure of a fibrillarin homologue from *Methanococcus jannaschii*, a hyperthermophile, at 1.6 Å resolution. *EMBO J.* **19,** 317–323.

Warner, J. R., and Knopf, P. M. (2002). The discovery of polyribosomes. *Trends Biochem. Sci.* **27,** 376–380.

Watkins, N. J., Leverette, R. D., Xia, L., Andrews, M. T., and Maxwell, E. S. (1996). Elements essential for processing intronic U14 snoRNA are located at the termini of the mature snoRNA sequence and include conserved nucleotide boxes C and D. *RNA* **2,** 118–133.

West, M., Hedges, J. B., Chen, A., and Johnson, A. W. (2005). Defining the order in which Nmd3p and Rpl10p load onto nascent 60S ribosomal subunits. *Mol. Cell Biol.* **25,** 3802–3813.

Williams, I., Richardson, J., Starkey, A., and Stansfield, I. (2004). Genome-wide prediction of stop codon readthrough during translation in the yeast *Saccharomyces cerevisiae*. *Nucleic Acids Res.* **32,** 6605–6616.

Wise, J. A. (1991). Preparation and analysis of low molecular weight RNAs and small ribonucleoproteins. *Methods Enzymol.* **194,** 405–415.

Yang, J. H., Zhang, X. C., Huang, Z. P., Zhou, H., Huang, M. B., Zhang, S., Chen, Y. Q., and Qu, L. H. (2006). snoSeeker: An advanced computational package for screening of guide and orphan snoRNA genes in the human genome. *Nucleic Acids Res.* **34,** 5112–5123.

Yu, Y.-T., Terns, R. M., and Terns, M. P. (2005). Mechanisms and functions of RNA-guided RNA modification. *In* "Fine-Tuning of RNA Modifications by Modification and Editing" (H. Grosjean, ed.), Vol. 12, pp. 223–262. Springer-Verlag, New York.

Yusupov, M. M., Yusupova, G. Z., Baucom, A., Lieberman, K., Earnest, T. N., Cate, J. H., and Noller, H. F. (2001). Crystal structure of the ribosome at 5.5 A resolution. *Science* **292,** 883–896.

Yusupova, G. Z., Yusupov, M. M., Cate, J. H., and Noller, H. F. (2001). The path of messenger RNA through the ribosome. *Cell* **106,** 233–241.

Zago, M. A., Dennis, P. P., and Omer, A. D. (2005). The expanding world of small RNAs in the hyperthermophilic archaeon Sulfolobus solfataricus. *Mol. Microbiol.* **55,** 1812–1828.

Zagorski, J., Tollervey, D., and Fournier, M. J. (1988). Characterization of an SNR gene locus in *Saccharomyces cerevisiae* that specifies both dispensable and essential small nuclear RNAs. *Mol. Cell Biol.* **8,** 3282–3290.

Zebarjadian, Y., King, T., Fournier, M. J., Clarke, L., and Carbon, J. (1999). Point mutations in yeast CBF5 can abolish *in vivo* pseudouridylation of rRNA. *Mol. Cell Biol.* **19,** 7461–7472.

Zhong, T., and Arndt, K. T. (1993). The yeast SIS1 protein, a DnaJ homolog, is required for the initiation of translation. *Cell* **73,** 1175–1186.

CHAPTER FOURTEEN

The U1 snRNA Hairpin II as a RNA Affinity Tag for Selecting snoRNP Complexes

Dorota Piekna-Przybylska,[*,1] Ben Liu,[†,1] *and* Maurille J. Fournier[*]

Contents

[*] Department of Biochemistry and Molecular Biology, University of Massachusetts, Amherst, Massachusetts
[†] Dana Farber Cancer Institute, Harvard Medical School, Boston, Massachusetts
[1] These authors contributed equally to this work

Methods in Enzymology, Volume 425
ISSN 0076-6879, DOI: 10.1016/S0076-6879(07)25014-1

Abstract

When isolating ribonucleoprotein (RNP) complexes by an affinity selection approach, tagging the RNA component can prove to be strategically important. This is especially true for purifying single types of snoRNPs, because in most cases the snoRNA is thought to be the only unique component. Here, we present a general strategy for selecting specific snoRNPs that features a high-affinity tag in the snoRNA and another in a snoRNP core protein. The RNA tag (called U1hpII) is a small (26 nt) stem-loop domain from human U1 snRNA. This structure binds with high affinity ($K_D = 10^{-11}$ M) to the RRM domain of the snRNP protein U1A. In our approach, the U1A protein contains a unique affinity tag and is coexpressed *in vivo* with the tagged snoRNA to yield snoRNP-U1A complexes with two unique protein tags—one in the bound U1A protein and the other in the snoRNP core protein. This scheme has been used effectively to select C/D and H/ACA snoRNPs, including both processing and modifying snoRNPs, and the snoRNA and core proteins are highly enriched. Depending on selection stringency other proteins are isolated as well, including an RNA helicase involved in snoRNP release from pre-rRNA and additional proteins that function in ribosome biogenesis. Tagging the snoRNA component alone is also effective when U1A is expressed with a myc-Tev-protein A fusion sequence. Combined with reduced stringency, enrichment of the U14 snoRNP with this latter system revealed potential interactions with two other snoRNPs, including one processing snoRNP involved in the same cleavages of pre-rRNA.

1. INTRODUCTION

Advances in affinity purification technologies have provided an enormous boost to detecting and isolating biochemical complexes in cells. Powerful new enrichment procedures and breakthroughs in mass spectral analysis have opened the way to purifying multimolecular complexes on a microscale and identifying the associated proteins by analyzing mass spectral data and genomic databases. In addition to characterizing free complexes, the new technologies have made it possible to examine complexes in different contexts, for example, snRNPs at different stages of mRNA splicing or rRNP intermediates in ribosome biogenesis. This last situation is of special interest to us, in particular the study of the scores of small nucleolar RNA–protein complexes (snoRNPs) that participate in ribosome synthesis.

Affinity strategies work best, of course, when the tagged molecules are unique to a complex of interest. In this regard, the small nucleolar RNA–protein complexes (snoRNPs) pose a special challenge, because, in most cases, the snoRNA molecule is the only distinguishing component; this is believed to be the case for nearly all modifying snoRNPs. Thus, isolating snoRNPs of these types by affinity tagging methods will require tagging the snoRNA in a fashion that does not interfere with its formation and function. Addressing this need is the thrust of the present chapter, and a strategy is described that is effective for several types of yeast snoRNPs. The scheme should have good usefulness for other RNP complexes as well.

1.1. Why isolate snoRNPs?

Eukaryotes contain three classes of snoRNP complexes, which are categorized by the RNA component. Two classes are large and defined by conserved snoRNA elements, boxes C and D in one case, and boxes H and ACA in the other (as reviewed in Bachellerie *et al.*, 2002; Bertrand and Fournier, 2004; Kiss, 2004; Meier, 2005; Yu *et al.*, 2005). The snoRNPs in these groupings each contain four core proteins that are also class-specific. Most snoRNPs mediate nucleotide modification, $2'$-O-methylation by the C/D species and pseudouridylation by the H/ACA species. A few snoRNPs in these two families function in processing (cleavage) of pre-rRNA, and one of these, the U3 snoRNP (a C/D species), has a special, early role in rRNA synthesis and contains additional proteins that presumably relate to its unique function. The third class of snoRNPs is defined by one species, the MRP snoRNP, which is required for a particular rRNA processing event. The snoRNA and two proteins in this complex are unique, although both the RNA structure and most protein components are related to those in the RNase P complex, which has been purified with excellent success (Chamberlain *et al.*, 1998; Salinas *et al.*, 2005). The population of snoRNPs in the yeast *Saccharomyces cerevisiae* is the best characterized to date and contains five species only known to participate in processing, 68 species only known to function in modification, and two species that participate in both processing and modification (Lowe and Eddy, 1999; Samarsky and Fournier, 1999; Schattner *et al.*, 2004; Torchet *et al.*, 2005; http://people.biochem.umass.edu/sfournier/fournierlab/snornadb/main.php).

Only a few reports exist on affinity isolation of snoRNPs, and these focus on the specialized U17, U3, and MRP processing snoRNPs (Dragon *et al.*, 2002; Lubben *et al.*, 1995; Salinas *et al.*, 2005; Watkins *et al.*, 2000). The isolation of the U17 RNP was performed by Me_3G immunoaffinity chromatography followed by Mono Q anion-exchange chromatography. Seven proteins were isolated, including Gar1 (identified) and three others corresponding in size to the remaining H/ACA core proteins, although these had not yet been identified (Lubben *et al.*, 1995). An early procedure designed for the U3 snoRNP featured an anticap–antibody column for the first enrichment step and then selection of the U3 snoRNP with the first tag reported for a specific snoRNA (Watkins *et al.*, 2000; and see later).

This latter study identified the four C/D core proteins and another that is specific for the U3 snoRNP. Our understanding about the full complement of proteins that associate with the U3 snoRNP came later, from a two-protein tagging approach that yielded a remarkable plethora of proteins that the U3 snoRNP encounters during rRNA synthesis and ribosome assembly (Bernstein *et al.*, 2004; Bleichert *et al.*, 2006; Dragon *et al.*, 2002; Gerczei and Correll, 2004; Hoang *et al.*, 2005; Wehner *et al.*, 2002).

The presence of unique proteins in the U3 and MRP snoRNPs suggests that other processing snoRNPs may also contain unique proteins that can be exploited in affinity selection. However, this has not been demonstrated, and the possibility remains open that the remaining processing snoRNPs do not contain any unique proteins. The same situation could well apply for the scores of modifying snoRNPs, leaving the snoRNA as the only distinguishing component. To our knowledge, no snoRNPs that function only in modification have been isolated and subjected to protein analysis. From a theoretical perspective, it seems highly unlikely that each modifying snoRNP carries a unique protein. If this were the case, such a protein(s) would most probably be specific to the snoRNA. Arguing against this possibility is the fact that archaeal snoRNP orthologs have been constituted *in vitro* and shown to be active in modifying assays (Charpentier *et al.*, 2005; Omer *et al.*, 2002; Tran *et al.*, 2003; and Chapter 16). These findings show that only the single-guide snoRNA and core proteins are required for site-specific binding and modification, suggesting that affinity selection of most snoRNPs will require tagging the RNA component. The selection strategy described here was developed for two general purposes: (1) to search for novel integral snoRNP proteins among the processing and modifying snoRNPs, and (2) to identify other proteins that associate with snoRNPs *in vivo* to gain new insights into how snoRNPs carry out their functions.

1.2. Affinity tags used in RNP isolations

Several excellent affinity tags exist for proteins including the following: c-myc, calmodulin binding peptide (CBP), FLAG, HIS, glutathione *S*-transferase (GST), protein A, and the TAP tag (Einhauer *et al.*, 2002; Hefti *et al.*, 2001; Lu *et al.*, 1997; Manstein *et al.*, 1995; Nilsson *et al.*, 1985; Rigaut *et al.*, 1999; Terpe, 2003; Vaillancourt *et al.*, 2000). The choices are sparser, however, for tagging RNA components in RNP complexes. Examples that have been used effectively include signals known as: StreptoTag, Streptavidin tag, and Sephadex tag, which are all RNA sequences defined by selection analysis from pools of random-sequence RNA molecules through *in vitro* selection, including SELEX (Bachler *et al.*, 1999; Ellington and Szostak, 1990; Srisawat and Engelke, 2001, 2002; Tuerk and Gold, 1990).

The StreptoTag is an RNA sequence (46nt) that binds to the antibiotic streptomycin with a dissociation constant (K_D) of approximately 1 μM (Bachler *et al.*, 1999). Used with yeast group II introns and viral RNA, this tag has been used to identify novel proteins that bind to these RNAs (Bock-Taferner and Wank, 2004; Dangerfield *et al.*, 2006). In those studies, the hybrid RNA was immobilized on a streptomycin affinity column, and the proteins were selected from cellular or nuclear extracts.

The Streptavidin tag (S1, 44nt) and Sephadex tag (D8, 33nt) are RNA sequences that bind to the protein streptavidin and to the dextran in Sephadex, respectively (Srisawat and Engelke, 2001, 2002). Both RNA motifs were incorporated into the RNA molecule of RNase P to select RNP complexes, with either streptavidin-agarose or Sephadex G-200 matrix. Two versions of the tagged RNA were stable in the cell. Although the protein pattern of the resulting RNP was not determined, the complex(es) was enzymatically active. The S1 tag binds to streptavidin with a $K_D = $ ~70 nM; however, its natural ligand biotin has a much higher affinity ($K_D = $ ~10^{-14} M), thus blocking biotin in crude cell lysates is recommended before applying the extract to a streptavidin column (Srisawat and Engelke, 2002). The D8 tag used with Sephadex has a lower affinity for Sephadex than the S1 tag has for streptavidin. Thus, extensive washing of beads with bound D8-tagged RNA leads to slow loss of this RNA (Srisawat and Engelke, 2002).

As with protein affinity tags, the main requirements for an effective RNA affinity tag are high specificity and the ability to place the tag into the RNA without compromising the production, stability, and function of the mature RNP complex. A tag that functions well in one RNA molecule may not work well in others. Thus, a diverse number of RNA affinity tags seems highly desirable, because identifying a suitable tag may be the only limitation in designing a successful isolation scheme.

1.3. The rationale of the present RNA tagging scheme

In the strategy described here, we tag the snoRNA with a small stem-loop structure from human U1 snRNA that binds tightly to the snRNP protein U1A (Scherly *et al.*, 1989). The U1A protein contains a c-myc tag for selection of the snoRNA-U1A complexes. Because the U1A protein is coexpressed with the tagged snoRNA, they can interact during snoRNP assembly. A core protein containing another tag is also coexpressed. Barring interference effects, the resulting snoRNP complex will include the four core proteins (one tagged), the tagged snoRNA, and the tagged U1A protein. The snoRNP is then enriched in two stages: (1) selection of the entire family of C/D or H/ACA snoRNPs by means of the tagged core protein, and (2) selection of the snoRNP of interest by means of the tagged U1A protein (Fig. 14.1).

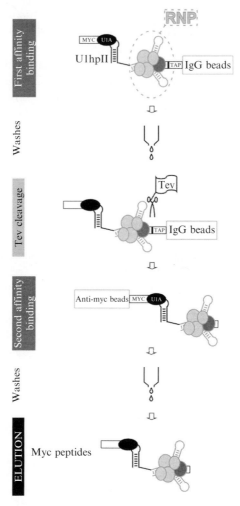

Figure 14.1   Overview of the selection strategy for snoRNP complexes that exploit the U1hpII-U1A couple. The snoRNA component contains an U1hpII domain ("the RNA-tag"), and the U1A protein bears a myc tag. Both are coexpressed in the same cells. The procedure involves two stages of enrichment. The first is based on selection of a tagged core protein common to all snoRNPs in the same family (selection here is with the TAP tag). After cleavage with Tev protease, the second stage of enrichment is based on selection of the snoRNA of interest, which is tagged with the U1hpII domain and complexed with the U1A protein (selection here is with the myc tag of the U1A protein).

Our tagging scheme for the snoRNA is based on remarkable properties described for the interaction of the human U1A protein with U1 snRNA and with its own pre-mRNA. The protein binds to both types of RNAs

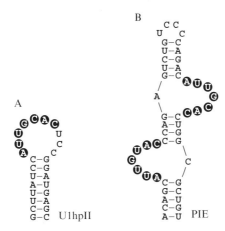

Figure 14.2   The human U1A protein binds with high selectivity and affinity to two related natural RNA motifs. (A) The U1hpII motif in human U1 snRNA. (B) The polyadenylation inhibition element (PIE) present in U1A pre-mRNA. The highlighted nucleotides (black background) in the loop regions of U1hpII and PIE interact tightly with the N-terminal domain of human U1A protein.

with unusually high selectivity and affinity. Binding occurs through similar sequence motifs; however, the motifs are embedded in different structures (Fig. 14.2) (Hall and Stump, 1992; van Gelder *et al.*, 1993). The motif featured by us is from the snRNA and is encoded in the loop portion of a 26-nucleotide stem-loop structure called U1hpII. The recognition sequence is seven nucleotides (AUUGCAC), and binding occurs through the RRM domains of the U1A protein (Scherly *et al.*, 1989; Tsai *et al.*, 1991). Binding with pre-mRNA occurs in a larger, highly folded structure known as PIE (polyadenylation inhibition element), which occurs naturally in the 3′-UTR of U1A pre-mRNA (Boelens *et al.*, 1993). The PIE domain contains two copies of the recognition motif, one in each of two adjacent bulges. One motif is the same as that in U1 RNA, and the other differs at one position; the PIE domain consists of 47 nucleotides (van Gelder *et al.*, 1993). Structural studies of each type of complex revealed the smaller U1hpII hairpin is "ready" for association of U1A protein, whereas binding of two U1A molecules to PIE involves a substantial bending of this RNA motif of approximately 100° (Allain *et al.*, 1997; Grainger *et al.*, 1999; Oubridge *et al.*, 1994; Varani *et al.*, 2000).

We elected to use the U1hpII domain in our strategy because of its small size, which we reasoned might have a lower potential for interfering with snoRNP synthesis or function. In the same context, we predicted this domain might be more amenable than the larger, more complicated PIE domain for use in repeat contexts, as a means of solving possible steric interference effects or to otherwise improve efficiency of U1A binding and snoRNP selection. We note that the PIE domain has been used with good success in an enrichment scheme for the yeast U3 snoRNP (Watkins *et al.*, 2000). The proteins in the final U3 RNP preparation included the Snu13, the three other C/D core proteins, and the U3-specific Rrp9 protein. Although these results demonstrate the feasibility of the use of the PIE domain for snoRNP isolation, we elected to feature the smaller U1hpII domain to minimize potential deleterious effects of tagging. In this regard, the U3 snoRNA is considerably larger (333nts) than most other yeast snoRNAs and contains many disposable segments; indeed, fully 50% of yeast U3 can be deleted without loss of rRNA processing function (Samarsky and Fournier, 1998). Other snoRNAs may be less forgiving than U3, especially the smaller snoRNAs. The yeast snoRNAs range in size from 78–1004nts, with the main distribution between 90 and 200nts. In any case, in the context of isolating the U3 snoRNP, the U1A-PIE combination yielded a preparation with five proteins that have also been identified in preparations selected without a RNA tag (Dragon *et al.*, 2002).

The results presented later show that the U1hpII domain and U1A protein can be used effectively to select both C/D and H/ACA snoRNPs (including small snoRNAs) with all four core proteins and, depending on the age of culture and the stringency of the wash conditions, various other proteins that associate with these complexes. We focused initially on two universal processing snoRNPs: the C/D box U14 snoRNP and the H/ACA box U17 snoRNP; the latter snoRNP is also called snR30 in yeast. Our longer-range intention was to identify proteins associated with these snoRNP complexes, either as integral components or partners in some stage of snoRNP synthesis or function.

1.4. The snoRNP substrates

To place the present work in the context of rRNA synthesis, the U14 and U17 snoRNPs are required for processing cleavages that create a 20S precursor to 18S rRNA, from a nearly full-length precursor that contains $(5' \rightarrow 3')$ the 18S, 5.8S, and 25S rRNAs. Stepwise, the earliest processing events that require snoRNPs include: (1) cleavage in the 5'-portion of 5'-external transcribed spacer (5' ETS) by a mechanism that requires the U3 snoRNP, followed by (2) the two cleavages that release the 20S precursor to 18S rRNA (Kressler *et al.*, 1999; Venema and Tollervey, 1999). These latter reactions are coupled, and both require the U3, U14, and U17 snoRNPs working in concert. In yeast, these processing events are followed by a cleavage requiring the MRP snoRNP that leads to generation of the mature 5' end of 5.8 S rRNA (Chu *et al.*, 1994; Clayton, 1994). In vertebrates, snoRNPs are known to be required for additional cleavages.

The substrates for the processing snoRNPs are precursor rRNP complexes in early stages of preribosome assembly. The various pre-rRNA molecules are associated with ribosomal proteins, assembly factors, and, in some cases, the RNA polymerase I transcription machinery. Indeed, characterization of proteins coselected with the U3 snoRNP has revealed associations with a substantial number of proteins involved in the transcription process (Gallagher et al., 2004). This is consistent with the knowledge that the specialized U3 snoRNP associates with new rRNA transcripts very early and is thought to be part of the terminal knobs visualized in actively transcribing complexes (Dragon et al., 2002; Mougey et al., 1993). It is not known when the other processing snoRNPs bind or act. Results from pulse-chase labeling studies of rRNA synthesis argue that processing occurs at the level of the full-length transcript; however, impressive images from electron microscopy suggest that a cleavage that separates the small and large rRNA segments can take place before transcription is complete (Osheim et al., 2004).

Most modified nucleotides created by snoRNPs are thought to be formed before the cleavage reactions take place. As in the case for the processing snoRNP, the modifying snoRNPs act on pre-rRNP complexes, not a naked transcript. It remains to be determined precisely when these snoRNPs bind, carry out modification reactions, and are released. Nonetheless, because of the nature of the pre-rRNP substrates on which snoRNPs act, a myriad of proteins could be present in preparations of affinity-selected snoRNPs; some could be relevant and others not. For us, identifying such associations and sorting through their relevance offered the promise of gaining valuable new insights into the mechanisms of snoRNP-mediated processing and modification.

We show here that our strategy is effective for enriching processing snoRNPs from both major snoRNP families. In addition to the expected core proteins, other proteins were coselected as well, depending on conditions, and several of these have links with ribosome synthesis. In work reported elsewhere by our laboratory, we show that a RNA helicase protein (Has1p) is coselected with the yeast U17/snR30 snoRNP, which was already known to be required for processing of 18S rRNA. In subsequent experiments, this helicase was found to associate with other snoRNPs as well (including U14) and to be needed for normal release of all snoRNPs tested from pre-rRNA (Liang and Fournier, 2006). That experience validates and highlights the potential of our enrichment approach. In addition to identifying relevant protein partners, we demonstrate here that the selection scheme can also be used to screen for interactions with other snoRNPs. The strategy and supporting results are described in the following sections, and full details of the method are provided in a subsequent final section, "Materials and Experimental Procedures."

2. METHODOLOGY

2.1. Overview of the method

In addition to tagging the snoRNA of interest (see later), a tag is incorporated into one of the snoRNP core proteins. The four core proteins in the C/D group include Snu13, Nop56, Nop58, and Nop1; and those in the H/ACA group are Nop10, Nhp2, Gar1, and Cbf5. Among these, Nop1 is the C/D methylase protein, and Cbf5 is the H/ACA pseudouridine synthase. Excellent progress is being made in defining the assembly pathways of these snoRNPs, both *in vitro* and *in vivo*, and in solving the structures of various snoRNP complexes by crystallography. These results will not be reviewed here, and readers are referred to Aittaleb *et al.* (2003); Ballarino *et al.* (2005); Dragon *et al.* (2000); Lafontaine and Tollervey (2000); Li and Ye (2006); Morlando *et al.* (2004); Preti *et al.* (2006); Watkins *et al.* (1998); and Yang *et al.* (2005). As progress continues, valuable new information about suitable RNA and protein sites to tag should become increasingly available.

As summarized in Fig. 14.1, the first of our two enrichment steps involves selection of a tag in a core protein, in particular the protein A moiety present in the TAP tag. For the C/D snoRNPs, the TAP tag is added to the C-terminal end of Nop58, and for the H/ACA snoRNPs, the tag is added to the C-terminal end of Gar1. The TAP sequence consists of two tandem IgG-binding domains of the *Staphylococcus aureus* protein A and a calmodulin-binding peptide, separated by a Tev protease cleavage site (Rigaut *et al.*, 1999). An IgG matrix is used in the first binding reaction, and, after washing, the selected material is released by cleavage with the Tev protease. The calmodulin-binding tag is not used in our procedure but is available for a situation where additional enrichment is indicated.

The second enrichment step relies on selection of a c-myc tag embedded in the human recombinant U1A protein, which is bound to the U1hpII RNA domain present in the snoRNA of interest (Fig. 14.1). Thus, after release of the snoRNPs from the first affinity column, the eluted material is incubated with anti-myc beads for the second affinity selection. After these beads are washed, the bound material is released with free myc peptides, which disrupt binding of the U1A protein to the column.

2.2. Tagging the snoRNP components

As with other *in vivo* tagging strategies, the key requirements for successfully tagging snoRNP components are to avoid interference effects on snoRNP synthesis and function and to achieve good enrichment with a suitable yield. In the case of the core proteins, results from other laboratories have shown that the TAP tag can be inserted into the Nop58 and Gar1 core proteins at the C-terminus without disrupting function or stability (Dragon *et al.*, 2002; Ganot *et al.*, 1997; Gavin *et al.*, 2002). In our case, identifying a suitable tag for several snoRNAs is the main challenge. Untoward effects here could impair processing of the snoRNA, assembly and localization of the snoRNP, and the stability and activity of the final snoRNP complex could be compromised.

We anticipated two special concerns with the indirect tagging strategy selected. One relates to the nature of the new RNA sequence added to the snoRNA and the other to binding of the U1A protein to the tagged snoRNA. In our scheme, the tagged snoRNA and U1A protein are coexpressed conditionally, from separate galactose-induced transcription units. The functionality of the tagged U14 and U17 snoRNAs was assessed by replacing the normal chromosomal coding sequences with the tagged version. Thus, the cells contain a single allele for these two snoRNAs. In some cases indicated later, the chromosomal allele of snoRNA gene is intact, and the tagged snoRNA is generated from a single-copy plasmid as a second allele.

Most of our tagging experience with the U1hpII domain has come from additions at the 5′ ends of the U14 and U17 and several modification guide snoRNAs from each snoRNA family. All these taggings have been successful. The RNA tag has also been inserted internally into the U14 species, and this is effective as well (see later). Placement at the 5′ end includes a short spacer (10–40nt) and one or more copies of the U1hpII sequence; vectors are available with multiple repeats of this U1A binding signal (see later). The spacer was included with the hope of avoiding interference between binding of U1A to the snoRNA and snoRNP assembly and function. Because these tagging schemes were successful, we have limited our tagging of other snoRNA sites to two internal sites in U14 snoRNA.

Our preliminary results revealed that U14 snoRNA with a single tag at the 5′ end was stable in the cell; however, up to 50% of the tag was lost by nucleolytic trimming, with most damage occurring during the snoRNP isolation procedure. With a view to enhancing the snoRNP yield, we increased the copy number of the tag, reasoning that a higher proportion of snoRNPs would retain the U1A binding signal over the course of the enrichment procedure. A set of expression vectors was created that can be used for producing snoRNAs with different numbers of the U1A protein binding signal at the 5′ end (Fig. 14.3A). In this series, the U1hpII signal corresponds to a stem of 8 base pairs capped with a loop of 10 nucleotides (Fig. 14.2A). Except where indicated, most of the results shown here are for snoRNA variants with 5′ extensions that contain nine binding motifs separated by six nucleotides and a 3′ spacer of 30nt for U14 and 32nt for U17.

These coding units are fused to the inducible *GAL1* promoter, and cloning is facilitated with a short multicloning site downstream of the

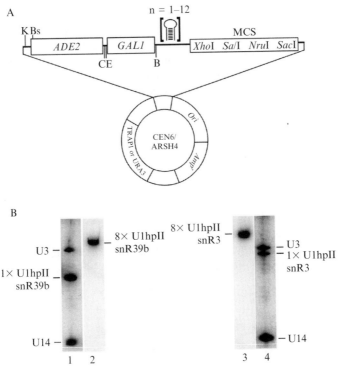

Figure 14.3 Tagging the snoRNA molecule. (A) Plasmid designed for expression of any RNA tagged at the 5' end with one and multiple copies of the U1hpII domain. (B) Tagged snoRNAs are stable. Northern results show accumulation of two snoRNAs tagged with single and eight U1hpII domains: snR39b (C/D class, lanes 1, 2) and snR3 (H/ACA class, lanes 3, 4). For reference the U3 and U14 snoRNAs are also shown. K, *Kpn*I; Bs, *Bst*BII; C, *Cla*I; E, *Eco*RI; B, *Bam*HI; MCS, multicloning site.

U1hpII tag(s). The expression vector also contains the selective marker gene *ADE2* upstream of the *GAL1* promoter. The vector has been designed so that the experimental snoRNA coding segment, and *GAL1* promoter can be readily substituted for the normal chromosomal coding unit. The sequence on the 5' side of this cassette can be introduced into *Kpn*I and *Bst*BII cloning sites, and the fused *Kpn*I/*Sac*I fragment that contains the entire expression cassette can be used to replace the wild-type snoRNA coding unit by homologous recombination. The accompanying *ADE2* marker is used to select transformants (Burke *et al.*, 2000).

The U1A protein is expressed from a plasmid, also from an inducible *GAL1* promoter. Because the N-terminal RRM domain of the human U1A protein is necessary and sufficient for interaction with the U1hpII domain, we use this segment only in our strategy. Here, the segment corresponds to the first 102 amino acids of the U1A protein, and this fragment is fused with one or more copies of the myc affinity tag. The recombinant U1A protein also contains a nuclear localization signal (NLS) (Bachler *et al.*, 1999). In the results that follow, the N-terminal RRM domain of U1A is fused with nine copies of the myc epitope; the 37 kDa product is designated U1A-myc9. However, we know that three or even one myc epitope is sufficient for stable association with the anti-myc beads used in the second affinity selection step (see section titled "Other Approaches for Enriching snoRNPs with the U1hpII tag/U1A Couple").

RNA patterns are shown in Fig. 14.3B for two tagged snoRNAs coexpressed with the U1A protein. The test snoRNAs are a C/D methylation guide snoRNA (snR39B) and an H/ACA pseudouridylation guide snoRNA (snR3); each was transcribed from an expression cassette on a plasmid that encodes one or eight U1A binding signals. Northern data show that each of the tagged snoRNAs accumulates well in the presence of the U1A-myc9 protein, relative to untagged U14 and U3 control snoRNAs. These results indicate that the tagging strategy does not disrupt snoRNA production or accumulation. Because snoRNA stability depends on snoRNP formation, the data also indicate that the tags and U1A protein do not disrupt snoRNP assembly.

2.3. Affinity enrichment of the processing snoRNPs U14 and U17

2.3.1. Experimental strains

Two strains were created with a constitutively expressed TAP-tagged core protein—Nop58 for the U14 snoRNP and Gar1 for the U17 snoRNP. In each strain, the natural chromosomal coding unit for the snoRNA has been replaced with a *GAL1*-dependent allele that expresses a tagged version of the snoRNA. Preliminary Northern data showed the tagged snoRNAs accumulate, as well as the wild-type snoRNAs; however, with 10–30% degradation when the culture density exceeded 1.85×10^7 cell/ml ($OD_{660} = 1$; data not shown). In another experiment with the U17 system degradation was even greater for cultures at OD_{660} values of ~2 (X. H. Liang, unpublished). Less damage (up to 5%) was observed with younger cells, and we elected to prepare snoRNPs from cultures with OD_{660} values of 0.5–0.8. Although the use of cells that are in a state of balanced physiological growth in early log phase seems best (for U14 and U17), this property must be balanced with yield considerations as well.

To maintain the plasmid that expresses the U1A-myc9 protein, cells are grown in selective galactose medium deficient in either uracil (U14) or tryptophan (U17). Although both the tagged snoRNA and U1A-myc9 protein are conditionally expressed from the *GAL1* promoter, other promoters can be also used (and an example is given in "Other Approaches for Enriching snoRNPs with the U1hpII tag/U1A Couple").

The main stages of the enrichment procedure are described here, and full details starting with cell culturing and extract preparation are provided in the "Materials and Experimental Procedures" section. Conditions for carrying out the two selection steps were established first with 1-liter cultures (minipreparations). Larger preparations to be used for mass spec protein identification were then derived from 10–15 liters of cells, but processed in 5-liter batches, which correspond to approximately 15 g of cells.

2.3.2. Core protein selection (step 1)

Extract is mixed with the IgG beads and incubated for 4 h at 4°, and all subsequent manipulations except the Tev cleavage reaction and final elution are carried out at this temperature. After an initial buffer wash with gravity settling, the beads are loaded onto a (Flex-6) column and subjected to further washing with a second buffer (60 vol). The beads are then incubated with Tev protease (and RNase inhibitor, at 16° for 2 h) to release the bound material.

2.3.3. RNA selection (step 2)

The material from the step 1 selection is incubated with anti–myc (agarose) beads (2 h, 4°) to select the U1A-myc9 protein. After loading on a (Flex-6) column, the beads are washed with buffer (60 vol), and the bound material is released with buffer containing c-myc peptide (3 small-volume elutions). Whereas step 1 selects the TAP-tagged Nop58 or Gar1 protein associated with many snoRNAs, step 2 selects the U1A-myc9 protein, which is stably associated with a single snoRNA. Eluted proteins are concentrated by precipitation with acetone or trichloroacetic acid (TCA).

2.3.4. RNA and protein analysis

RNA and protein enrichment and quality are evaluated by Northern and Western analysis by use of small portions of the various fractions (indicated in the figures). Protein patterns are examined by SDS-PAGE (on 10 or 13.5% gels) and visualized by silver staining. Preparations from 5 liters (~15 g of cells) are used for small diagnostic gels, and at least twice this amount to prepare protein bands for mass spec analyses. Aliquots are sampled and frozen throughout the procedure to evaluate each step.

The effectiveness of the 2-step strategy for preparing the U14 and U17 snoRNPs is shown in Fig. 14.4, with enrichment results for both the U1A protein and specific snoRNAs. Western data track the U1A protein (myc epitope) over the course of the second stage of enrichment and show it is present at a modest level in the final eluted material (panel A, U14-left and U17-right). An enrichment of ~50% is achieved after step 1 (Fig. 14.4A). The protein A moiety of the TAP-tagged core proteins was also probed but is only apparent in the initial extract, because protein A is removed by

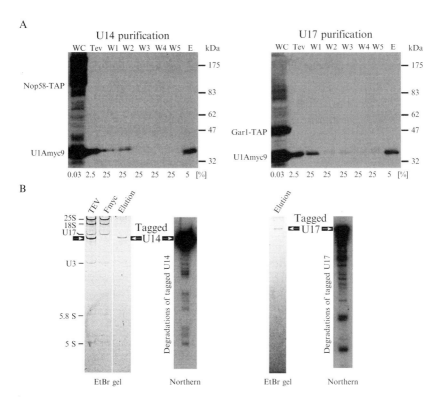

Figure 14.4   Second-stage enrichment of the U14 and U17 snoRNPs. (A) The U1A-myc9 protein is present in the eluted snoRNP complexes. The initial selections were for TAP-tagged core proteins, Nop58 for U14 (left) and Gar1 for U17 (right). The U1A-myc9 protein is selected by the second affinity column (anti-myc beads) and persists through the washing procedure. The presence and stability of the U1A-myc9, Nop58, and Gar1 proteins was monitored by Western analysis (with anti-myc and PAP antibodies). Nop58 and Gar1 are not visible in the second enrichment stage because of loss of the protein A moiety after Tev protease cleavage. WC, whole cell extract; Tev, material released from the IgG column by the Tev protease; W1–W5, wash fractions from the anti-myc beads (second affinity column); E, elution. The numbers at the bottom represent the percentage of the total volume analyzed at each step. (B) RNA enrichment in the second stage. The RNAs were fractionated on polyacrylamide-urea gels, and the patterns for U14 (left pair) and U17 (right pair) were examined by ethidium bromide staining (left panels, EtBr gel) and Northern analysis (right panels), respectively. Tagged U14 released from the IgG beads (Tev) was selected by anti-myc beads through the U1A-myc9 protein. In the final elution step, only one band of RNA is observed by ethidium bromide staining, which corresponds in size to tagged U14. Similarly, ethidium bromide staining shows a dominant band of tagged U17 in the eluted complex. Northern results confirm the identities of the U14 and U17 snoRNAs in the final step. Some degradation is also evident, slight for U14 and more substantial for U17; Tev, material (5%) released from the IgG beads; Fmyc, flow-through material (5%) after binding with anti-myc beads; elution, the final fraction (5%) of enriched snoRNPs.

the Tev cleavage reaction. Silver staining of total protein in the final preparations shows all core proteins to be present (see later).

RNA screening after each selection stage shows good enrichment results as determined by ethidium bromide staining (left lanes) and Northern probing (right lanes) (Fig. 14.4B). In particular, the data show that the tagged snoRNAs are efficiently selected by the second affinity column (Fig. 14.4B; the Fmyc lane shows the flow-through). Importantly, when a control isolation was carried out with untagged snoRNA, no RNA was selected by the U1A–myc9 protein, indicating that the U1hpII/U1A selection scheme is very specific (data not shown). Ethidium bromide staining shows that the final U14 and U17 snoRNP fractions contain a single band that corresponds in size to the tagged snoRNA. Northern data confirm these bands to be the U14 and U17 snoRNAs and reveal the occurrence of some degradation, slight for U14 and more substantial for U17 (rightmost lane). Most of the damage to U17 occurs during the isolation procedure (data not shown). Despite the damage to the U17 RNA, we reasoned that both preparations should be suitable for identifying proteins associated with the complexes. Consistent with this view, preparations made in this way contain all of the expected snoRNP core proteins plus others related to snoRNP function.

After concentrating the final fractions, the protein patterns were examined by electrophoresis on a gel with silver staining. To obtain gel bands for mass spec, we analyzed the yields from a minimum of 30 g of cells (Fig. 14.5). Patterns are shown for both snoRNP types with cells grown to different densities and with different selection stringencies. These variables were assessed, but not in detail, with the aim of obtaining pattern differences that could be helpful in considering the relevance of novel noncore proteins. For both types of snoRNPs all four core proteins are clearly visible, as well as others, and the identities were determined by mass spectrometry analysis (LC/MS/MS). The noncore proteins included both natural yeast proteins and proteins that are part of the isolation strategy (i.e., the U1A–myc9 protein, the Tev protease, and the RNase inhibitor) added in the protease cleavage reaction step.

Comparing the two sets of U14 preparations shown, the tagged Nop58 protein is readily apparent in one but not in the other (compare left panels in Figs. 14.5A,B), and variations have been seen for other preparations as well. This variation in yield could be due to interference by one or both affinity tags that might, for example, reduce Nop58 binding affinity or increase susceptibility to damage by protease. As noted, culture age and wash volumes differed in the second enrichment stage, and from other experience, it is clear that culture stage is a key variable in such preparations. We observed some degradation of the TAP tagged Nop58 when the culture density exceeded 1.85×10^7 cells/ml ($OD_{660} = 1$; data not shown). The preparation depleted of Nop58 (Fig. 14.5B) received fewer washes

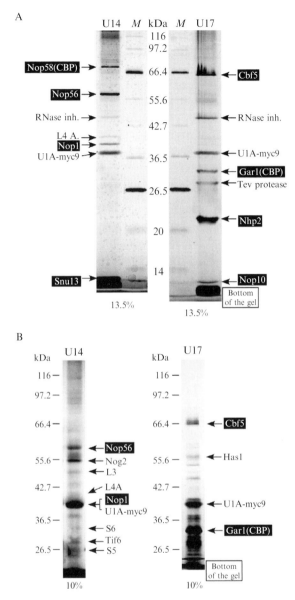

Figure 14.5 Protein patterns of the enriched U14 and U17 complexes. The proteins were concentrated by acetone precipitation and resolved in 13.5% or 10% SDS-PAGE (left and right pairs, respectively). Bands were visualized by silver staining and identified by mass spectrometry analysis (LC/MS/MS), direct in the case of the U14 results shown and by reference to another, more complex gel from our laboratory in the case of U17 (Liang and Fournier, 2006). Key variables here are cell culture stage (early-log and mid-log, panels A and B) and wash stringency (60 vols and 40 vols, panels A and B). Core proteins are highlighted with a black background. All four are evident for both

(40 volumes vs 60) but was derived from an older culture than that in Fig. 14.5A ($OD_{660} = 1.5$ vs 0.5–0.8). Experience with several other U14 preparations argues that snoRNP quality is best with cells harvested in early log phase, where quality is judged by the pattern of core proteins. Consistent with a more stringent wash condition, fewer noncore proteins are apparent in the U14 preparation subjected to more washing, and this situation has been seen in other cases in which greater wash volumes were used (Fig. 14.5A). The less-washed preparation contains more proteins (Fig. 14.5B), and among these six species were identified. Two are involved in maturation of ribosome precursors (Nog2, Tif6) and four are ribosomal proteins (L3, L4, S5, S6). The presence of these various proteins is consistent with association of the U14 snoRNP with pre-rRNP complexes but, of course, could also reflect spurious, nonspecific binding by RNA or protein–protein interactions.

Tif6 and Nog2 are known to be involved in biogenesis of the 60S subunit (Basu *et al.*, 2001; Saveanu *et al.*, 2001) and have been detected in some early, large subunit precursors formed after cleavage at the A2 site in pre-rRNA, which separates the maturation complexes of the small and large subunits (Gavin *et al.*, 2002; Ho *et al.*, 2002; Milkereit *et al.*, 2003). Nog2 and the C/D core protein Nop1 were also found in the pre-60S complex, where Tif6 was used as bait (Gavin *et al.*, 2002). Many other factors isolated with tagged Tif6 are involved in very early stages of pre-60S rRNP maturation (Milkereit *et al.*, 2003). Interestingly, Tif6 was also found in one early pre-40S complex (Ho *et al.*, 2002). These correlations strongly suggest that associations of Tif6 and Nog2 with the U14 snoRNA are relevant.

In the same context, the U3 snoRNA and its specific protein Rrp9 have also been found in a few early pre-60S complexes, although like U14, the U3 snoRNP is only known to act on upstream pre-rRNA species (Harnpicharnchai *et al.*, 2001; Nissan *et al.*, 2002; Oeffinger and Tollervey, 2003). Studies by others on protein and RNA components associated with pre-40S and pre-60S complexes showed that some factors are present in both complexes and thus cannot be classified as specific to 40S or 60S subunit maturation. Such factors have been proposed to link 60S and 40S subunit synthesis, and some are already present in 90S preribosome

---

snoRNPs in the higher percent gel (panel A); however, the smallest core proteins migrated out of the lower percent gel (panel B). CBP, calmodulin peptide attached to the C-end of Nop58 and Gar1. Detection of some ribosomal proteins and the Tif6, Nog2, and Has1 proteins suggests that some snoRNPs are isolated from preribosomal complexes. Tif6 and Nog2 are involved in biogenesis of the 60S subunit and have been detected in early pre-60S complexes formed shortly after cleavage at the A2 site in pre-rRNA, which separates the small and large ribosomal subunit (see text). Has1 is a helicase involved in release of U17 and other snoRNPs from large preribosomal RNP complexes (see text).

complexes before processing occurs at the A2 site of pre-rRNA (Takahashi *et al.*, 2003). These latter findings suggest that large subunit maturation factors and ribosomal proteins that coenrich with the U14 complex could arise from relevant interactions of these types *in vivo*.

Comparison of protein patterns for various U17 snoRNP preparations did not reveal any special loss of the tagged core protein, as seen for the U14 snoRNP (Fig. 14.5). We note that the two smallest core proteins (Nhp2 and Nop10) are absent in the second gel pattern shown here, because they were allowed to migrate out of a lower density gel (10 vs 13.5%; Fig. 14.5B). Importantly, all core proteins are consistently present in these preparations. Under the higher-volume wash condition, no other abundant proteins are evident, whereas several additional bands are apparent in the lower-volume condition (Fig. 14.5B). One of the more abundant proteins present is the RNA helicase Has1, which is required for normal and timely production of the small and large subunits (Emery *et al.*, 2004). After identifying Has1 in our U17 preparations, our group discovered that it is associated with other snoRNPs as well, including the U14 species, and that Has1 is required for normal release of snoRNPs from the largest pre-rRNA species. This effect was demonstrated for eight snoRNPs tested, including both processing and modifying snoRNPs (Liang and Fournier, 2006). Although these findings seem to validate the present enrichment strategy, we do not know whether Has1 associates with the free U17 snoRNP or at the level of a pre-rRNA–U17 complex.

2.4. Other approaches for enriching snoRNPs with the U1hpII tag/U1A couple

Our strategy can also be used in other formats to isolate snoRNPs and, we believe, other types of RNP complexes as well. In this section, we demonstrate the feasibility of: (1) inserting the RNA tag into an internal site of a target snoRNA (U14), and (2) the use of the U1A protein as the only bait protein, that is, with no core protein tagged. The small size of the RNA tag (26nts) provides optimism that it can be embedded within an RNA molecule without disrupting its function. More worrisome is the prospect of interference caused by binding of the larger U1A protein partner (102 amino acids, without any protein affinity tag).

2.4.1. Tagging a snoRNA internally is effective too

We tested two internal tagging schemes for the U14 snoRNA, avoiding several segments in the 126nt molecule that are essential for its synthesis or function (Fig. 14.6A). These include sequences of 13 and 14nts that bind to 18S rRNA for pairing and methylation guide function (domains A and B), a 32-nt highly folded structure essential for rRNA activity processing (the Y-domain), and the box C and D elements and adjoining helical segments

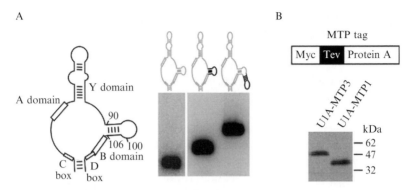

Figure 14.6 Insertion of the RNA tag into the interior of the U14 snoRNA does not interfere with snoRNP accumulation and function. (A) The major features of the U14 RNA are labeled. A single U1hpII tag (26nts) was inserted into U14 in two contexts: between nts A100 and G101 and to replace the small stem-loop between positions G90 and C106. Both U14 variants support growth (not shown), and Northern results show that stability and accumulation are comparable to natural U14. The Northern samples are identified here with U14 structures, with the tagged portions highlighted in black. (B) Protein A was provided with a novel tag (MTP) to allow it to be used as the sole bait in affinity selection. The tag includes 1 or 3 myc epitopes, a Tev cleavage site, and a protein A sequence. Western results (with PAP antibody) show that the U1A-MTP fusion protein expressed from the *NOP1* promoter is stable. The numbers of myc epitopes in the U1A variants examined are coded in the tag name (i.e., MTP1 and MTP3).

required for snoRNP assembly, stability, and nucleolar localization (Dunbar and Baserga, 1998; Huang *et al.*, 1992; Jarmolowski *et al.*, 1990; Li and Fournier, 1992; Liang and Fournier, 1995; Samarsky *et al.*, 1996, 1998). We elected to place the tag within the (unnamed) stem-loop structure located immediately upstream of domain B.

Two tagged U14 constructs were evaluated, one with the U1hpII tag inserted between positions A100 and G101 of the loop, and a second in which the entire stem-loop domain was replaced with the U1hpII sequence (Fig. 14.6A). Each U14 variant was expressed from a plasmid that contains a normal chromosomal segment that encodes U14 and the snR190 snoRNA, which are cotranscribed from the natural promoter. The plasmids were introduced into a yeast strain in which U14 is expressed conditionally from the chromosome with a *GAL1* promoter. The test strain also constitutively expresses a new tagged variant of the U1A protein from the *NOP1* promoter (see later). Both versions of the tagged U14 are metabolically stable and accumulate at levels similar to untagged U14 snoRNA (Fig. 14.6A); importantly, both versions support normal growth (data not shown). Thus, production and activity of the U14 snoRNP are not impaired by the presence of the new RNA tag (and the new version of the U1A protein described next).

2.4.2. U1A protein as the only bait

To determine whether the U1A protein can be used as the only bait for affinity selection, we designed a new tag segment that combines the two protein tags used earlier. This segment contains one or three copies of the myc epitope and the protein A sequence, separated by the Tev protease cleavage site (Fig. 14.6B). We call the new tag MTP (*Myc*, *Tev*, *Protein A*), and code the number of myc epitopes as MTP1 and MTP3. Plasmid-encoded sequences for both versions of the U1A-MTP protein were constructed, with expression driven by the *NOP1* promoter. Western analysis showed that both proteins have good stability and accumulation properties (Fig. 14.6B). One-liter isolations of the U14 snoRNP were performed with both internally tagged versions of U14 snoRNA and both versions of the U1A protein with the same two-stage conditions described previously.

Northern data show that both tagged variants of U14 are enriched equally well with U1A-MTP1 (Fig. 14.7A), and similar results were obtained with the U1A-MTP3 (data not shown). Staining of an RNA gel with ethidium bromide shows only one U14 RNA species that corresponds to the tagged variant (Fig. 14.7B). Data are shown here for one tagged U14 species only, and only for the U1A-MTP1 bait. No differences in U14 RNA enrichment were apparent for the two schemes used to tag the snoRNA and with the two different U1A tagging schemes with MTP.

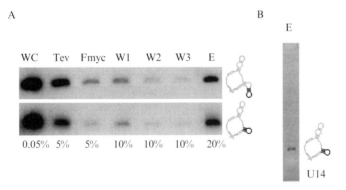

Figure 14.7 Two-step selection of internally tagged U14 that uses U1A-MTP1 as the only bait protein. (A) Northern results showing selection of two internally tagged U14 variants from the second step of isolation and stability of RNA on the anti-myc column. Both tagged versions of U14 were released from IgG beads (Tev) and then selected by anti-myc beads through the U1A protein containing one myc epitope (Fmyc). WC, whole cell extract; Tev, material after release from the IgG column by Tev protease; Fmyc, flow though after binding with anti-myc beads; W1–3, washes of anti-myc beads (second affinity column); E, RNA elution. (B) In the final elution only a single RNA band corresponding in size to the internally tagged U14 was observed on an ethidium bromide-stained gel.

The results show that one U1hpII hairpin is sufficient for stable interaction with the U1A protein and that one myc epitope in the MTP tag is sufficient for stable association with the anti-myc beads. Thus, this alternate, single-bait U1A protein scheme can also be applied for enriching RNP complexes. Although the results are for an internally tagged snoRNA, we suggest that the same approach could be effective for a snoRNA bearing a tag at one end, although an extension may be necessary in some cases.

2.5. Screening for interactions between snoRNPs

In principle, the approaches described here can be used to screen for interactions between different snoRNPs (or a snoRNP and another RNP), by determining whether an additional RNA species copurifies with the tagged snoRNP of interest. Coenrichment would require an association that is sufficiently stable to persist through the selection procedure. If detected, the presence of another snoRNA could reflect a relevant *in vivo* interaction or a selective artifact. Authentic multi-snoRNP complexes may not coenrich, of course, if the interactions are weak or short lived. In such cases, genetic screening *in vivo* might be more effective, by use of conventional two- and three-hybrid approaches or a new "RNA-hybrid" system (Fashena *et al.*, 2000; Hook *et al.*, 2005; Piganeau *et al.*, 2006). In the following, we show that our selection scheme can, indeed, enrich other snoRNPs, when selection is for the tagged U14 snoRNP.

Because of the specificity of the U1A-MTP tagging system, only one snoRNA should be enriched in the first selection step in contrast to using a core protein as bait. Thus, if additional snoRNAs are enriched, this would infer that the test snoRNP interacts directly or indirectly with other snoRNPs. In the initial two-step enrichment procedure, the additional snoRNAs were sequentially washed out in the second step of purification. Although most of the minor snoRNAs are likely due to contamination, others may be in common complexes with the tagged snoRNP and, thus, be of special interest. Such species could be lost with the additional manipulations. For this reason we decided to reduce the affinity selection to a single step and to monitor the RNA content of the wash carried out after selection by the IgG beads.

In this approach, the cell extract was prepared as previously, except that we included in all buffers the detergent NP40 (0.1%) with the aim of making the selection condition more stringent. After incubation with the IgG beads (4 h), the beads were loaded into a column and subjected to washing with buffer (500 bed volumes; 50 ml/0.1 ml beads). After release of the selected material by Tev cleavage, the RNA composition was examined by gel electrophoresis, with detection by ethidium bromide staining or Northern analysis (Fig. 14.8). Patterns are shown for the U14 snoRNP

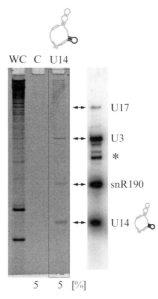

Figure 14.8 The U1hpII/U1A-MTP pair can be used as bait to identify other interacting RNA molecules. Enrichment of two other snoRNPs with the tagged U14 complex after the first selection step (IgG beads) suggests direct interactions between the snoRNPs or presence in the same complex. The ethidium bromide–staining pattern is for the material (5%) released from the IgG beads (lane: U14) after intensive washing (500 vol.). No RNA was selected in the control experiment (lane: C) where no tagged RNA was expressed. The bands above the tagged U14 snoRNA were identified by Northern analysis as the snR190 and U3 snoRNAs. The signals (*) below the U3 band were identified as degradation products of this snoRNA. The weak U17 signal in the Northern pattern suggests that this snoRNP is not selected with the tagged U14, but is a background RNA. The minor snoRNPs are washed out in the two-step selection procedure (see Fig. 14.7B).

preparation and a control sample from cells in which the U1A-MTP1 bait protein is expressed, but no tagged snoRNA is present.

Strikingly, three highly enriched RNA bands are apparent in the material released from the beads, whereas no such bands were detected in the control experiment. The Northern results show that these bands correspond to the tagged U14 snoRNA, as expected, and the snoRNA species U3 and snR190. A signal was also obtained for U17 snoRNA, but this is much weaker (Fig. 14.8). The bands located just below the U3 signal were confirmed to be U3 degradation products. If we consider U17 an internal control, the data indicate that the U3 and snR190 species are enriched at a higher level—on roughly the same scale as the tagged U14. This high level of enrichment suggests that the U3 and snR190 snoRNPs are not likely contaminants but, rather, are present in one or more complexes with the U14 snoRNP. Both snoRNAs were washed out in the second step of

U14 selection, indicating that the interactions with the U14 snoRNP are not stable.

Although additional work is needed to understand this phenomenon, these early results could reflect U14 relationships that are already inferred in the literature. In particular, U14 and snR190 are cotranscribed from the same transcription unit and are believed to be derived from the same pre-snoRNA maturation complex (Chanfreau *et al.*, 1998). In the case of U3 and U14, both are required (with U17) for the same two cleavages (A1 and A2) that create the 20S precursor to 18S rRNA. This requirement is consistent with the possibility that U3 and U14 (and perhaps U17) are in a common complex(es), most likely in the 90S pre-ribosome that contains full-length pre-rRNA. The reduced yield of U17 relative to U3 could reflect selective loss of this snoRNP from the complex or involvement in different, but related, complexes. By use of computational techniques, we have searched for potential interactions between these snoRNAs and found only short (6nt) base-pairing possibilities. Thus, any direct interactions between these snoRNPs are likely to involve proteins as well.

2.6. Potential limitations of these approaches

As with other tagging approaches, applying the strategies described here must be done on an empirical basis. The first concern, of course, will be to discover a suitable site(s) for tagging the snoRNA or other RNA that does not create interference effects. We show that the U1hpII tag can be used effectively when placed slightly upstream (10–40nt) of the 5′ end or internally in a target RNA. Some degradation of the RNA was observed with the 5′ location, which was only slight for U14, but substantial for U17. This damage could be triggered by the small RNA tag or from binding of the much larger U1A protein to the tag. We do not know whether the tagged U17 (or U14) snoRNA actually supports growth or whether the requirement is satisfied by trimmed variants lacking the tag. Importantly, however, it is clear that U17 snoRNA tagged at the 5′ end assembles with the four expected core proteins and another linked to ribosome synthesis.

As noted, we subsequently discovered that a protein coselected with the U17 snoRNP, Has1, is required for release of this snoRNP and others from pre-rRNA complexes (Liang and Fournier, 2006). This discovery seems to validate the selection strategy presented, and additional support comes from the fact that all five noncore proteins coselected by the U14 snoRNP are either involved in ribosome maturation or are ribosomal proteins. Although these correlations suggest that the snoRNPs and noncore proteins identified are present in the same pre-rRNP complexes, it does not follow that they interact directly. Furthermore, as with any selection scheme, spurious associations are always possible. The relevance of each association must be determined independently by complementary biochemical or genetic approaches. An important adjunct for the present strategy would be to carry out new affinity isolations with the newly identified protein (or RNA) as bait to determine whether any of the same components are selected. Corroborating results would provide evidence of relevance, but not proof, whereas negative results would be just that.

For some work, the solution conditions used in our method may prove to be a limitation. The interaction between the U1hpII domain and the U1A protein is sensitive to ionic strength and studies by others indicate that $MgCl_2$ and NaCl concentrations should not exceed 5 mM and 250 mM respectively (Hall and Stump, 1992).

2.7. Other applications

On the basis of our experience with the snoRNP complexes, we are confident that the tagging strategies described will be useful for isolating other snoRNPs and other types of RNP complexes as well. For the snoRNPs, affinity selection provides the best direct approach to identifying proteins that associate with particular species, and thereby gain valuable clues about partner molecules involved in snoRNP synthesis, trafficking, and function. The obvious caveat is that enriched proteins and RNAs may be coselected for totally spurious reasons.

In time, we may learn that all modifying snoRNPs act in the same way with the same facilitating partner molecules, despite different RNA substrate specificities, but it is too early to rule out special situations. For example, the pseudouridine guide snoRNA in yeast called snR86 is unusually long with 1004nt, whereas most other H/ACA snoRNAs are much smaller, approximately 190nt (Torchet et al., 2005). This difference could reflect the presence of specialized, noncanonical sites of binding or function. Alternately, the snoRNAs and snRNAs in yeast are generally larger than in other organisms (especially vertebrates) and may simply indicate an atypical evolution. The same reasoning applies to the processing snoRNAs, where it is clear already that the U3 snoRNA binds additional, noncore proteins in an essential 5' domain. Especially intriguing are sno-like RNAs that have been referred to as "orphan" snoRNAs, which could have functions other than guiding modification (Bachellerie et al., 2002). Many such species identified in mouse brain have the key features of guide snoRNAs; however, candidate RNA targets have not yet been identified in all cases (Huttenhofer et al., 2001).

Finally, the new snoRNP selecting schemes could be valuable for discovering interacting partners for other RNAs as well, such as guide RNAs used in classical RNA editing and for the rapidly growing number of other noncoding small RNAs.

3. Materials and Experimental Procedures

3.1. Yeast strains

The following strains were used in these studies.

YS153 (MATa ura3, his3, trp1, HIS3::GAL1::U14) (Jarmolowski *et al.*, 1990); YBH20 (MATa ura3, ade2, trp1, his3, leu2, GAR1-TAP::URA3); YBH23 (MATa ura3, ade2, trp1, his3, leu2, nop58::HIS3 [pBL277-NOP58-TAP, TRP1]); YBH25 (MATa ura3, ade2, trp1, his3, leu2, GAR1-TAP::URA3, [pBL276-GAL1 NLS U1A-myc9 TRP1]); YBH26 (MATa ura3, ade2, trp1, his3, leu2, nop58::HIS3 [pBL277-NOP58-TAP, TRP1], [pBL272-GAL1 NLS U1A-myc9 URA3]); YBH70 (MATa ura3, ade2, trp1, his3, leu2, GAR1-TAP::URA3, ADE2::GAL1::9× U1hpII SNR30 [pBL276-GAL1 NLS U1A-myc9 TRP1]); YBH71 (MATa ura3, ade2, trp1, his3, leu2, nop58::HIS3 [pBL277-NOP58-TAP, TRP1], ADE2:: GAL1::9× U1hpII SNR128 [pBL272-GAL1 NLS U1A-myc9 URA3]); YDP-C (MATa ura3, his3, trp1, HIS3::GAL1::U14, [pDKG4-SNR190-U14 TRP1], [pDK104-pNOP1 NLS U1A-MTP1 URA3]); YDP70 (MATa ura3, his3, trp1, HIS3::GAL1::U14, [pDK90-SNR190-U14 U1hpII in A100-C101 TRP1], [pDK103-pNOP1 NLS U1A-MTP3 URA3]); YDP71 (MATa ura3, his3, trp1, HIS3::GAL1::U14, [pDK90-SNR190-U14 U1hpII in A100-C101 TRP1] [pDK104-pNOP1 NLS U1A-MTP1 URA3]); YDP73 (MATa ura3, his3, trp1, HIS3::GAL1::U14, [pDK91-SNR190-U14 U1hpII in G90-C106 TRP1] [pDK104-pNOP1 NLS U1A-MTP1 URA3]).

3.2. TAP tagging of core proteins

TAP tagging of genes *GAR1* and *NOP58* was performed as described elsewhere (Puig *et al.*, 2001).

3.3. Plasmid constructs

Plasmids that can be used for tagging the 5′ end of any RNA molecule with the human U1hpII hairpin were created as follows. The *ADE2* gene was amplified by PCR (oligos 1 and 2, see later) from genomic DNA and subcloned into *Kpn*I and *Cla*I sites of pBL143, resulting in pBL175. Plasmid pBL143 is a pRS314 derivative that contains an snR38-derived snoRNA gene under control of the *GAL1-GAL10* dual promoter (Liu and Fournier, 2004; Sikorski and Hieter, 1989). The first snoRNAs tagged with one U1hpII tag were U14 and U17. For 1× U1hpII-U14, the genomic fragment was generated by PCR (oligos 3 and 4) and inserted into *Bam*HI/*Sac*I of pBL175, resulting in pBL257. For 1× U1hpII-U17, the genomic fragment was generated by PCR (oligos 5/6), cut with *Bam*HI, and inserted in *Bam*HI and *Sac*I (blunted with mung bean nuclease) of pBL175, resulting in pBL259.

To create a plasmid with multiple cloning sites for tagging any RNA with 1× U1hpII, a PCR fragment (oligos 7/8, template pBL257) was inserted into *Bam*HI/*Sac*I of pBL257, producing pBL279. The series of plasmids containing a multicopy U1hpII tag were generated by PCR with oligos 9 and 10 (the PCR product defined by oligos 11/12 was used as a template) and subcloned into *Bam*HI/*Xho*I of pBL279, resulting in pBL291 (2× U1hpII), pBL292 (3× U1hpII), and pBL293 (4× U1hpII). Two 3× U1hpII fragments were used to generate a plasmid with a 6× U1hpII tag. One 3× U1hpII fragment (cut from pBL292 with *Xho*I, then blunted with mung bean nuclease and cut with *Bam*HI) and another 3× U1hpII fragment (cut from pBL292 with *Bam*HI, then blunted with mung bean nuclease and cut with *Xho*I) were ligated and subcloned into *Bam*HI and *Xho*I sites of pBL279, resulting in pBL296 (6× U1hpII). The plasmid with the 8× U1hpII tag was generated from two 4× U1hpII fragments (pBL293) similar to that used for pBL296, resulting in pBL297. The set of plasmids pBL294, pBL295, pBL298, and pBL299 are derived from pRS316 (*URA3*) with the same expression cassettes as pBL291, pBL292, pBL296, and pBL297, respectively. The plasmid with the 12× U1hpII tag was generated from two 6× U1hpII fragments (pBL299) in similar way as used for pBL297, resulting in pBL300.

To maintain the genotypes of yeast strains with TAP-tagged core proteins and a plasmid-encoded U1A-myc9 (YBH25, YBH26), tagged C/D and H/ACA snoRNAs were expressed from a second allele on a different parental plasmid. The expression cassette *ADE2-GAL1*-nxU1hpII-snoRNA (flanked by *Kpn*I and *Sac*I) should be in a plasmid with the *TRP1* marker (pRS314) in the case of a tagged C/D snoRNA and in a plasmid with the *URA3* marker (pRS316) in the case of a tagged H/ACA snoRNA.

Expression cassettes encoding tagged U14 and U17 snoRNA genes were prepared to replace the natural chromosomal coding sequences. For tagging U14 with multiple copies of U1hpII, the 1× U1hpII U14 fragment of pBL257 (cleaved with *Bam*HI, then blunted with mung bean nuclease and cut with *Sac*I) was subcloned into the *Nru*I/*Sac*I sites of pBL297 (8× U1hpII). To facilitate recombination with the U14 construct, a fragment of ~800nt containing the coding sequence for snR190 (upstream of the U14 sequence) was amplified by PCR (oligos 13/14) and subcloned into *Kpn*I/*Bst*BI sites of a plasmid with tagged U14 DNA, resulting in pBL303. The *Kpn*I/*Sac*I fragment of this plasmid (snR190-*ADE2-GAL1*-9× U1hpII-U14) was used for yeast transformation (YBH23). The accompanying *ADE2* marker was used to select transformants on plates containing galactose.

For tagging U17 snoRNA with multiple copies of U1hpII, the *Eco*RI/*Nru*I fragment (GAL1-8× U1hpII) of pBL297 was inserted into the *Eco*RI and *Bam*HI sites (blunted with mung bean nuclease) of pBL259, resulting in pBL301. The entire expression cassette of this plasmid (*ADE2-GAL1*-9× U1hpII-U17) was amplified by PCR (oligos 15/6; oligo 15 includes a 40nt upstream coding sequence of the U17 snoRNA for homologous recombination), and this PCR product was used for yeast transformation (YBH20).

Two plasmids encoding different versions of internally tagged U14 snoRNA were prepared (pDK90 and pDK91). In both cases, the tagged sequences are expressed in the context of the natural transcription unit that also specifies the snoRNA snR190, both under control of the natural promoter. Stepwise, a genomic fragment that defines an expression cassette for snR190 and U14 was amplified by PCR (oligos 13/4) and subcloned into *KpnI*/*SacI* sites of pRS314, resulting in pDKG4. One tagged version (encoded in pDK90) contains U14 with a single U1hpII sequence inserted between nucleotides A100 and G101. In the other version (encoded in pDK91), the U1hpII tag is substituted for the segment between G90 and C106, creating a deletion of 15nt in this U14 variant. To create these constructs, the PCR fragments of genomic DNA were amplified with oligos 13/16 and 13/17, digested with *KpnI*/*AvrII*, and subcloned into the corresponding sites of pDKG4 to yield pDK90 and pDK91, respectively.

The plasmid used to express U1A-myc9 (with a nuclear localization signal) was created by moving a *SacI*/*KpnI* fragment (*GAL1-NLS-U1A-myc9*) of the vector p415 GAL1 NLS-U1A-myc9 (Bachler *et al.*, 1999) into corresponding sites in pRS314 and pRS316 to yield pBL276 and pBL272, respectively. The construct that expresses U1A-MTP from the promoter for Nop1 was prepared as follows. A 350nt fragment of genomic DNA with the *NOP1* promoter was amplified with oligos 18/19 and subcloned into the *XhoI*/*HindIII* sites of pRS415 (creating pDK81). The *SacII*/*SacI* sites of this plasmid were used to insert the *URA3* gene amplified with oligos 20/21 from pBS1539 (Puig *et al.*, 2001). Next, the cloning sites *PstI* and *XbaI* were used to insert a segment of the TAP sequence that includes the Tev protease cleavage site and the protein A sequence. This product was generated from pBS1539 by amplification with oligos 22/23 and yielded pDK98. Fragments of U1A-myc (with 1 and 3 myc epitopes) were amplified from pBL272 (with oligos 24/25), and subcloned into *HindIII*/*PstI* sites in pDK98, resulting in pDK103 (U1A-MTP3) and pDK104 (U1A-MTP1).

3.4. DNA oligonucleotides

1. GGGGTACCCAATGTGTCCATCTGACATTACTATTTTGC;
2. GGATCGATTAAGCGTTGATTTCTATGTATGAAGTCCAC;
3. ACGTGGATCCAGCTTATCCATTGCACTCCGGATGAGCT-
   GAGGAGACAGATAATATATATAATTTATG;
4. TGCAGAGCTCAACCCACATCAACACTCCACCGTGAC;

5. ACGTGGATCCAGCTTATCCATTGCACTCCGGATGAGCT-
   GTTTCTCTTTGCTTATTTGTAAAAAAAAAAG;
6. GGAATTAAGAGAGCGGACAAGTCGAATG;
7. ACCTCTATACTTTAACGTCAAGGAG;
8. AAAATGCAGAGCTCTCGCGAGTCGACCTCGAGGCTCAT-
   CCGGAGTGCAATGGATAAGC;
9. ACGTGGATCCAGCTTATCCATTGCACTCCGGATGAGCA-
   AAATTAGCTTATCCATTGCACTCCGGATGAGCAA;
10. GCCGCTCGAGGCTCATCCGGAGTGCAATGGATAAGCTT-
    TTTAAGCTCATCCGGAGTGCAATGGATAAGCTAA;
11. ACGTGGATCCAGCTTATCCATTGCACTCCGGATGAGCA-
    AAAAAAGCTTATCCATTGCACTCCGGATGAGC;
12. GCCGCTCGAGGCTCATCCGGAGTGCAATGGATAAGCTT-
    TTTTTGCTCATCCGGAGTGCAATGGATAAGCT;
13. CGGGGTACCGCTCACTGCTCACTCGATCACAGAC;
14. AATTTTCGAATTACCCACATTTGATAATCTCATGAG;
15. AACCATAGTCTCGTGCTAGTTCGGTACTATACAGGGAA-
    GGCAATGTGTCCATCTGACATTACTATTTTGC;
16. AGAGCTCATCACTCAGACATCCTAGGAAGGTCTCAGCT-
    CATCCGGAGTGCAATGGATAAGCTTAAAGAAGAGCGGT-
    CACC;
17. TGCAGAGCTCATCACTCAGACATCCTAGGAAGTGCTCAT-
    CCGGAGTGCAATGGATAAGCACGGTCACCGAGAGTACT-
    AAC;
18. CCGCTCGAGCTGTCCTCCGTTCTGTAAAA;
19. CCCAAGCTTAGTGGATCCATGGTACTGTTTTAG;
20. TCCCCGCGGTTGTTGTTCCTTACCATTAAG;
21. CGAGCTCGGAGACAATCATATGGGAGAAGC;
22. GCTCTAGATCAGGTTGACTTCCCCGCGGAATTC;
23. AAAACTGCAGGATTATGATATTCCAACTACTG;
24. CCCAAGCTTTCTAGAACTAGTATGCCAA;
25. AAAACTGCAGCAAGTATTCCTCGGAGATTAGC;
26. AGGAGAGCTCTGGTTAACTTGTCAGACTGCCATTTGTAC-
    CCACCCATAG;
27. CCGTGGAAACTGCGAATGTTAAGGAACCAG;
28. AGATGTCTGCAGTATGGTTTTACCCAAATG;
29. AGCTCATCCGGAGTGCAATGGATAAGCT;
30. GGCTCAGATCTGCATGTGTTGTAT;

3.5. Growth conditions

Yeast culturing is performed at 30° with shaking at 250 rpm. Yeast strains YBH70 and YBH71 are grown in galactose-containing medium deficient in adenine, uracil, and tryptophan. Strains YBH25 and YBH26 are used as controls (no tagged RNA) and are grown in galactose medium deficient in tryptophan and uracil. Strains YDP70, YDP71, YDP73, and YDP-C (control—no tagged RNA) are grown in glucose-containing medium deficient in uracil and tryptophan.

3.6. Affinity purification of snoRNPs

The snoRNP selection strategy described in the following is for cultures of 5 liters. For mass spectrometry analysis of the proteins, two or three such preparations can be combined as needed to achieve suitable amounts of protein. One-liter cultures were used to evaluate and adjust the enrichment conditions and for the experiment with the internally tagged U14 snoRNA. All steps were done as described for 5-liter cultures, but with proportionally less IgG beads, anti-myc beads, and volumes of buffers and washing.

All enrichment steps, except the Tev cleavage reaction and final elution, were performed at $4°$. To monitor the elution of the tagged RNAs and tagged proteins, samples taken from the various purification steps were frozen at $-80°$ and examined by Northern and Western analyses.

Cultures are harvested at early-log phase ($OD_{660} = 0.5$–0.8) by centrifugation ($4000g$), and the cell pellet is washed twice with cold water and resuspended in one volume of buffer A (25 mM Tris-HCl, pH 7.6, 5 mM MgCl$_2$, 150 mM KCl, 0.5 mM DTT, 10% glycerol) containing Complete^{TM} protease inhibitors (one tablet/50 ml; Roche Molecular Biochemicals). Typically, a 5-liter culture yields 15 g of cells, which are resuspended in 15 ml of A buffer. The cells are lysed by vortex mixing with glass beads. An amount of acid-washed glass beads (425–600 microns, Sigma) equal in volume to the cell pellet is added, and the cells are broken by ten 30-sec cycles of robust mixing, with an interval of 30 sec on ice between each cycle. One volume (15 ml) of buffer A is added to the lysate, and it is then cleared by two centrifugations, first at $4000g$ for 10 min and next at $15,000g$ for 30 min.

Before the initial binding step, the IgG beads (IgG Sepharose 6 Fast Flow; Amersham Pharmacia Biotech) are washed with buffer A. The clarified extract is mixed with 500 μl of the IgG beads (corresponding to a total suspension of 750 μl) and incubated with gentle mixing for 4 h at $4°$. The washing steps are performed using a Flex-6 column (1.0×5 cm; Kontes). First, the mixture is left on ice for 10 min to allow the beads to settle. The supernatant is removed, and the beads are resuspended in 20 (bed) volumes of buffer A. After 10 min on ice, the supernatant is removed, the beads are resuspended in 20 vol of buffer A and applied to the column. The wash rate is controlled (at 1 ml/min) and continued with 60 vol of buffer B (25 mM Tris-HCl, pH 7.6, 5 mM MgCl$_2$, 150 mM KCl, 0.5 mM DTT, 0.5 mM EDTA, 10% glycerol).

Next, 300 U of Tev protease (Invitrogen) is mixed with 600 U of RNase inhibitor (Promega) in 3 ml buffer C (25 mM Tris-HCl, pH 8.0, 5 mM MgCl$_2$, 150 mM KCl, 1 mM DTT, 0.5 mM EDTA), added to the beads, and the mixture is transferred to a 15-ml Falcon tube. The Tev cleavage reaction is performed at 16° for 2 h with gentle mixing. To remove the beads, the mixture is applied to the column. The flow-through solution is then incubated with 150 μl anti-myc beads (prewashed with buffer A) at 4° for 2 h with gentle mixing (Anti-c-Myc Agarose, Sigma). The beads are loaded onto a Flex-6 column and washed at a rate of 1 ml/min with 60 vol of buffer A. For elution of selected material, the beads are mixed with 0.25 ml of buffer A containing myc peptide (0.5 mg/ml; Sigma) and transferred into a 1.5-ml Eppendorf tube. The elution is performed with three cycles of incubation, each for 20 min at 10° with gentle mixing. The liquid phases from each cycle are combined and subjected for protein precipitation.

The proteins are precipitated with acetone (3 volumes and overnight storage at −20°) or TCA (15%, on ice for a few hours), followed by centrifugation and washing of the pellet with acetone (twice for TCA). Pellets are then resuspended in electrophoresis loading buffer (60 mM Tris-HCl, pH 6.8, 2% SDS, 150 mM mercaptoethanol, 5 mg/ml Bromophenol Blue, 10% glycerol), incubated at 95° for 5 min, and cooled to room temperature. Samples from multiple preparations were combined (from a total of approximately 30 g of yeast cells) and fractionated by 10% or 13.5% SDS-PAGE and visualized by silver staining (Bio-Rad). Proteins in individual bands were identified by mass spectrometry analysis (LC/MS/MS).

3.7. Screening for coselecting RNAs

The test for snoRNAs that might copurify with the U14 snoRNP was done with yeast strains YDP73 and YDP-C as the control strain. Preparation of cell extract and the binding and washing procedures were done as previously except that buffer A also contained 0.1% Nonidet (NP40). The IgG beads were washed with 500 vol of buffer A (with 0.1% NP40), and the material selected was eluted by Tev protease cleavage as described previously.

3.8. RNA isolation and Northern analysis

Total yeast RNA is prepared as described elsewhere (Balakin *et al.*, 1993). During the enrichment procedures, RNA is isolated from fractions collected at each step as follows. A sample volume is adjusted to 0.5 ml and extracted twice, first with 1 vol of phenol and next with 1 vol of phenol-chloroform 5:1 (Sigma). To facilitate RNA precipitation, 10 μg of glycogen is added to each sample before precipitation. RNA is separated on a gel of

8% polyacrylamide-7 M urea, stained with ethidium bromide (0.1–0.5 μg/ml), and then subjected to Northern analysis (e.g., King *et al.*, 2003). The snoRNAs are detected with the following oligonucleotide probes: U3 (26); U14 (27); U17 (28); tagged snR3 and snR39b (29), and snR190 (30).

3.9. Western analysis

To track the tagged proteins during the snoRNP isolations, samples were taken at different stages of enrichment, and the proteins were precipitated with TCA. The proteins are fractionated by 10% SDS-PAGE and subjected to Western blotting by use of, in order, anti-c-myc antibody (9E10, 1:5000; Roche Diagnostics); and PAP reagent (1:500; Sigma), which recognizes the protein A moiety of the TAP sequence. For detection, ECL blotting detection reagents (Amersham Pharmacia Biotech) are used.

ACKNOWLEDGMENTS

We are grateful to Uwe von Ahsen for providing a plasmid that encodes the NLS U1A-myc9 protein (p415 GAL1 NLS-U1A-myc9). Mass spectrometry services were provided by the Lerner Research Institute Mass Spectrometry Laboratory for Protein Sequencing at the Cleveland Clinic Foundation and the Laboratory for Proteomic Mass Spectrometry, University of Massachusetts Medical School. We are also grateful to Aswini Panigrahi and Kenneth Stuart at the Seattle Biomedical Research Institute for generous and helpful attempts to catalog by mass spectrometry all proteins in stage-2 snoRNP preparations, which had not been prefractionated by PAGE. This work was supported by a grant from NIH (GM19351) to M. J. F.

REFERENCES

Aittaleb, M., Rashid, R., Chen, Q., Palmer, J. R., Daniels, C. J., and Li, H. (2003). Structure and function of archaeal box C/D sRNP core proteins. *Nat. Struct. Biol.* **10,** 256–263.

Allain, F. H., Howe, P. W., Neuhaus, D., and Varani, G. (1997). Structural basis of the RNA-binding specificity of human U1A protein. *EMBO J.* **16,** 5764–5772.

Bachellerie, J. P., Cavaille, J., and Huttenhofer, A. (2002). The expanding snoRNA world. *Biochimie* **84,** 775–790.

Bachler, M., Schroeder, R., and von Ahsen, U. (1999). StreptoTag: A novel method for the isolation of RNA-binding proteins. *RNA* **5,** 1509–1516.

Balakin, A. G., Schneider, G. S., Corbett, M. S., Ni, J., and Fournier, M. J. (1993). SnR31, snR32, and snR33: Three novel, non-essential snRNAs from *Saccharomyces cerevisiae*. *Nucleic Acids Res.* **21,** 5391–5397.

Ballarino, M., Morlando, M., Pagano, F., Fatica, A., and Bozzoni, I. (2005). The cotranscriptional assembly of snoRNPs controls the biosynthesis of H/ACA snoRNAs in *Saccharomyces cerevisiae*. *Mol. Cell Biol.* **25,** 5396–5403.

Basu, U., Si, K., Warner, J. R., and Maitra, U. (2001). The *Saccharomyces cerevisiae* TIF6 gene encoding translation initiation factor 6 is required for 60S ribosomal subunit biogenesis. *Mol. Cell Biol.* **21,** 1453–1462.

Bernstein, K. A., Gallagher, J. E., Mitchell, B. M., Granneman, S., and Baserga, S. J. (2004). The small-subunit processome is a ribosome assembly intermediate. *Eukaryot. Cell* **3**, 1619–1626.

Bertrand, E., and Fournier, M. J. (2004). The snoRNPs and related machines: Ancient devices that mediate maturation of rRNA and other RNAs. *In* "The Nucleolus" (M. O. J. Olson, ed.), pp. 225–261. Landes Bioscience Publishing, Georgetown, TX; New York, NY.

Bleichert, F., Granneman, S., Osheim, Y. N., Beyer, A. L., and Baserga, S. J. (2006). The PINc domain protein Utp24, a putative nuclease, is required for the early cleavage steps in 18S rRNA maturation. *Proc. Natl. Acad. Sci. USA* **103**, 9464–9469.

Bock-Taferner, P., and Wank, H. (2004). GAPDH enhances group II intron splicing *in vitro*. *Biol. Chem.* **385**, 615–621.

Boelens, W. C., Jansen, E. J., van Venrooij, W. J., Stripecke, R., Mattaj, I. W., and Gunderson, S. I. (1993). The human U1 snRNP-specific U1A protein inhibits poly-adenylation of its own pre-mRNA. *Cell* **72**, 881–892.

Burke, D., Dawson, D., and Stearns, T. (2000). Methods in yeast genetics: A Cold Spring Harbor Laboratory Course Manual 2000 Edition. Cold Spring Harbor Laboratory Press, Cold Spring Harbor, New York.

Chamberlain, J. R., Lee, Y., Lane, W. S., and Engelke, D. R. (1998). Purification and characterization of the nuclear RNase P holoenzyme complex reveals extensive subunit overlap with RNase MRP. *Genes Dev.* **12**, 1678–1690.

Chanfreau, G., Rotondo, G., Legrain, P., and Jacquier, A. (1998). Processing of a dicistronic small nucleolar RNA precursor by the RNA endonuclease Rnt1. *EMBO J.* **17**, 3726–3737.

Charpentier, B., Muller, S., and Branlant, C. (2005). Reconstitution of archaeal H/ACA small ribonucleoprotein complexes active in pseudouridylation. *Nucleic Acids Res.* **33**, 3133–3144.

Chu, S., Archer, R. H., Zengel, J. M., and Lindahl, L. (1994). The RNA of RNase MRP is required for normal processing of ribosomal RNA. *Proc. Natl. Acad. Sci. USA* **91**, 659–663.

Clayton, D. A. (1994). A nuclear function for RNase MRP. *Proc. Natl. Acad. Sci. USA* **91**, 4615–4617.

Dangerfield, J. A., Windbichler, N., Salmons, B., Gunzburg, W. H., and Schroder, R. (2006). Enhancement of the StreptoTag method for isolation of endogenously expressed proteins with complex RNA binding targets. *Electrophoresis* **27**, 1874–1877.

Dragon, F., Gallagher, J. E., Compagnone-Post, P. A., Mitchell, B. M., Porwancher, K. A., Wehner, K. A., Wormsley, S., Settlage, R. E., Shabanowitz, J., Osheim, Y., Beyer, A. L., Hunt, D. F., and Baserga, S. J. (2002). A large nucleolar U3 ribonucleoprotein required for 18S ribosomal RNA biogenesis. *Nature* **417**, 967–970.

Dragon, F., Pogacic, V., and Filipowicz, W. (2000). *In vitro* assembly of human H/ACA small nucleolar RNPs reveals unique features of U17 and telomerase RNAs. *Mol. Cell Biol.* **20**, 3037–3048.

Dunbar, D. A., and Baserga, S. J. (1998). The U14 snoRNA is required for 2′-O-methyla-tion of the pre-18S rRNA in *Xenopus* oocytes. *RNA* **4**, 195–204.

Einhauer, A., Schuster, M., Wasserbauer, E., and Jungbauer, A. (2002). Expression and purification of homogenous proteins in *Saccharomyces cerevisiae* based on ubiquitin-FLAG fusion. *Protein Expr. Purif.* **24**, 497–504.

Ellington, A. D., and Szostak, J. W. (1990). *In vitro* selection of RNA molecules that bind specific ligands. *Nature* **346**, 818–822.

Emery, B., de la Cruz, J., Rocak, S., Deloche, O., and Linder, P. (2004). Has1p, a member of the DEAD-box family, is required for 40S ribosomal subunit biogenesis in *Saccharomyces cerevisiae*. *Mol. Microbiol.* **52**, 141–158.

Fashena, S. J., Serebriiskii, I., and Golemis, E. A. (2000). The continued evolution of two-hybrid screening approaches in yeast: How to outwit different preys with different baits. *Gene* **250,** 1–14.

Gallagher, J. E., Dunbar, D. A., Granneman, S., Mitchell, B. M., Osheim, Y., Beyer, A. L., and Baserga, S. J. (2004). RNA polymerase I transcription and pre-rRNA processing are linked by specific SSU processome components. *Genes Dev.* **18,** 2506–2517.

Ganot, P., Caizergues-Ferrer, M., and Kiss, T. (1997). The family of box ACA small nucleolar RNAs is defined by an evolutionarily conserved secondary structure and ubiquitous sequence elements essential for RNA accumulation. *Genes Dev.* **11,** 941–956.

Gavin, A. C., Bosche, M., Krause, R., Grandi, P., Marzioch, M., Bauer, A., Schultz, J., Rick, J. M., Michon, A. M., Cruciat, C. M., Remor, M., Hofert, C., *et al.* (2002). Functional organization of the yeast proteome by systematic analysis of protein complexes. *Nature* **415,** 141–147.

Gerczei, T., and Correll, C. C. (2004). Imp3p and Imp4p mediate formation of essential U3-precursor rRNA (pre-rRNA) duplexes, possibly to recruit the small subunit processome to the pre-rRNA. *Proc. Natl. Acad. Sci. USA* **101,** 15301–15306.

Grainger, R. J., Norman, D. G., and Lilley, D. M. (1999). Binding of U1A protein to the 3′ untranslated region of its pre-mRNA. *J. Mol. Biol.* **288,** 585–594.

Hall, K. B., and Stump, W. T. (1992). Interaction of N-terminal domain of U1A protein with an RNA stem/loop. *Nucleic Acids Res.* **20,** 4283–4290.

Harnpicharnchai, P., Jakovljevic, J., Horsey, E., Miles, T., Roman, J., Rout, M., Meagher, D., Imai, B., Guo, Y., Brame, C. J., Shabanowitz, J., Hunt, D. F., and Woolford, J. L., Jr. (2001). Composition and functional characterization of yeast 66S ribosome assembly intermediates. *Mol. Cell* **8,** 505–515.

Hefti, M. H., Van Vugt-Van der Toorn, C. J., Dixon, R., and Vervoort, J. (2001). A novel purification method for histidine-tagged proteins containing a thrombin cleavage site. *Anal. Biochem.* **295,** 180–185.

Ho, Y., Gruhler, A., Heilbut, A., Bader, G. D., Moore, L., Adams, S. L., Millar, A., Taylor, P., Bennett, K., Boutilier, K., Yang, L., Wolting, C., *et al.* (2002). Systematic identification of protein complexes in *Saccharomyces cerevisiae* by mass spectrometry. *Nature* **415,** 180–183.

Hoang, T., Peng, W. T., Vanrobays, E., Krogan, N., Hiley, S., Beyer, A. L., Osheim, Y. N., Greenblatt, J., Hughes, T. R., and Lafontaine, D. L. (2005). Esf2p, a U3-associated factor required for small-subunit processome assembly and compaction. *Mol. Cell Biol.* **25,** 5523–5534.

Hook, B., Bernstein, D., Zhang, B., and Wickens, M. (2005). RNA-protein interactions in the yeast three-hybrid system: Affinity, sensitivity, and enhanced library screening. *RNA* **11,** 227–233.

Huang, G. M., Jarmolowski, A., Struck, J. C., and Fournier, M. J. (1992). Accumulation of U14 small nuclear RNA in *Saccharomyces cerevisiae* requires box C, box D, and a 5′, 3′ terminal stem. *Mol. Cell Biol.* **12,** 4456–4463.

Huttenhofer, A., Kiefmann, M., Meier-Ewert, S., O'Brien, J., Lehrach, H., Bachellerie, J. P., and Brosius, J. (2001). RNomics: An experimental approach that identifies 201 candidates for novel, small, non-messenger RNAs in mouse. *EMBO J.* **20,** 2943–2953.

Jarmolowski, A., Zagorski, J., Li, H. V., and Fournier, M. J. (1990). Identification of essential elements in U14 RNA of *Saccharomyces cerevisiae*. *EMBO J.* **9,** 4503–4509.

King, T. H., Liu, B., McCully, R. R., and Fournier, M. J. (2003). Ribosome structure and activity are altered in cells lacking snoRNPs that form pseudouridines in the peptidyl transferase center. *Mol. Cell* **11,** 425–435.

Kiss, T. (2004). Biogenesis of small nuclear RNPs. *J. Cell Sci.* **117,** 5949–5951.

Kressler, D., Linder, P., and de La Cruz, J. (1999). Protein trans-acting factors involved in ribosome biogenesis in *Saccharomyces cerevisiae*. *Mol. Cell Biol.* **19,** 7897–7912.

Lafontaine, D. L., and Tollervey, D. (2000). Synthesis and assembly of the box C + D small nucleolar RNPs. *Mol. Cell Biol.* **20,** 2650–2659.

Li, D., and Fournier, M. J. (1992). U14 function in *Saccharomyces cerevisiae* can be provided by large deletion variants of yeast U14 and hybrid mouse-yeast U14 RNAs. *EMBO J.* **11,** 683–689.

Li, L., and Ye, K. (2006). Crystal structure of an H/ACA box ribonucleoprotein particle. *Nature* **443,** 302–307.

Liang, W. Q., and Fournier, M. J. (1995). U14 base-pairs with 18S rRNA: A novel snoRNA interaction required for rRNA processing. *Genes Dev.* **9,** 2433–2443.

Liang, X. H., and Fournier, M. J. (2006). The helicase Has1p is required for snoRNA release from pre-rRNA. *Mol. Cell Biol.* **26,** 7437–7450.

Liu, B., and Fournier, M. J. (2004). Interference probing of rRNA with snoRNPs: A novel approach for functional mapping of RNA *in vivo*. *RNA* **10,** 1130–1141.

Lowe, T. M., and Eddy, S. R. (1999). A computational screen for methylation guide snoRNAs in yeast. *Science* **283,** 1168–1171.

Lu, Q., Bauer, J. C., and Greener, A. (1997). Using *Schizosaccharomyces pombe* as a host for expression and purification of eukaryotic proteins. *Gene* **200,** 135–144.

Lubben, B., Fabrizio, P., Kastner, B., and Luhrmann, R. (1995). Isolation and characterization of the small nucleolar ribonucleoprotein particle snR30 from *Saccharomyces cerevisiae*. *J. Biol. Chem.* **270,** 11549–11554.

Manstein, D. J., Schuster, H. P., Morandini, P., and Hunt, D. M. (1995). Cloning vectors for the production of proteins in *Dictyostelium discoideum*. *Gene* **162,** 129–134.

Meier, U. T. (2005). The many facets of H/ACA ribonucleoproteins. *Chromosoma* **114,** 1–14.

Milkereit, P., Kuhn, H., Gas, N., and Tschochner, H. (2003). The pre-ribosomal network. *Nucleic Acids Res.* **31,** 799–804.

Morlando, M., Ballarino, M., Greco, P., Caffarelli, E., Dichtl, B., and Bozzoni, I. (2004). Coupling between snoRNP assembly and 3′ processing controls box C/D snoRNA biosynthesis in yeast. *EMBO J.* **23,** 2392–2401.

Mougey, E. B., O'Reilly, M., Osheim, Y., Miller, O. L., Jr., Beyer, A., and Sollner-Webb, B. (1993). The terminal balls characteristic of eukaryotic rRNA transcription units in chromatin spreads are rRNA processing complexes. *Genes Dev.* **7,** 1609–1619.

Nilsson, B., Abrahmsen, L., and Uhlen, M. (1985). Immobilization and purification of enzymes with staphylococcal protein A gene fusion vectors. *EMBO J.* **4,** 1075–1080.

Nissan, T. A., Bassler, J., Petfalski, E., Tollervey, D., and Hurt, E. (2002). 60S pre-ribosome formation viewed from assembly in the nucleolus until export to the cytoplasm. *EMBO J.* **21,** 5539–5547.

Oeffinger, M., and Tollervey, D. (2003). Yeast Nop15p is an RNA-binding protein required for pre-rRNA processing and cytokinesis. *EMBO J.* **22,** 6573–6583.

Omer, A. D., Ziesche, S., Ebhardt, H., and Dennis, P. P. (2002). *In vitro* reconstitution and activity of a C/D box methylation guide ribonucleoprotein complex. *Proc. Natl. Acad. Sci. USA* **99,** 5289–5294.

Osheim, Y. N., French, S. L., Keck, K. M., Champion, E. A., Spasov, K., Dragon, F., Baserga, S. J., and Beyer, A. L. (2004). Pre-18S ribosomal RNA is structurally compacted into the SSU processome before being cleaved from nascent transcripts in *Saccharomyces cerevisiae*. *Mol. Cell* **16,** 943–954.

Oubridge, C., Ito, N., Evans, P. R., Teo, C. H., and Nagai, K. (1994). Crystal structure at 1.92 Å resolution of the RNA-binding domain of the U1A spliceosomal protein complexed with an RNA hairpin. *Nature* **372,** 432–438.

Piganeau, N., Schauer, U. E., and Schroeder, R. (2006). A yeast RNA-hybrid system for the detection of RNA-RNA interactions *in vivo*. *RNA* **12,** 177–184.

Preti, M., Guffanti, E., Valitutto, E., and Dieci, G. (2006). Assembly into snoRNP controls 5′-end maturation of a box C/D snoRNA in *Saccharomyces cerevisiae*. *Biochem. Biophys. Res. Commun.* **351,** 468–473.

Puig, O., Caspary, F., Rigaut, G., Rutz, B., Bouveret, E., Bragado-Nilsson, E., Wilm, M., and Seraphin, B. (2001). The tandem affinity purification (TAP) method: A general procedure of protein complex purification. *Methods* **24,** 218–229.

Rigaut, G., Shevchenko, A., Rutz, B., Wilm, M., Mann, M., and Seraphin, B. (1999). A generic protein purification method for protein complex characterization and proteome exploration. *Nat. Biotechnol.* **17,** 1030–1032.

Salinas, K., Wierzbicki, S., Zhou, L., and Schmitt, M. E. (2005). Characterization and purification of *Saccharomyces cerevisiae* RNase MRP reveals a new unique protein component. *J. Biol. Chem.* **280,** 11352–11360.

Samarsky, D. A., and Fournier, M. J. (1998). Functional mapping of the U3 small nucleolar RNA from the yeast *Saccharomyces cerevisiae. Mol. Cell Biol.* **18,** 3431–3444.

Samarsky, D. A., and Fournier, M. J. (1999). A comprehensive database for the small nucleolar RNAs from *Saccharomyces cerevisiae. Nucleic Acids Res.* **27,** 161–164.

Samarsky, D. A., Fournier, M. J., Singer, R. H., and Bertrand, E. (1998). The snoRNA box C/D motif directs nucleolar targeting and also couples snoRNA synthesis and localization. *EMBO J.* **17,** 3747–3757.

Samarsky, D. A., Schneider, G. S., and Fournier, M. J. (1996). An essential domain in *Saccharomyces cerevisiae* U14 snoRNA is absent in vertebrates, but conserved in other yeasts. *Nucleic Acids Res.* **24,** 2059–2066.

Saveanu, C., Bienvenu, D., Namane, A., Gleizes, P. E., Gas, N., Jacquier, A., and Fromont-Racine, M. (2001). Nog2p, a putative GTPase associated with pre-60S subunits and required for late 60S maturation steps. *EMBO J.* **20,** 6475–6484.

Schattner, P., Decatur, W. A., Davis, C. A., Ares, M., Jr., Fournier, M. J., and Lowe, T. M. (2004). Genome-wide searching for pseudouridylation guide snoRNAs: Analysis of the *Saccharomyces cerevisiae* genome. *Nucleic Acids Res.* **32,** 4281–4296.

Scherly, D., Boelens, W., van Venrooij, W. J., Dathan, N. A., Hamm, J., and Mattaj, I. W. (1989). Identification of the RNA binding segment of human U1 A protein and definition of its binding site on U1 snRNA. *EMBO J.* **8,** 4163–4170.

Sikorski, R. S., and Hieter, P. (1989). A system of shuttle vectors and yeast host strains designed for efficient manipulation of DNA in *Saccharomyces cerevisiae. Genetics* **122,** 19–27.

Srisawat, C., and Engelke, D. R. (2001). Streptavidin aptamers: Affinity tags for the study of RNAs and ribonucleoproteins. *RNA* **7,** 632–641.

Srisawat, C., and Engelke, D. R. (2002). RNA affinity tags for purification of RNAs and ribonucleoprotein complexes. *Methods* **26,** 156–161.

Takahashi, N., Yanagida, M., Fujiyama, S., Hayano, T., and Isobe, T. (2003). Proteomic snapshot analyses of preribosomal ribonucleoprotein complexes formed at various stages of ribosome biogenesis in yeast and mammalian cells. *Mass Spectrom. Rev.* **22,** 287–317.

Terpe, K. (2003). Overview of tag protein fusions: From molecular and biochemical fundamentals to commercial systems. *Appl. Microbiol. Biotechnol.* **60,** 523–533.

Torchet, C., Badis, G., Devaux, F., Costanzo, G., Werner, M., and Jacquier, A. (2005). The complete set of H/ACA snoRNAs that guide rRNA pseudouridylations in *Saccharomyces cerevisiae. RNA* **11,** 928–938.

Tran, E. J., Zhang, X., and Maxwell, E. S. (2003). Efficient RNA 2′-O-methylation requires juxtaposed and symmetrically assembled archaeal box C/D and C′/D′ RNPs. *EMBO J.* **22,** 3930–3940.

Tsai, D. E., Harper, D. S., and Keene, J. D. (1991). U1-snRNP-A protein selects a ten nucleotide consensus sequence from a degenerate RNA pool presented in various structural contexts. *Nucleic Acids Res.* **19,** 4931–4936.

Tuerk, C., and Gold, L. (1990). Systematic evolution of ligands by exponential enrichment: RNA ligands to bacteriophage T4 DNA polymerase. *Science* **249,** 505–510.

Vaillancourt, P., Zheng, C. F., Hoang, D. Q., and Breister, L. (2000). Affinity purification of recombinant proteins fused to calmodulin or to calmodulin-binding peptides. *Methods Enzymol.* **326,** 340–362.

van Gelder, C. W., Gunderson, S. I., Jansen, E. J., Boelens, W. C., Polycarpou-Schwarz, M., Mattaj, I. W., and van Venrooij, W. J. (1993). A complex secondary structure in U1A pre-mRNA that binds two molecules of U1A protein is required for regulation of polyadenylation. *EMBO J.* **12,** 5191–5200.

Varani, L., Gunderson, S. I., Mattaj, I. W., Kay, L. E., Neuhaus, D., and Varani, G. (2000). The NMR structure of the 38 kDa U1A protein-PIE RNA complex reveals the basis of cooperativity in regulation of polyadenylation by human U1A protein. *Nat. Struct. Biol.* **7,** 329–335.

Venema, J., and Tollervey, D. (1999). Ribosome synthesis in *Saccharomyces cerevisiae. Annu. Rev. Genet.* **33,** 261–311.

Watkins, N. J., Newman, D. R., Kuhn, J. F., and Maxwell, E. S. (1998). *In vitro* assembly of the mouse U14 snoRNP core complex and identification of a 65-kDa box C/D-binding protein. *RNA* **4,** 582–593.

Watkins, N. J., Segault, V., Charpentier, B., Nottrott, S., Fabrizio, P., Bachi, A., Wilm, M., Rosbash, M., Branlant, C., and Luhrmann, R. (2000). A common core RNP structure shared between the small nucleolar box C/D RNPs and the spliceosomal U4 snRNP. *Cell* **103,** 457–466.

Wehner, K. A., Gallagher, J. E., and Baserga, S. J. (2002). Components of an interdependent unit within the SSU processome regulate and mediate its activity. *Mol. Cell Biol.* **22,** 7258–7267.

Yang, P. K., Hoareau, C., Froment, C., Monsarrat, B., Henry, Y., and Chanfreau, G. (2005). Cotranscriptional recruitment of the pseudouridylsynthetase Cbf5p and of the RNA binding protein Naf1p during H/ACA snoRNP assembly. *Mol. Cell Biol.* **25,** 3295–3304.

Yu, Y.-T., Terns, R. M., and Terns, M. P. (2005). Mechanisms and functions of RNA-guided RNA modification. *In* "Fine-Tuning of RNA Modifications by Modification and Editing" (H. Grosjean, ed.), Vol. 12, pp. 223–262. Springer-Verlag, New York.

CHAPTER FIFTEEN

A Dedicated Computational Approach for the Identification of Archaeal H/ACA sRNAs

Sébastien Muller, Bruno Charpentier, Christiane Branlant, *and* Fabrice Leclerc

Contents

Abstract

Whereas dedicated computational approaches have been developed for the search of C/D sRNAs and snoRNAs, as yet no dedicated computational approach has been developed for the search of archaeal H/ACA sRNAs. Here we describe a computational approach allowing a fast and selective identification of H/ACA sRNAs in archaeal genomes. It is easy to use, even for biologists having no special expertise in computational biology. This approach is a stepwise knowledge-based approach, combining the search for common structural features of H/ACA motifs and the search for their putative target sequences. The first step is based on the ERPIN software. It depends on the establishment of a secondary structure-based "profile." We explain how this profile is built and how to use ERPIN to optimize the search for H/ACA motifs. Several examples of applications are given to illustrate how powerful the method is, its limits, and how the results can be evaluated. Then, the possible target rRNA sequences corresponding to the identified H/ACA motifs are searched by use of a descriptor-based method (RNAMOT). The principles and the practical aspects of this method are also explained, and several examples are given here as well to help users in the interpretation of the results.

Laboratoire de Maturation des ARN et Enzymologie Moléculaire, Nancy Université, Faculté des Sciences et Techniques, Vandoeuvre-les-Nancy, France

Methods in Enzymology, Volume 425
ISSN 0076-6879, DOI: 10.1016/S0076-6879(07)25015-3

1. INTRODUCTION

Pseudouridine (Ψ) is one of the most abundant posttranscriptionally modified nucleotides found in tRNAs, rRNAs, and UsnRNAs (Rozenski *et al.*, 1999). Uridine to pseudouridine conversion can be catalyzed by a single protein with RNA recognition capacity and RNA/Ψ-synthase activity. It can also be catalyzed by H/ACA RiboNucleoProtein complexes (H/ACA sRNPs) containing a guide RNA with H/ACA boxes, called snoRNAs and scaRNAs in Eukarya and sRNAs in Archaea. These RNAs are associated with a set of proteins, Nhp2p, Nop10p, Cbf5, and Gar1p in yeast, NHP2, NOP10, Dyskerin (or NAP57), and GAR1 in human (Henras *et al.*, 1998; Khanna *et al.*, 2006; Lafontaine *et al.*, 1998; Meier and Blobel, 1994; Wang and Meier, 2004; Watkins *et al.*, 1998), and their homologs L7Ae, aNOP10, aCBF5, and aGAR1 in Archaea (Baker *et al.*, 2005; Charpentier *et al.*, 2005; Rozhdestvensky *et al.*, 2003). Proteins aCBF5 and Cbf5/Dyskerin belong to the TruB family of RNA/Ψ-synthases (Hamma *et al.*, 2005; Lafontaine *et al.*, 1998; Manival *et al.*, 2006; Meier and Blobel, 1994; Rashid *et al.*, 2006; Wang and Meier, 2004).

The guiding properties of H/ACA RNAs are based on the formation of a complex structure: two sequences from the RNA substrate, that are separated by a 5′-UN-3′ dinucleotide, base pair with the two strands of an internal loop of the guide RNA (pseudouridylation pocket). The U residue in the 5′-UN-3′ dinucleotide is the target site of the reaction (Balakin *et al.*, 1996; Ganot *et al.*, 1997a,b; Ni *et al.*, 1997).

Several studies combining computational and experimental approaches have been dedicated to the identification of eukaryal H/ACA snoRNAs in human, mouse, *D. melanogaster*, *S. cerevisiae*, and *S. pombe* (Huang *et al.*, 2004, 2005; Kiss *et al.*, 2004; Li *et al.*, 2005; Schattner *et al.*, 2004, 2006; Torchet *et al.*, 2005; Yuan *et al.*, 2003). Most of these RNAs were found to have a two stem-loop structure. Most generally, each of the two stem loops contains a pseudouridylation pocket. These stem loops are linked by a single-stranded sequence containing the ANANNA H box and the 3′ stem loop is flanked by a single-stranded element containing an ANA trinucleotide. Whereas the global architecture of eukaryal snoRNAs is highly conserved, the sizes and base compositions of their stem-loop structures are highly variable.

Knowledge regarding archaeal H/ACA sRNAs is more limited, because they have only been characterized in *Archaeoglobus fulgidus* (Tang *et al.*, 2002a), *Pyrococcus* species (Baker *et al.*, 2005; Charpentier *et al.*, 2005; Rozhdestvensky *et al.*, 2003), and *Methanocaldococcus jannaschii* (Thebault *et al.*, 2006). Identified Ψ residues in archaeal rRNAs are also very limited: 4, 6, and 3 Ψ residues were detected in the 23S rRNA of *Halobacterium halobium, Sulfolobus acidocaldarius,* and *Haloarcula marismortui,* respectively (Del Campo *et al.,* 2005; Massenet *et al.,* 1999; Ofengand and Bakin, 1997). However, the available data reveal some specific features of archaeal H/ACA sRNAs compared with their eukaryal counterparts. Indeed, although the target U residue is identified by the same base-pairing rules between the guide RNA and the targeted sequence, the global architecture of H/ACA sRNAs is more variable. One, two, or three contiguous stem-loop structures containing a pseudouridylation pocket can be present. They may even contain an additional stem-loop structure without a pseu-douridylation pocket (see Pf6, Rozhdestvensky *et al.,* 2003). In addition, compared with their eukaryal counterparts, they are structurally more con-strained, with a strong conservation of the relative positions, sizes, and base compositions of their structural elements. In each stem-loop structure, the stems delineating the pseudouridylation pocket are highly enriched in G-C and C-G pairs. Each stem loop also displays a K-turn motif or a K-loop motif (Charpentier *et al.,* 2005; Charron *et al.,* 2004; Hamma *et al.,* 2005; Klein *et al.,* 2001; Li and Ye, 2006; Rozhdestvensky *et al.,* 2003; Vidovic *et al.,* 2000) in its apical part. K-turn and K-loop motifs are characterized by the presence of an A·G and G·A sheared pair flanked by a U residue. Each stem-loop structure is tailed by an ANA box that is more frequently an ACA triplet. By inspection of the secondary structure of the identified H/ACA sRNAs, we observed that the K-turn or K-loop motif on the one hand and the ACA box on the other hand are located at a well-defined distance from the targeted U residue in the pseudouridylation pocket.

Therefore, although not unique, the characteristic features of the H/ACA sRNA stem-loop structures, flanked by their associated ANA box, are specific enough to represent a signature that can be used for H/ACA sRNA gene identification by computational search in sequence databases. For simplification, we will designate these modular elements as H/ACA motifs. On the basis of the peculiar structure formed by the H/ACA sRNA and the RNA substrate, the finding by computational analysis of putative target sequences can also be used as a complementary source of information to discriminate true H/ACA sRNA genes from false-positive DNA sequences.

Taking advantage of the specificity of archaeal H/ACA sRNAs, we developed a knowledge-based approach that combines the consecutive searches for sequences coding for H/ACA motifs in complete sequences of archaeal genomes and their complementary RNA targets in rRNA sequences. The identification of H/ACA motifs is performed by use of a profile-based approach, taking advantage of the present knowledge on H/ACA sRNAs. The knowledge used includes data from comparative sRNA sequence analysis (Baker *et al.,* 2005; Charpentier *et al.,* 2005; Rozhdestvensky *et al.,* 2003; Tang *et al.,* 2002a; Thebault *et al.,* 2006), *in vitro* reconstitution of active H/ACA sRNPs by use of various WT and mutated H/ACA sRNAs and various RNA substrates (Charpentier *et al.,* 2005; S. Muller *et al.,* unpublished results) and the recent determination of the 3D structure of a complete H/ACA sRNP particle (Li and Ye, 2006). Then, the identification of possible RNA targets associated with the identified H/ACA motifs is performed by use of a descriptor-based approach, taking into account the rules of complementarities that were also inferred from the data from *in vitro* reconstitution (S. Muller *et al.*, unpublished data). Because it is a knowledge-based approach, each unknown H/ACA motif associated with an identified target can, after an appropriate validation, contribute to increase and refine our knowledge on the H/ACA structural features and their target recognition. Thus, the general performance of the approach may be improved.

This chapter is dedicated to biologists who have no special expertise in computational methods. Our goal is to teach how to use the proposed strategy to get an H/ACA sRNA gene identification that will be as extensive as possible with a minimized number of false-positive results. The use of a computational approach for the prediction of H/ACA sRNA genes in Archaea and the prediction of their target sites are as important as *in vitro* reconstitution assays that are available to validate the results (see Chapter 16 by Charpentier *et al.*). The combined use of computational predictions and *in vitro* tests should increase considerably our knowledge on archaeal sRNAs in the near future.

2. METHOD

Our strategy for the detection of H/ACA sRNAs follows a stepwise and iterative procedure in which the first step is the search for H/ACA-like motifs through archaeal genomes, and the second step is the determination of the associated RNA target(s) in rRNAs (Fig. 15.1). In the first step (Fig. 15.1, step 1), H/ACA-like motifs are detected by use of the profile-based ERPIN program (Gautheret and Lambert, 2001). This program has been applied to the search of a wide range of RNA motifs (Lambert *et al.*, 2002, 2004; Legendre *et al.*, 2005). Once H/ACA-like motifs are identified, their putative target(s) in rRNAs are searched (Fig. 15.1, step 2) by use of the descriptor-based RNAMOT program. This program has also been extensively applied to the identification of several kinds of RNA motifs, and the data obtained have been experimentally validated (Bourdeau *et al.*, 1999; Laferriere *et al.*, 1994; Lescure *et al.*, 1999, 2002). In the present strategy, the use of RNAMOT is atypical, because this program is applied to the search of RNA motifs formed by partial base pairing of the internal loops of the H/ACA motifs with ribosomal RNAs. Because of the peculiar application of both ERPIN and RNAMOT in the present strategy, we had to create an appropriate H/ACA profile for ERPIN and several RNA target descriptors for RNAMOT. We will describe the rationale used for their generation

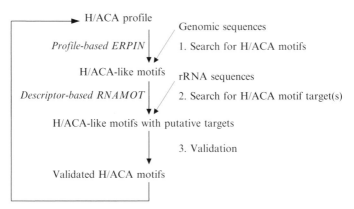

Figure 15.1 The multistep strategy proposed for the identification of H/ACA motifs. A description of the three steps leading to the identification of new H/ACA sRNA candidates: first an ERPIN-based step is used for the search of H/ACA-like motifs, then, a descriptor-based step is used for the search of their targets. Each of them includes tests of the validity of the results and a final global validation step is performed, taking into account the present knowledge on H/ACA sRNAs and rRNA pseudouridylation. The sequences of the new validated H/ACA motifs can be integrated in the H/ACA profile, and the search can be repeated on the same genome or run on other genomes.

and how to use them to achieve maximal efficiency and specificity of the delivered data.

By this procedure, H/ACA-like motifs that exhibit at least one putative RNA target are identified. They are then subjected to an evaluation step (Fig. 15.1, step 3). This evaluation step is based on: (1) current knowledge on Ψ positions in archaeal rRNA, (2) comparison with already identified archaeal H/ACA sRNAs and, (3) experimental data obtained from structure–function analysis of H/ACA sRNA by use of the *in vitro* reconstituted system. Obviously, the final proof of the validity of the prediction requires experimental tests of the proposed guiding property by use of the *in vitro* assembly procedure (see Chapter 16 by Charpentier *et al.*) and the identification of a Ψ residue at the targeted position in the rRNA by the CMCT approach (see Chapter 2). When the demonstration is completed, the newly identified H/ACA motifs can be used to enrich the ERPIN profile.

2.1. Search for H/ACA-like motifs

The search for H/ACA-like motifs requires the establishment of a profile, as implemented in the ERPIN program. This H/ACA profile is based on our present understanding of the common structural features of archaeal H/ACA motifs. Once the profile is built, the search can be started. The basic ERPIN command, parameters, and options and the present stage of their optimization are explained by use of various examples of searches of H/ACA motifs in archaeal genomes. Through the proposed interpretations of the data obtained and by giving some tips, we intend to help users speed up their search, analyze and evaluate the results, and improve the initial profile.

2.1.1. Requirements

2.1.1.1. Hardware and software The ERPIN program used in this approach runs and has been tested under most UNIX platforms (for more details about compatible operating systems, see references). The ERPIN program is available as a stand-alone version that can be obtained from D. Gautheret (Gautheret and Lambert, 2001) and as a web server (http://tagc.univ-mrs.fr/erpin/) (Lambert *et al.*, 2004). The results presented here were obtained with the most recent release of ERPIN (version 5.5).

2.1.1.2. Sequence data The sequence format used for ERPIN applications is FASTA. Already published archaeal genomic sequences can be downloaded from the NCBI ftp site: ftp://ftp.ncbi.nih.gov/genomes/Bacteria. The sequences of the 26 already identified H/ACA sRNAs sequences were used to establish the H/ACA profile (Baker *et al.*, 2005; Charpentier *et al.*, 2005; Rozhdestvensky *et al.*, 2003; Tang *et al.*, 2002a). This profile is available at http://tagc.univ-mrs.fr/asterix/erpin/.

2.1.2. Establishment of the ERPIN profile

The H/ACA profile is established on the basis of a sequence alignment of all the H/ACA motifs of the already identified H/ACA sRNAs (Baker *et al.*, 2005; Charpentier *et al.*, 2005; Rozhdestvensky *et al.*, 2003; Tang *et al.*, 2002a). It is based on and includes secondary structure information, and its correct establishment is critical for the sensitivity and specificity of the ERPIN search.

To build this profile, the sequences aligned are considered as a succession of single-stranded (sx) and double-stranded (Hx) elements, x is a number defined according to the position of the element in the H/ACA sequence (5′ to 3′). Each nucleotide in a defined structural element carries the number attributed to this element (Fig. 15.2B). These numbers correspond to the first lines of the profile (Fig. 15.2E). Segments known to be single-stranded, like the ACA triplet or the two guide sequences, or segments that are single-stranded or double-stranded, depending on the considered H/ACA motif, are each defined by a unique stretch of a given number. In contrast, obligatory helices are identified by the presence of two stretches of residues that are all assigned to the same number and have identical lengths. When two stretches with an identical numbering are defined in the profile, the RNA sequences that do not contain the corresponding base-pair interaction are not selected in the search.

Constitution of the profile is complicated by the fact that, despite a strong conservation, H/ACA motifs show some structural variations. For

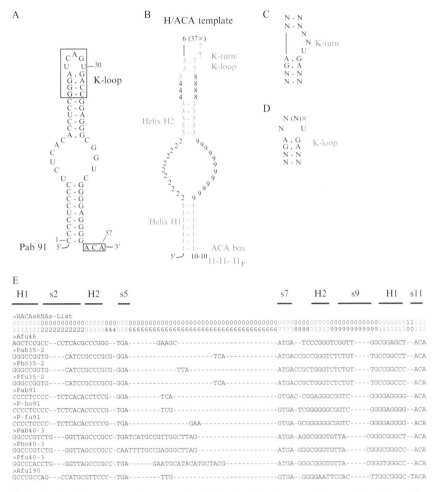

Figure 15.2 The structural elements of H/ACA sRNAs taken into consideration in the ERPIN H/ACA profile and an example of H/ACA profile. (A) Secondary structure proposed for the *P. abyssi* Pab91 sRNA (Charpentier *et al.*, 2005). The ACA triplet at the 3′ end of the motif, the G/C rich helices 1 and 2, and the K-loop structure on top of helix 2 are shown. (B) The structural elements taken into consideration to build the H/ACA profile are shown, with their respective numbers (1–11), which are defined by reference to the 5′ to 3′ orientation of the sequence. (C) and (D) The K-turn and K-loop motifs used in the search are shown, respectively. (E) A portion of the sequence alignment of the H/ACA profile is given. The two stretches of identical numbers correspond to the helical elements H1 and H2. As the total number of defined structural elements exceeds 9, the digits are presented in two lines. The annotations on the top of the alignment indicates the various elements taken into consideration to build the profile, with the associated names: the basal helix (H1), the apical helix (H2), the K-turn motif 5′ and 3′ elements (s5 and s7), the ACA box (s11) and the 5′ and 3′ guide sequences (s2 and s9, respectively). The buffer elements are not indicated. For each of the two priority orders used in the search, only the H1, H2, s5, s7 and s11 elements were unmasked step by step.

instance, the sizes of helices 1 and 2 slightly vary from one motif to the other, and a limited number of bulge residues or mismatches can be present, in particular in helix 2. An extreme case of variation is exemplified in the second motif of the *P. furiosus* Pf7 sRNA, where the 5′ strand of the pseudouridylation pocket is separated from helix 1 by an additional stem-loop structure (see Pf7, Rozhdestvensky *et al.*, 2003). Finally, the presence of either an apical K-loop or an apical K-turn motif with a terminal stem loop increases the diversity of the H/ACA motifs. For an optimized prediction of H/ACA motifs, we had to take this variability into consideration. This was of high importance, because ERPIN does not select sequences that contain bulge residues and internal loops in the double-stranded regions identified by the H/ACA profile. To overcome this program limitation, two base-paired elements, H1 and H2, were defined, corresponding to the estimated minimal regular succession of base pairs present in helices 1 and 2 (7 bps for H1 and 5 bps for H2, respectively). Furthermore, to accommodate the size variations of helices 1 and 2, segments that can be either single stranded or double stranded were included in the profile ("buffer" segments). A maximal size was defined for each "buffer" segment, by taking into account constant distances in H/ACA motifs (distances between the ACA trinucleotide and the pseudouridylation pocket and between the K-turn/K-loop motif and this pocket, respectively), as well as the overall lengths of known H/ACA motifs.

To be able to select RNA with either a K-loop or a K-turn motif, we defined these elements in the H/ACA profile by their common structural features: namely, two A·G and G·A sheared pairs with sequence constraints at the position 5′ to the GA sequence in the 5′ strand and at the two successive positions 5′ to the GA sequence in its 3′ strand (Fig. 15.2). Therefore, the two possible motifs are defined by the presence of a BGA (where B is any nucleotide except A) and RUGA elements that are separated by a loop. The maximal size of this loop was fixed to 37nts.

The 41 H/ACA motifs of the 26 identified H/ACA sRNAs were manually aligned, taking into account the following structural elements (Fig. 15.2B and E): a basal helical element H1 (No. 1), the 5′ and 3′ strands of the pseudouridylation pocket (Nos. 2 and 9, maximum lengths 11 and 12nts, respectively), an upper helical element H2 (No. 3), one K-turn or K-loop motif defined by a BGA element in the 5′ strand, a RUGA element in the 3′ strand and an apical loop (Nos. 5, 7, and 6, respectively), the ANA triplet (No. 11) and 3 "buffer" elements (Nos. 4, 8, and 10) that can be single or double stranded and are used for flexibility. Only three of the conserved structural elements have a sequence imposed by the presence of conserved residues in all the aligned RNA sequences of the profile, namely, the 5′ and 3′ elements of the K-turn/K-loop motif and the ANA box (Nos. 5, 7, and 11, respectively). The "buffer" elements numbered 4, 8, and 10 can be considered as gaps, which can be filled or not. The maximal size of elements 4, 8, and 10 were fixed to 3, 4, and 2nts, respectively.

The preceding description of the rationale for generation of the ERPIN profile will help the user of this approach to introduce new H/ACA sequences in the profile and will also facilitate the interpretation of the data obtained by use of this profile.

2.1.3. Practical procedure

The ERPIN searches based on the defined profile are performed by use of a local version of the program or the ERPIN web server version accessible at http://tagc.univ-mrs.fr/asterix/erpin/. Note that the current version of the ERPIN server does not allow the modifications of the profile and optional parameters. However, this server access provides a 2D structure representation of the selected putative H/ACA motif.

First, one needs to specify the genome sequences and the H/ACA profile to be used. To this end, the "compulsory" commands <profile> and <genome> allow the user to enter the names and locations of the ERPIN profile and sequence database, respectively. By default, the search is performed on both strands of the genome. Then, various options are available to finely tune the ERPIN search.

When activating the "no mask" option, the genomic screening will involve the simultaneous search of all the structural elements defined in the profile. We do not recommend this possibility; because of the great number of elements defined in the H/ACA profile, the computer search will be very slow. To increase the performance of the search (both speed and efficiency) screening for the presence of some selected structural elements can be done with a priority order. This order is defined by the use of successive masks. The first mask restricts the search to one or more selected element(s) (the unmasked elements, command "umask"). The masked elements will influence the search by their delimited sizes. The sequences of the unmasked elements of the H/ACA motifs, which are aligned in the profile, are compared with the inspected genomic sequences. Then, the remaining selected masked elements are unmasked step by step ("-add" option). The first step is fast, because only a few elements are searched through the entire genome. In the second step, the added elements are only searched within the sequence portions of the genome identified in the first step and so on in the following steps.

In this priority order strategy, a cutoff value can be specified to determine how stringent the search is at each of the steps defined by the successive masks (i.e., how similar the elements have to be in comparison with those included in the profile to be identified as hits). Different cutoff values can be defined for the different steps of the selection, depending on the relative importance of the elements specified in the mask. The definition of the masks, their order of use, and the cutoff values used at each step have an impact on the results.

To explain how these cutoff values are defined, we have to introduce the notion of lod-scores and scores. By comparison of one residue (single-stranded element) or a pair of residues (double-stranded elements) in a given element of one aligned H/ACA motif in the profile with the corresponding residues, or pairs of residues, in the other aligned motifs, ERPIN can establish a score of similarity (lod-score) for individual residues or pairs of residues. On the basis of the sum of the lod-scores of all the residues or pairs of residues in a given element, ERPIN defines a score for this element. By extension, when a mask is used, a score can be established for the overall unmasked regions of each of the H/ACA motifs aligned in the profile. The cutoff value for the step of the search where this mask is used will be defined by reference to the score values established for all the motifs aligned in the profile.

The cutoff values can be defined as absolute values or as percentages. We will use percentages. They indicate the minimal percentage of sequences in the profile that are captured as hits on the basis of their calculated scores. When a 100% or higher percentage is used as cutoff (100% is the default cutoff value if no specification is given), all the sequences aligned in the profile will be captured as hits. Lower percentages indicate that not all the sequences in the profile are selected, thus, the search in the analyzed genome will be highly stringent. On the other hand, a higher cutoff value (>100%) indicates that sequences with some divergence relative to the aligned motifs of the profile can be selected. Note that the possibility of defining cutoff values >100% is available only for releases of ERPIN >5.3.

The two highly discriminating structural elements in the H/ACA motifs are the helical elements H1 and H2 and the K-turn/K-loop structure, respectively. Searches can be done by giving prevalence to one or the other of these two elements. In the following applications, we illustrate the relative efficiencies of two ordered series of three successive masks. In order 1, preference is given to the selection of the helical elements H1 and H2. Indeed, the presence of these two elements H1 (No. 1) and H2 (No. 3) is first tested, then the 5' and 3' strands of the K-turn/K-loop (Nos. 5 and 7) are searched, and finally, the presence of a putative ACA motif (No. 11) is investigated. When the second series of masks is used, designated as order 2, the first two steps are inverted, so that priority is given to the two strands of the K-turn/K-loop in the search. In these two procedures, the elements 2, 4, 6, 8, 9, and 10, namely the buffer elements, the apical loop and the two strands of the internal loop, are always masked.

2.1.4. Illustrative tests performed on archaeal genomes whose H/ACA sRNAs have been characterized

We will first present various tests run on genomes for which H/ACA sRNAs were characterized (Baker et al., 2005; Charpentier et al., 2005; Rozhdestvensky et al., 2003; Tang et al., 2002a; Thebault et al., 2006). These tests illustrate the influence of the priority order and the cutoff values on the detection of H/ACA-like motifs. They also show how the ERPIN parameters are optimized, taking into consideration the present state of the profile. In the blind tests presented in Table 15.1 (columns denoted blind tests), no prior knowledge on the H/ACA sRNAs identified in the analyzed species or in the genus, in the case of the *Pyrococcus* species, was included. To this end, the *A. fulgidus* genome was subjected to an ERPIN search by use of a profile in which the five identified *A. fulgidus* H/ACA motifs were eliminated. Similarly, the *M. jannaschii* genome was subjected to an ERPIN search by use of a profile in which the six identified *M. jannaschii* H/ACA motifs were eliminated. Finally, for the searches performed on each of the *Pyrococcus* genomes, the 30 motifs identified in *P. abyssi, P. furiosus,* and *P. horikoshii* were eliminated from the profile. In parallel, we evaluated the effect of the presence in the H/ACA profile of motifs belonging to three species of the same genus (*Pyrococcus*) on searches performed on the *A. fulgidus* and *M. jannaschii* genomes. To this end, we performed blind tests on the *A. fulgidus* and *M. jannaschii* genomes by use of an H/ACA profile containing motifs from only one of the *Pyrococcus* species (*P. abyssi*) (Table 15.1B). For each of these blind tests, the two priority orders (1 and 2) were applied, as well as three different cutoff values (95%, 100%, and 110%). The putative H/ACA motifs selected in the assay are next compared with the already identified H/ACA motifs. "Positive results" in Table 15.1 correspond to motifs already identified, whereas the other ones are denoted "likely false-positive results."

As illustrated in Table 15.1A, priority order 1 always allows the identification of the larger number of validated H/ACA motifs. The detection of 60–90% of the known H/ACA motifs with a cutoff value of 110% demonstrates the performance of the approach. The number of likely false-positive results increases with this cutoff value. However, as will be explained later, the inspection of the motifs and their location in the genome allows an efficient discrimination of likely false-positive results.

Some interesting observations can be made by inspection of Table 15.A and B: the presence of motifs from three species of the *Pyrococcus* genus in the profile did not bias the search. On the contrary, it has a marked positive effect on the selection of true H/ACA motifs in both *A. fulgidus* and *M. jannaschii*. The presence of all the H/ACA motifs from *Pyrococcus* is even required for an efficient selection of the H/ACA motifs from *A. fulgidus*.

For training in data interpretation and to demonstrate that secondary structure analysis of the candidate motifs helps discriminate the false-positive results, we provide the potential secondary structures of the candidate motifs selected for *M. jannaschii* in the blind tests presented in Table 15.1A (Fig. 15.3), and we comment on these results. As evidenced in Fig. 15.3 (panel B1), all of the four selected motifs corresponding to validated H/ACA motifs (denoted FW1, FW3, RC1, and RC3 in the search compilation) have an ACA triplet and a canonical K-turn structure, whereas the two likely false-positive sequences (FW2 and RC2 motifs, respectively) contain a CCA and an ACG motif instead of the ACA triplet (panel B2). Neither of them

Table 15.1A Tests of the effects of priority order, cutoff values and the use of a second step of selection in ERPIN searches

	Number of H/ACA sRNA motifs already identified	Cutoff	Blind Tests				Second Step Tests			
			Order (1)		Order (2)		Order (1)		Order (2)	
			Number of positive results	Number of likely false-positive results	Number of positive results	Number of likely false-positive results	Number of positive results	Number of likely false-positive results	Number of positive results	Number of likely false-positive results
A. fulgidus	5 [1]	95%	1	0	1	0	1	0	1	0
		100%	2	1	2	1	2	0	2	0
		110%	3	3	3	3	4	3	3	2
P. abyssi	10 [2, 3, 4]	95%	4	0	2	0	9	1	5	0
		100%	4	0	2	0	9	1	5	0
		110%	9	1	4	1	9	2	5	2
P. furiosus	10 [2, 3, 4]	95%	5	0	3	0	9	0	7	0
		100%	5	0	3	0	9	0	7	0
		110%	7	1	4	1	9	1	7	0
P. horikoshii	10 [2, 3, 4]	95%	3	0	2	0	9	1	6	0
		100%	3	0	2	0	9	0	6	0
		110%	6	2	4	0	9	0	6	0
M. jannaschii	6 [5]	95%	3	1	3	0	3	1	3	0
		100%	3	1	3	0	3	1	3	0
		110%	4	2	4	1	4	1	5	0

[1] Tang *et al.*, 2002.
[2] Rozhdestvensky *et al.*, 2003.
[3] Charpentier *et al.*, 2005.
[4] Baker *et al.*, 2005.
[5] Thébault *et al.*, 2006.

Table 15.1B Tests of the effects of priority order, cutoff values and the use of a second step of selection in ERPIN searches

Number of H/ACA sRNA motifs already identified	Cutoff	Blind Tests			
		Order (1)		Order (2)	
		Number of positive results	Number of likely false-positive results	Number of positive results	Number of likely false-positive results
A. fulgidus 5 [1]	95%	1	0	1	0
	100%	1	0	1	0
	110%	1	1	1	1
M. jannaschii 6 [5]	95%	3	1	3	0
	100%	3	1	3	0
	110%	3	1	3	1

(A) Blind tests performed on each of the archaeal genomes whose H/ACA sRNAs have been studied (A. fulgidus, P. abyssi, P. furiosus, P. horikoshii and M. jannaschii). In the first step of these blind tests, for each of the studied genomes, the known H/ACA motifs of this species or of species of the same genus were removed for the H/ACA profile used for the ERPIN search. Searches were run with the two orders of priority and with three different cutoff values (95, 100 and 110%). In a given test, identical cutoff values were used for each of the masks. The name of the studied genome is given in the first column. The number of H/ACA motifs already identified in this species is indicated in the second column. The number of true H/ACA motifs (positives results) and likely false positive H/ACA motifs found in the ERPIN search are given for each assay. Results obtained with each of the priority orders are shown in parallel. In the second step of these blind tests, the true H/ACA motifs that were found in the first step were included in the H/ACA profiles. The tests use the same parameters as in the first step.
(B) Similar blind tests were performed on the A. fulgidus and M. jannaschii genomes with a profile lacking the P. furiosus and P. horikoshii H/ACA motifs.

A

```
>Methanocaldococcus jannaschii DSM2661 complete genome
FW 1 216280..216350 35.25 4.17e-06
CTCCCCAgtgggttag----CCCTCt--TGA.cgtagcagagctaaagg----ATGAta--GAGGGtgttaaac----TGGGGAG--ACA
FW 2 864064..864141 30.10 1.20e-04
GCCCTGGtggtgta------GCCCGgc-CTA.tcatacgggactgtcactcccGTGAct--CGGGTtcaaatcccgg-CCAGGGCg-CCA
FW 3 986084..986151 32.88 2.12e-05
CCCCCTAcggccca------CGCTCt--CGA.cgtggcagagccaagg-----ATGAga--GAGGGtaggata-----TAGGGGGg-ACA
>Methanocaldococcus jannaschii DSM2661 complete genome
RC 1 118060..118133 18.34 3.75e-02
CTACCCAcaagggcg-----CCGTGc--CGA.gtagccgttatggcttca---ATGAaggcCACGGtttttcca-----AGGGTAGatACA
RC 2 1150252..1150326 18.71 3.21e-02
GCGCCGGccgggatttgaacCCGGGt--CGC.tggcttggaaggccaga----GTGAta--CCAGGctacacca----CCGGCGC--ATG
RC 3 1659450..1659520 42.10 1.34e-08
GCCCCGGgaaaccgc-----GGGGGa--TGA.gcgacagcccggcaagct---GTGAgt--CCCCTttgctccc----CCGGGGC--ACA
```

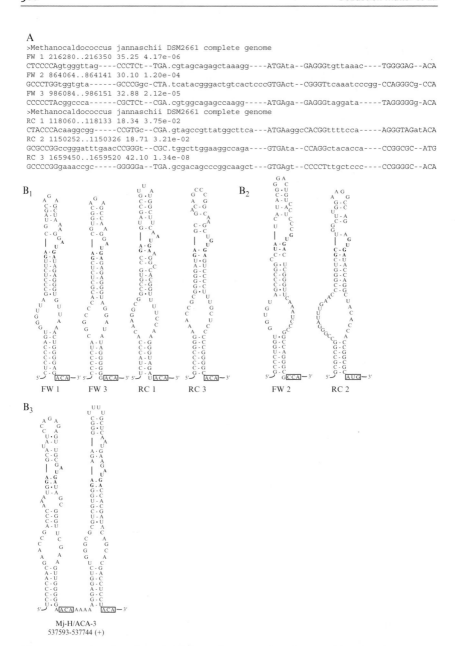

Figure 15.3   Results of an ERPIN search run on the *M. jannaschii* genomic sequence and the secondary structures proposed for the identified candidates. (A) The results of the ERPIN program, as displayed in output, are shown. The ERPIN program was run on the *M. jannaschii* genome sequence, using as a profile an alignment of the known H/ACA motifs, except for the ones identified in *M. jannaschii* (Thebault *et al.*, 2006) (see Table 15.1). The names of the candidates are defined by reference to the DNA strand screened and the order of appearance. ERPIN gives the positions relative to the forward

contain the canonical A·G and G·A sheared pairs tandem. In addition, they correspond to tRNA sequences, tRNA^{Asp}(GUC) and tRNA^{Gly}(UCC), respectively. Therefore, these likely false-positive H/ACA-like motifs can be unambiguously discarded from the ERPIN results. Note that most tRNA sequences can be folded into a stem-loop structure, explaining their detection with the profile that we used. The undetected motifs correspond to two motifs present in a unique sRNA (Mj-H/ACA-3, Thebault et al., 2006). The 5′ motif has probably been discarded because of the internal loop present in helix 2. The second one probably differs by the sequences of helices 1 and 2 compared with the H/ACA motifs of the alignment.

Interestingly, when the four validated H/ACA motifs found in the blind test (FW1, FW3, RC1, and RC3) are included in the profile, the 3′-terminal motif of RNA Mj-H/ACA-3 is found in the search by use of the priority order 2 and a cutoff value of 110% (Table 15.1A). Hence, the use of both priority orders may be interesting in some specific cases.

More generally, a stepwise investigation including a second run after the inclusion in the profile of the validated H/ACA motifs selected in a first run, strongly improves the number of selected motifs and decreases the number of likely false-positive results (Table 15.1A). When the priority order 1 is used, only one of the previously identified H/ACA motifs is not found in the three *Pyrococcus* species, the one proposed to guide U to Ψ conversion at position 2575 in the *P. furiosus* LSU rRNA (see Pf3, stem I, Rozhdestvensky et al., 2003).

2.1.5. Example of the search of H/ACA-like motifs in a yet unexplored genome

The genome from *Thermococcus kodakarensis*, a species belonging to the same order as species of the *Pyrococcus* genus, was used in this teaching example. The profile used contained the 41 known archaeal H/ACA motifs. The search was first performed with priority order 1, and the cutoff value defined for each step of the selection was 110%. Hence, the following ERPIN command was used:

```
erpin<profile><genome>−1,11−umask 1 3−add 5 7−add 11−cutoff 110%110%110%
```

"1,11" indicates the beginning and the end of the chain of structural elements considered in the profile. In this case, all the elements of the profile are considered.

strand for each selected H/ACA-like motif. (B) A secondary structure is proposed for each of the candidates. The motifs in Panel B1 were retained after inspection of their structures, whereas those in panel B2 were discarded. Panel B3 represents the proposed secondary structures of the two known H/ACA motifs that were not detected in the first step of the blind search.

The search time was short: less than 3 min of CPU time on a 3.2-Ghz P IV. Sixteen candidate motifs were identified. Six of them were encoded by the plus strand and 10 by the minus strand. At the last step of the selection, ERPIN computes an e-value reflecting the statistical significance of the selected H/ACA-like motifs. This e-value represents the probability of encountering this motif at random (Lambert *et al.*, 2005). Hence, it depends on the scores calculated for all the structural elements and on the size of the database. The output file displays the candidate sequences that satisfy the cutoff values defined at each selection step and the e-values of these sequences (Fig. 15.4). Only candidates with a negative e-value deserve further analysis.

In the search performed on the *Thermococcus kodakarensis* genome, nine candidates had negative e-values between 1e-06 and 4e-02, which was an indication for a good fit with the H/ACA motifs in the profile. In contrast, seven candidates had e-values between 1e-01 and 4, suggesting that they may correspond to false-positive results. The possible secondary structure of each candidate is shown in Fig. 15.4 (panels B1 and B2). The nine candidates with satisfying e-values displayed all the expected structural elements (panel B1). Only the FW3 candidate has an AUA motif instead of the ACA motif. Despite its high e-value, the FW1 motif has a canonical K-loop, an ACA motif, and helices 1 and 2 with correct lengths. The high e-value observed may be due to the presence of two G·U pairs in helix 1. Furthermore, motifs FW1, FW2 and FW3 may belong to the same H/ACA sRNA,

A
```
>Thermococcus-kodakarensis
FW   1   462059..462114    21.90   7.56e-01
TGCCCCTgcgcgagga----CCCCG---GGGaga--------------------ATGAac--CGGGGggcgatgc----GGGGGCA--ACA
FW   2   462115..462195    33.44   1.32e-04
GCCCGGCctcagcgaggtccCCGCG---GGAggggccttccgcgtcccggagcacg--ATGAc---CGCGGgaaacccag---GCCGGGC--ACA
FW   3   462201..462272    27.43   3.49e-02
GGCCCGCcagggttag----CCCGCc--CAAgggtggcgtacgccttcg--------ATGAgg--GCGGGagttaccg----GCGGGCC--ATA
FW   4   899083..899151    32.54   3.74e-04
GCCCCCGggaaaccgc----GGGGGa--TGAgcgcctcggcgcgagcc----------GTGAtt--CCCCTtcgctccc----CGGGGGC--ACA
FW   5  1545183..1545239   35.55   7.44e-06
GCCCTCCcctctcacac---CCCCG---GGAgccg---------------------GTGAc---CGGGGggcggtcgg---GGAGGGC--ACA
FW   6  1769611..1769691   19.95   1.64e+00
CGCCCGGgtggtgta-----GCCCGgc-CCAtcatacgggactgtcactccc-----GTGAcc--CGGGTtcaaatcccggcCCGGGCGccATA
>Thermococcus-kodakarensis
RC   1   47849..47907     35.19   1.28e-05
GGCCTGGcgtcccgccct--CCCCGg--GGAaac--------------------GTGAac--CGGGGcttcctg-----CCAGGCCt-ACA
RC   2   465056..465126    34.46   3.57e-05
GGGCCCGgtctccgga----CCGCT---GGAgggggtctctggccgtca--------ATGAg---AGCGGgggatcagc---CGGGCCC--ACA
RC   3   465127..465197    33.20   1.77e-04
CCGCCCGgtggcccgt----GTCCC---GGAgggggccgaggcccaca--------GTGAa---GGGATgagacgatgc--CGGGCGG--ACA
RC   4   865615..865690    19.53   1.91e+00
GGGAGCAtcatgac------TCCGT---CAActccagccctgacgaccttcctaatgATGAtcc-TCGGGttaacg------TGCTCCC--AGA
RC   5   986951..987023    19.50   1.93e+00
AGCCACGctggagttcatg-CCCAGctcGGAaaagagttc----------------GTGAagacCGGGGaaatgcccg---CGAGGCTttACA
RC   6  1107274..1107352   34.45   3.64e-05
GGGCCGGgagcatccac---CCGCG---GGAgca---------------------GTGAc---CGCGGgcctctgtac---CCGGCCC--ACA
RC   7  1107332..1107400   30.26   3.64e-03
GGGCTCGgtacaaccgc---CTCCG---CAAggtatcgggttcc-----------GTGAg---CGGAGcgtgctcacgc-CGAGCCC--ACA
RC   8  1371594..1371659   18.26   2.94e+00
GTCCTCTacctgaaatg---GGCT----CAAggtgtccattc--------------GTGAg---AAAGTtagaagcct---GGAGGATg-AAA
RC   9  1404046..1404129   21.71   8.22e-01
CGGCGCCtattcctgcgc--TGGCGa--TGAtaccctgaggtatagctgagagaccgGTGAggc-CAACGgtaagacc----GGCGCCGa-ATA
RC  10  1460837..1460911   17.72   3.52e+00
TCGAGGGtctgaaagt----CCAGAa--TGAaacggaagacagt------------ATGAcagtCTTGGttaagtttaaaaCCCTCGGcgAGA
```

Figure 15.4 *(continued)*

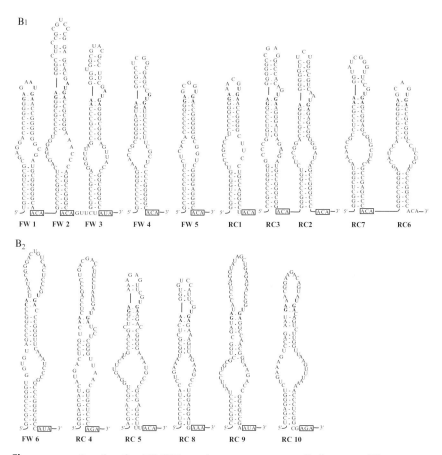

Figure 15.4   Results of an ERPIN search run on a yet-unstudied genome, *Thermococcus kodakarensis*, and proposed secondary structures of the identified candidates. Same legend as in Fig. 15.3. Panel (A) also indicates the absolute score and the e-value of the candidate. Panel (B1) displays the candidates retained after inspection of their structures, and panel (B2) displays the discarded candidates.

because their coding sequences are adjacent in the genome. Two pairs of motifs with a negative e-value are also encoded by adjacent sequences (RC3 and RC2 on the one hand, and RC7 and RC6 on the other hand). Therefore, they likely belong to two H/ACA sRNAs, each containing two H/ACA motifs. In addition, the coding sequences of the 10 motifs represented in panel B1 are all located between ORFs.

In contrast, the six candidate motifs with an e-value between 1e-01 and 4 (panel B2) can be discarded for several reasons: (1) except for motif RC5, they do not contain an ACA triplet; (2) most of them contain bulge residues or an internal loop in helix 1 or in helix 2; (3) the large terminal loops of motifs FW6, RC4, RC9, and RC10 are not structured. Finally, the candidate motif FW 6 corresponds to a tRNA^{Asp}(GUC), and the coding sequences of the five other motifs are located within ORFs. On the basis of our present knowledge, sRNA genes may partially overlap ORFs but are not included in ORFs. A strong argument for the validation of the 10 H/ACA motifs in *T. kodakarensis* (Fig. 15.4, panel B1) is their homology with the 10 H/ACA sRNAs found in the *Pyrococcus* species. Hence, we predict that the 10 motifs detected in *T. kodakarensis* are the homologs of the 10 experimentally demonstrated *Pyrococcus* H/ACA motifs.

Interestingly, no new putative motifs with a satisfying negative e-value were found when a second screening was performed after insertion of the 10 validated motifs in the profile by use of the same parameters as in the first screening step (order 1 and cutoff values of 110%). Only the number of motifs with a high e-value increased, suggesting that all of the true H/ACA motifs were detected in the first search. Therefore, the 41 known H/ACA motifs aligned in the profile seem to provide enough information for an exhaustive selection of the *T. kodakarensis* motifs in a single run.

We next tested whether introduction of the probable *T. kodakarensis* H/ACA motifs in the profile would increase the selection of the *Pyrococcus* species known H/ACA motifs in blind tests (Table 15.2). This is, indeed, the case. When a blind ERPIN search is performed on the *P. abyssi* genome with a profile including the H/ACA motifs from *A. fulgidus, M. jannaschii, P. furiosus*, and *P. horikoshii*, nine of the known H/ACA motifs are detected with the priority order 1 and a 95% cutoff value (Table 15.2). No false-positive result is found with a 95% cutoff value. When the *T. kodakarensis* motifs are also added in the alignment, the 10 *P. abyssi* motifs are found when a cutoff value of 95% is used. However, the number of false-positive results increases in this case (Table 15.2).

2.1.6. Recommendations

On the basis of the data presented in Tables 15.1 and 15.2, the cutoff values used should be defined, taking into account the number and origin of the H/ACA motifs aligned in the profile. If they belong to species that are phylogenetically closely related to the studied species, small cutoff values can be used. They are expected to allow a selection of the true motifs with a minimum of noise. When a phylogenetically distant species is studied, we recommend the use of a cutoff value of 110% for each step to select true motifs that may exhibit divergences relative to the canonical structure. Hence, the closer the species, the smaller the cutoff values can be.

As in most cases, use of priority order 1 gives better results; we recommend the use of this order and only exceptionally, when needed, order 2.

As illustrated by the blind tests in Table 15.1, to complete the identification of H/ACA motifs in a new species, it may be worthwhile performing a second screening, after inclusion in the profile of the H/ACA-like motifs that were identified in the first step and contain all the needed characteristic structural features.

Table 15.2 Tests of the effect of the integration of the *T. kodakarensis* H/ACA motifs identified in this work, on blind tests performed on the *P. abyssi* genome

Number of H/ACA sRNA motifs already identified	Cutoff	Blind Tests performed with the profile A[a]				Bind Tests performed with the profile B[a]			
		Order (1)		Order (2)		Order (1)		Order (2)	
		Number of positive results	Number of likely false-positive results	Number of positive results	Number of likely false-positive results	Number of positive results	Number of likely false-positive results	Number of positive results	Number of likely false-positive results
P. abyssi 10 [2,3,4]	95%	9	0	7	0	10	1	7	1
	100%	10	1	8	0	10	2	8	2
	110%	10	1	8	1	10	4	8	5

[a] The profile A contains the H/ACA motifs of *A. fulgidus*, *M. jannaschii*, *P. horikoshii* and *P. furiosus*. The profile B corresponds to the profile A with, in addition, the *T. kodakarensis* H/ACA motifs.

The H/ACA motifs detected in *T. kodakarensis* are included in the H/ACA profile available at http://tagc.univ-mrs.fr/asterix/erpin/. We are currently analyzing the H/ACA motifs of all completely sequenced archaeal genomes, and an updated H/ACA profile including not experimentally verified motifs will soon be available at the same URL.

Concerning the interpretation of the data, we recommend the selection of H/ACA-like motifs with negative e-values for the subsequent analysis (screening for the target sequence). Motifs with e-values between 1 and 1e-02 can be maintained if they contain almost all the needed structural elements; a default in one of the structural elements may be accepted at this stage of the selection, because screening for target sequences will also help in discriminating true candidates.

Note that one strong discriminatory feature of ERPIN results is the overlap of candidate coding sequences with sequences coding for tRNAs, rRNAs, or proteins. Note also that, as observed in *T. kodakarensis*, when the apical loop closing the K-turn motif is long enough, an important criterion for the validation of the H/ACA motif is the possibility of forming a stem-loop structure in this apical region.

Finally, the newly identified sRNAs have to be experimentally confirmed with the approaches described in the third section of this chapter.

2.2. Search for targets of the H/ACA-like motifs

The identification of a putative RNA target for a given H/ACA motif is initially based on the search for a dinucleotide containing a uridine residue at the 5′ position. This dinucleotide should be flanked by two sequences complementary to the two strands of the expected pseudouridylation pocket. To this end, one has to first establish the most likely secondary structure of the H/ACA motif, and we will comment in the following on the difficulties that may be encountered at this step. Once the 2D structure is established, a series of RNA descriptors, describing the various possible base-paired structures that can be formed by the target RNA and the two guiding sequences, have to be settled (Fig. 15.5). The different steps in target identification are explained in the following.

2.2.1. Requirements

2.2.1.1. *Hardware and software* The RNAMOT program (Gautheret *et al.*, 1993) version 2.1 can be run on most UNIX platforms. The source and a tutorial are available at http://pages-perso.esil.univ-mrs.fr/~dgaut/download/. Any equivalent simple descriptor-based software such as RNAMotif (Macke *et al.*, 2001) can also be used. We selected the RNAMOT program because it is simple and easy to use.

2.2.1.2. *Sequence data*   The sequence format used for RNAMOT is FASTA. The ribosomal RNA sequences of each of the studied archaeal species were extracted from their genomic sequence subsets, including all the noncoding RNAs (file with extension .frn obtained at ftp://ftp.ncbi.nih.gov/genomes/Bacteria).

2.2.2. Procedure to build the RNAMOT descriptors and run the program

2.2.2.1. *Construction of the descriptors*   On the basis of our present knowledge of the archaeal H/ACA system, there is no requirement concerning the identity of the residue at the $3'$ position in the single-stranded dinucleotide (Fig. 15.5). Altogether, the two base-paired regions most generally contain at least nine base pairs, and one wobble pair can be included in these nine pairs.

In Fig. 15.5, we illustrate the various descriptors that have to be built to look for the possible targets of the H/ACA motif Pab91 of *P. abyssi*. A roughly equivalent number of base pairs in the two helical regions is expected to be the most frequent situation (4 and 5 base pairs or 5 and 4 base pairs). However, one cannot exclude the possibility of having highly asymmetrical base-pair interactions (1 + 8, 2 + 7, 3 + 6, 6 + 3, 7 + 2, and 8 + 1).

We will explain how one of the descriptors needed to look for one of the base-paired configurations is built; all of the other descriptors are built in the same way. The descriptor used as an example in Fig. 15.5 (panel C) is devoted to the search of a target sequence forming base-pair interactions including 4 and 5 residues of the $3'$ and $5'$ strands of the pseudouridylation pocket, respectively (descriptor designated as 4 + 5). The descriptor file includes two lines: the first one identifies the searched sequence, s1 in this example. The second line defines the sequence requirements.

s1

s1 11 : 11 YUYYUNRGGYU

The two numbers separated by a colon correspond to the minimum and maximum nucleotide lengths of the searched target sequence; Y and R correspond to pyrimidine and purine residues, respectively.

If no positive result is obtained when the first series of descriptors corresponding to a total of nine base pairs is used, it is recommended to look for possible targets able to form only eight or seven base pairs. The use of this low number of base pairs may allow the selection of interactions that include mismatches. Indeed, such interactions cannot be selected directly by the RNAMOT descriptors described previously.

In targets that include mismatches, an extension of the possible base-pair interactions should be tested by inspection of the guide and target

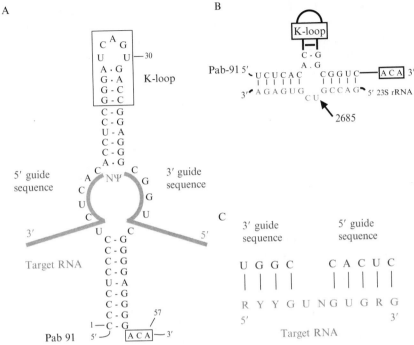

Figure 15.5 The strategy used for the identification of the targets of the H/ACA-like motifs. (A) The secondary structure of the *P. abyssi* Pab91 sRNA is shown with a schematic representation of the target RNA base pair interactions formed with the 5′ and 3′ guide sequences of the H/ACA motif (Charpentier *et al.*, 2005). The ΨN dinucleotide obtained after U to Ψ conversion is shown. (B) The known interaction formed between the *P. abyssi* Pab91 sRNA and its known target sequence in *P. abyssi* 23S rRNA. The targeted U2685 residue is indicated (Charpentier *et al.*, 2005). (C) As an example, we show the structure that would be used to define a 5 + 4 descriptor for the selection of target sequences of RNA Pab91. Once the 5′ and 3′ guide sequences are specified, the possible complementary sequences are written, taking into account possible wobble G·U base-pair interactions (Y = pyrimidines, R = purines) and the chosen numbers of base pairs to be formed with the 5′ and the 3′ guide sequence, respectively. The rRNA sequences are then screened with a descriptor of these possible complementary sequences by use of RNAMOT.

sequences. Even when a putative interaction including nine base pairs is found, we recommend looking for possible extensions of the base-pair interactions by inspection of the flanking sequences.

2.2.2.2. How to start the search? The program RNAMOT is started using the following command line:

rnamot -s -s <rRNA sequence > -d<descriptor > -o<output > -t

The name of the executable is "rnamot." The first "-s" specifies the mode of use (-s for search, by opposition to -a for alignment); the second "-s" is used to provide the name and location of the file containing the rRNA sequences that will be screened, and "-d" the name and location of the descriptor file. Finally, the "-o" option specifies the file in which the results are written. The "-t" option activates the display of the search stage and the number of positive results obtained. A search has to be executed for each of the descriptors. However, whatever descriptor is used, the search time does not exceed a few seconds.

2.2.3. Example of application: search for the target sequences of the *M. jannaschii* FW1 motif

In the *M. jannaschii* motif FW1, helices 1 and 2 contain nine and six successive Watson–Crick base pairs, respectively (Fig. 15.6, panel A). In addition, one wobble G·U pair can be present at the extremity of helix 2. The formation or the absence of this wobble pair modifies the pseudouridylation pocket and, therefore, the putative target sequences. Hence, two series of descriptors were built as described earlier. In the first series, the G·U pair at the extremity of helix 2 was considered to be formed. In the second series, it was considered to be opened. No putative target sequence was detected for this second series of descriptors. In contrast, two putative target sequences were detected with the first series, suggesting the formation of the G·U pair in the guide sRNA. One of the target sequences (Fig. 15.6, panel B, hit sequence 1) was obtained by use of a descriptor designed for the search of sequences forming three and six base pairs with the $3'$ and $5'$ guiding sequences, respectively ($3 + 6$ descriptor). The second target sequence (panel B, hit sequence 2) was selected with a $4 + 5$ descriptor. Hit sequence 1 belongs to 16S rRNA, and residue U1261 would be the target. Hit sequence 2 belongs to 23S rRNA, and residue U2015 would be the target.

Inspection of both rRNA flanking sequences reveals the possible extension by 2 (G–C and G·U) and 1 (G–C) base pairs of the interactions that can be formed with the $5'$ guide sequences for hit sequences 1 and 2, respectively. Therefore, the base-pair interaction formed for hit sequence 1 involves two stretches of three and eight base pairs. However, each of them contains two wobble pairs. Thus far, such a high number of G·U pairs has never been described for the previously studied H/ACA sRNAs and snoRNAs. Therefore, hit sequence 1 may correspond to a false-positive result. In contrast, hit sequence 2 forms two stretches of four and six uninterrupted Watson–Crick base pairs. A strong argument in favor of the validity of hit sequence 2 is the observation of H/ACA motifs proposed to target similar positions in the 23S rRNAs of *P. furiosus* (H/ACA motif Pf 7) and *A. fulgidus* (H/ACA motif Afu4) (Rozhdestvensky *et al.*, 2003) (Fig. 15.6, panels C and D).

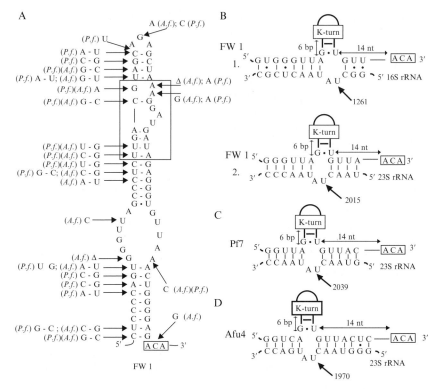

Figure 15.6 Target sequences found for the *M. jannaschii* FW1 motif by an RNAMOT search. (A) Secondary structure proposed for the *M. jannaschii* FW1 candidate. The compensatory mutations and single base substitutions, insertions, or deletions found in the *A. fulgidus* Afu4 (*A.f.*) and *P. furiosus* Pf7 (*P.f.*) motifs compared with the *M. jannaschii* FW1 motif are indicated. Δ indicates one missing nucleotide. (B) The possible base–pair interactions between the *M. jannaschii* FW1 motif and the two putative targeted sequences identified for this motif by the RNAMOT search (hit sequences 1 and 2) are shown. The distances between the basal extremity of helix 2 and either the K-loop or the ACA box are indicated. The target sequences proposed for the third H/ACA motif of the *P. furiosus* Pf7 sRNA (C) and the third H/ACA motif of the *A. fulgidus* Afu4 sRNA (Rozhdestvensky *et al.*, 2003) (D), and the interactions that they can form with the guide RNA are represented.

2.2.4. Application to the search of target sequences for the 10 H/ACA-like motifs found in *T. kodakarensis*

As described previously, we screened the *T. kodakarensis* genome for H/ACA-like motifs. We will describe how we looked for their possible target sequences with RNAMOT. For each of the 16 hit H/ACA motifs obtained by the ERPIN search, different series of descriptors were built to take into consideration the slight possible length variations of helix 2. For example, in the FW1 motif, a noncanonical G·A pair may be included at the basal extremity of helix 2 or may be opened so that its two nucleotides can be included in the guide sequences (Fig. 15.7, panel A). At least two alternative lengths of helix 2 are possible for almost all the selected motifs of *T. kodakarensis* (Fig. 15.7).

When the various series of descriptors built for each of the motifs are used, no putative target sequence was found for the six motifs that we discarded after inspection of their primary and secondary structures. In contrast, up to 3 putative target sequences were found for each of the 10 motifs that we retained after this inspection. In Fig. 15.7, we present the corresponding base-pair interactions by order of stability for each of the motifs. The order of presentation also takes into account the fact that some of the H/ACA motifs are expected to belong to the same sRNA (FW1, FW2, and FW3; RC2 and RC3; RC6 and RC7).

Most generally, when two or three possible target sequences were detected for a given motif, they differ by their stabilities and/or by the distance between helix 2 and the ACA trinucleotide. For instance, among the two hit sequences found for each of the FW2 and RC1 motifs, the hit sequences numbered 2 form shorter base-pair interactions with the guide sequences compared with the hit sequences numbered 1. In cases in which three putative target sequences were found (motifs RC2, RC6 and RC7), most frequently, one of them forms significantly more stable interactions than the two other ones. Therefore, in each case, we prefer the hit sequence forming the more stable interaction (hits numbered 1). However, peculiar results were obtained for the RC2 motif: both hits 1 and 2 form stable interactions (9 Watson–Crick base pairs and a G·U wobble pair for hit sequence 2 and 19 Watson–Crick pairs and two G·U pairs for hit sequence 1). By use of hit sequence 2, the K-turn structure and the ACA triplet are located at correct distances from the basal extremity of helix 2 (Fig. 15.7E). Thus, despite the lower stability of the interaction formed by hit sequence 2 compared with hit sequence 1, we have no criterion to eliminate this hit sequence. Furthermore, it should be noted that the interactions that can be formed with hit sequence 1 are unusually long. Formation of these interactions would result in the disruption of a large part of helix 1. Because helix 1 is expected to be important for the folding and the activity of the H/ACA sRNP complex, the possible interaction between the hit sequence 1 and motif RC2 may have another function than RNA pseudouridylation. Interestingly, this hit sequence is located 3 nts downstream from the 16S rRNA 5' extremity. We may imagine an RNA chaperone role of motif RC2, in addition to its possible role in pseudouridylation.

Note that no Watson–Crick base pair in helix 1 is opened in any of the interactions formed with the other H/ACA motifs and hit sequences. Only the G·U pair at the extremity of helix 1 in motif RC1 may be formed or opened, depending on whether a G·U pair is included or not in the interaction established between hit sequence 1 and the 5' strand of the guide sequence (Fig. 15.7D).

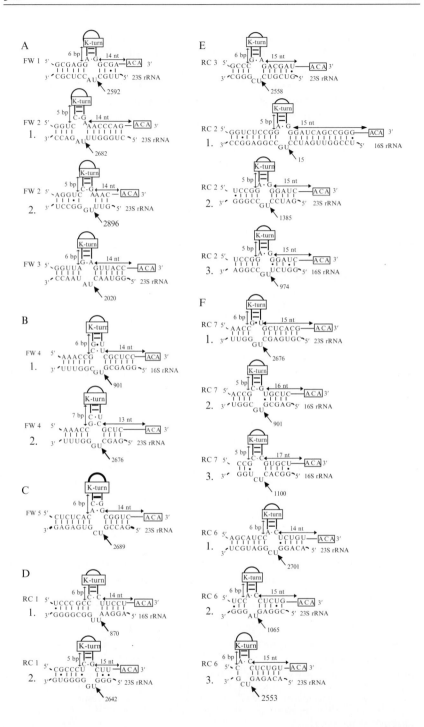

On the basis of the data presented in Fig. 15.7, the selected hit sequences can form from 4 up to 7 base pairs with the 5′ and 3′ guide sequences. Interestingly, for most of the more favorable hit sequences, a noncanonical pair is present at the basal extremity of helix 2, which may be required to ensure enough flexibility in the cruciform structure.

2.2.5. Tips for searching and analyzing H/ACA targets

As evidenced by the preceding examples, a major difficulty in the prediction of target sequences comes from the necessity of establishing the correct secondary structure of the putative H/ACA-like motif and to define the basal extremity of helix 2. Because of the irregularity and size variations of helix 2, several possible conformations can sometimes be proposed for a given motif. In contrast, helix 1, which consists in a regular stretch of Watson–Crick pairs, is generally easier to predict.

In addition, even when the correct secondary structure has been established, the size variability of the base-pair interactions formed with the 5′ and 3′ guide sequences increases the number of RNAMOT searches that have to be run for a given motif.

As illustrated by the *T. kodakarensis* example given previously, we recommend the initial use of a series of descriptors corresponding to interactions including 9 base pairs. A reduction of this number of base pairs in the first round of the selection would unnecessarily increase the number of false-positive results and limit the chance of obtaining positive results. Hits with a lower level of complementarity can be searched in a second step, if no hit is obtained with the 9 base pair descriptors. Restriction of the starting descriptors to interactions only including Watson–Crick pairs would avoid the selection of interactions containing one wobble G·U pair such as the one found for the *T. kodakarensis* motif FW1.

For interpretation of the results note that: (1) a non-Watson–Crick base pair is often present at the basal extremity of helix 2, (2) the distances between the basal extremity of helix 2 and the K-turn/K-loop sequence on the one hand, and the ACA triplet on the other hand, are most often 5–6 bps and 14–15nts, respectively, (3) the base-pair interactions with both guide sequences include most frequently 4–7 residues; however, more asymmetrical base-pair interactions cannot be excluded.

Figure 15.7 Putative target sequences identified for the 10 H/ACA motifs of *T. kodakar-ensis* by an RNAMOT search. Panels A–F display the base-pair interactions that can be formed between the putative target sequences identified by use of RNAMOT and each of the 10 H/ACA motifs identified for *T. kodakarensis*. Each panel corresponds to a given H/ACA sRNA. For each motif the possible base-pair interactions are represented according to their stability.

As illustrated by the *T. kodakarensis* example used in this chapter, the search of putative target sequences in rRNAs by RNAMOT can confirm the discrimination of true H/ACA motifs made by inspection of the H/ACA-like motifs obtained in the ERPIN search. However, if an H/ACA motif bears all the characteristic features of archaeal H/ACA motifs and if no putative target sequence is found in rRNAs, its target sequence may be contained in another kind of RNA. In this case, the search for possible target sequences in other RNAs will be useful.

2.3. Phylogenetic and experimental validation of the results

2.3.1. Phylogenetic validation

When new H/ACA motifs and their putative targets have been selected by the proposed approach, several comparisons with known data on archaeal H/ACA motifs may be helpful for their validation. This was illustrated in the interpretation of the RNAMOT hit sequences found for the FW1 motif of *M. jannaschii*, where three homologous H/ACA motifs from *P. furiosus, A. fulgidus,* and *M. jannaschii,* that differ by a limited number of base substitutions, insertions or deletions (Fig. 15.6), guide modifications at similar positions in 23S rRNA. Therefore, as a first step in the validation of the data, we recommend the comparison of the putative H/ACA motifs obtained for a new archaeal genome with the already identified H/ACA motifs. However, note that phylogenetic comparisons are limited by the rapid evolution of sRNAs.

Although pseudouridylation sites in archaeal rRNAs have only been identified in a limited number of species and are not strongly conserved from one species to the other (Del Campo *et al.,* 2005; Ofengand, 2002), a comparison of the putative target sites detected by RNAMOT with known pseudouridylation sites in archaeal rRNAs may also be useful, especially when Ψ residues have been experimentally identified in rRNAs from phylogenetically related species.

We will again use the *M. jannaschii* FW1 motif to illustrate this point. One of the two putative target sites found by the RNAMOT (U2015 in 23S rRNA) was experimentally found to be pseudouridylated in the 23S rRNA of archaeal species (Ofengand and Bakin, 1997; and for review, Ofengand, 2002; corrected in Del Campo *et al.,* 2005). The corresponding positions in the yeast and human large subunit rRNAs are also pseudouridylated, and the H/ACA snoRNAs snR191 in yeast and hU19 in human are, respectively, proposed to guide the modifications (Badis *et al.,* 2003; Bortolin and Kiss, 1998).

On the contrary, no counterpart of the second putative target residue (U1261) of motif FW1 was found to be pseudouridylated, neither in archaea nor in eukarya. Therefore, taken together, the FW1 motif of *M. jannaschii* can reasonably be considered as guiding U to Ψ conversion at position U2015 in 23S rRNA.

2.3.2. Experimental validations

The situation described previously for the FW1 motif of *M. jannaschii* is a highly favorable one in terms of validation. It is not always possible to find homologous H/ACA motifs in other species and/or the presence of a Ψ residue at the targeted position in another archaeal species. When this is not the case, direct experimental proof is particularly important to validate the data.

A first simple and important experiment to do is a Northern blot or primer extension analysis on total RNA extracted from the studied archaea. This will allow verification of the presence of the proposed H/ACA sRNA. Radiolabeled primers complementary to the H/ACA sRNAs candidates can be used for this purpose. In addition, these kinds of experiments will give information on the length of the sRNAs because, as explained before, archaeal sRNAs have variable lengths. A protocol for primer extension analysis is proposed in Chapter 2 in this volume by Motorine *et al.*

Another validation approach involves localizing the pseudouridylation positions in rRNAs by use of the CMCT-RT approach as described by Motorine *et al.* in Chapter 2. This is not an easy technique to handle. However, a detailed protocol and tips are given by Motorine *et al.*

Finally, the complete verification of the proposed target sequences can be obtained by *in vitro* reconstitution of an active H/ACA sRNP by use of *in vitro* transcribed guide RNA and target sequence and recombinant aCBF5, aNOP10, L7Ae, and aGAR1 proteins, as described in Chapter 16 by Charpentier *et al.*, in this volume.

3. CONCLUSIONS

The knowledge-based approach described herein, which combines the search for H/ACA motifs and their respective target(s), is an efficient approach as illustrated in the numerous examples given in this chapter. Its efficiency will be further improved, on enrichment of the H/ACA profile by inclusion of new validated motifs. One does not need much expertise in computing to use the ERPIN and RNAMOT softwares, which are easy to use. Moreover, a web server version of ERPIN is available for the user to become familiar with the method. No data other than the structural or functional features of H/ACA sRNAs, which are described in this chapter, is necessary.

Compared with the results recently obtained by application of the MilPat tool to the identification of H/ACA-like motifs in the genomes from *M. jannaschii* and three *Pyrococcus* species (Thebault *et al.*, 2006), the approach proposed in this chapter is more directed and, therefore, gives a very limited number of false-positive results. By the use of the H/ACA profile enriched by the H/ACA motifs of *T. kodakarensis* identified in this work, we detected, in the blind test presented in Table 15.2, the 10 H/ACA motifs present in *P. abyssi, P. furiosus,* and *P. horikoshii.* When the MilPat approach is used, among the 89 to 148 candidates H/ACA motifs found for the different *Pyrococcus* species studied, only 6 to 7 of the known motifs were detected, depending on the species. We think that after inclusion of the H/ACA motifs from a limited number of archaeal species belonging to different archaeal orders, which is currently being done by our team, we will be able to propose soon (at site http://tagc.univ-mrs.fr/asterix/erpin/), a highly powerful tool for the search of H/ACA sRNAs in any archaeal species. The strength of our approach is the coupling of the search of H/ACA-like motifs to the search of their target sequences. As evidenced in the given examples, both steps in this strategy participate in the selection of the true H/ACA motifs among the identified hit motifs.

We would like to point out that the strategy proposed in this chapter is well suited to the search of archaeal H/ACA motifs because of their structural specificities (presence of a K-turn or a K loop and of G-C rich helices). This strategy would be much less efficient if applied to the search of snoRNA coding sequences. However, a specific computational approach dedicated to eukaryal H/ACA snoRNAs has been developed (Schattner *et al.*, 2006).

REFERENCES

Badis, G., Fromont-Racine, M., and Jacquier, A. (2003). A snoRNA that guides the two most conserved pseudouridine modifications within rRNA confers a growth advantage in yeast. *RNA* **9,** 771–779.

Baker, D. L., Youssef, O. A., Chastkofsky, M. I., Dy, D. A., Terns, R. M., and Terns, M. P. (2005). RNA-guided RNA modification: Functional organization of the archaeal H/ACA RNP. *Genes Dev.* **19,** 1238–1248.

Balakin, A. G., Smith, L., and Fournier, M. J. (1996). The RNA world of the nucleolus: Two major families of small RNAs defined by different box elements with related functions. *Cell* **86,** 823–834.

Bortolin, M. L., and Kiss, T. (1998). Human U19 intron-encoded snoRNA is processed from a long primary transcript that possesses little potential for protein coding. *RNA* **4,** 445–454.

Bourdeau, V., Ferbeyre, G., Pageau, M., Paquin, B., and Cedergren, R. (1999). The distribution of RNA motifs in natural sequences. *Nucleic Acids Res.* **27,** 4457–4467.

Charpentier, B., Muller, S., and Branlant, C. (2005). Reconstitution of archaeal H/ACA small ribonucleoprotein complexes active in pseudouridylation. *Nucleic Acids Res.* **33,** 3133–3144.

Charron, C., Manival, X., Clery, A., Senty-Segault, V., Charpentier, B., Marmier-Gourrier, N., Branlant, C., and Aubry, A. (2004). The archaeal sRNA binding protein L7Ae has a 3D structure very similar to that of its eukaryal counterpart while having a broader RNA-binding specificity. *J. Mol. Biol.* **342,** 757–773.

Del Campo, M., Recinos, C., Yanez, G., Pomerantz, S. C., Guymon, R., Crain, P. F., McCloskey, J. A., and Ofengand, J. (2005). Number, position, and significance of the pseudouridines in the large subunit ribosomal RNA of Haloarcula marismortui and Deinococcus radiodurans. *RNA* **11,** 210–219.

Ganot, P., Bortolin, M. L., and Kiss, T. (1997a). Site-specific pseudouridine formation in preribosomal RNA is guided by small nucleolar RNAs. *Cell* **89,** 799–809.

Ganot, P., Caizergues-Ferrer, M., and Kiss, T. (1997b). The family of box ACA small nucleolar RNAs is defined by an evolutionarily conserved secondary structure and ubiquitous sequence elements essential for RNA accumulation. *Genes Dev.* **11,** 941–956.

Gautheret, D., and Lambert, A. (2001). Direct RNA motif definition and identification from multiple sequence alignments using secondary structure profiles. *J. Mol. Biol.* **313,** 1003–1011.

Gautheret, D., Major, F., and Cedergren, R. (1993). Modeling the three-dimensional structure of RNA using discrete nucleotide conformational sets. *J. Mol. Biol.* **229,** 1049–1064.

Hamma, T., Reichow, S. L., Varani, G., and Ferre-D'Amare, A. R. (2005). The Cbf5-Nop10 complex is a molecular bracket that organizes box H/ACA RNPs. *Nat. Struct. Mol. Biol.* **12,** 1101–1107.

Henras, A., Henry, Y., Bousquet-Antonelli, C., Noaillac-Depeyre, J., Gelugne, J. P., and Caizergues-Ferrer, M. (1998). Nhp2p and Nop10p are essential for the function of H/ACA snoRNPs. *EMBO J.* **17,** 7078–7090.

Huang, Z. P., Zhou, H., He, H. L., Chen, C. L., Liang, D., and Qu, L. H. (2005). Genomewide analyses of two families of snoRNA genes from *Drosophila melanogaster,* demonstrating the extensive utilization of introns for coding of snoRNAs. *RNA* **11,** 1303–1316.

Huang, Z. P., Zhou, H., Liang, D., and Qu, L. H. (2004). Different expression strategy: Multiple intronic gene clusters of box H/ACA snoRNA in *Drosophila melanogaster.* *J. Mol. Biol.* **341,** 669–683.

Khanna, M., Wu, H., Johansson, C., Caizergues-Ferrer, M., and Feigon, J. (2006). Structural study of the H/ACA snoRNP components Nop10p and the 3′ hairpin of U65 snoRNA. *RNA* **12,** 40–52.

Kiss, A. M., Jady, B. E., Bertrand, E., and Kiss, T. (2004). Human box H/ACA pseudouridylation guide RNA machinery. *Mol. Cell Biol.* **24,** 5797–5807.

Klein, D. J., Schmeing, T. M., Moore, P. B., and Steitz, T. A. (2001). The kink-turn: A new RNA secondary structure motif. *EMBO J.* **20,** 4214–4221.

Laferriere, A., Gautheret, D., and Cedergren, R. (1994). An RNA pattern matching program with enhanced performance and portability. *Comput. Appl. Biosci.* **10,** 211–212.

Lafontaine, D. L., Bousquet-Antonelli, C., Henry, Y., Caizergues-Ferrer, M., and Tollervey, D. (1998). The box H + ACA snoRNAs carry Cbf5p, the putative rRNA pseudouridine synthase. *Genes Dev.* **12,** 527–537.

Lambert, A., Legendre, M., Fontaine, J. F., and Gautheret, D. (2005). Computing expectation values for RNA motifs using discrete convolutions. *BMC Bioinformatics* **6,** 118.

Lambert, A., Fontaine, J. F., Legendre, M., Leclerc, F., Permal, E., Major, F., Putzer, H., Delfour, O., Michot, B., and Gautheret, D. (2004). The ERPIN server: An interface to profile-based RNA motif identification. *Nucleic Acids Res.* **32,** W160–W165.

Lambert, A., Lescure, A., and Gautheret, D. (2002). A survey of metazoan selenocysteine insertion sequences. *Biochimie* **84,** 953–959.

Legendre, M., Lambert, A., and Gautheret, D. (2005). Profile-based detection of microRNA precursors in animal genomes. *Bioinformatics* **21,** 841–845.

Lescure, A., Gautheret, D., Carbon, P., and Krol, A. (1999). Novel selenoproteins identified *in silico* and *in vivo* by using a conserved RNA structural motif. *J. Biol. Chem.* **274,** 38147–38154.

Lescure, A., Gautheret, D., and Krol, A. (2002). Novel selenoproteins identified from genomic sequence data. *Methods Enzymol.* **347,** 57–70.

Li, L., and Ye, K. (2006). Crystal structure of an H/ACA box ribonucleoprotein particle. *Nature* **443,** 302–307.

Li, S. G., Zhou, H., Luo, Y. P., Zhang, P., and Qu, L. H. (2005). Identification and functional analysis of 20 Box H/ACA small nucleolar RNAs (snoRNAs) from *Schizosaccharomyces pombe. J. Biol. Chem.* **280,** 16446–16455.

Macke, T. J., Ecker, D. J., Gutell, R. R., Gautheret, D., Case, D. A., and Sampath, R. (2001). RNAMotif, an RNA secondary structure definition and search algorithm. *Nucleic Acids Res.* **29,** 4724–4735.

Manival, X., Charron, C., Fourmann, J. B., Godard, F., Charpentier, B., and Branlant, C. (2006). Crystal structure determination and site-directed mutagenesis of the *Pyrococcus abyssi* aCBF5-aNOP10 complex reveal crucial roles of the C-terminal domains of both proteins in H/ACA sRNP activity. *Nucleic Acids Res.* **34,** 826–839.

Massenet, S., Motorin, Y., Lafontaine, D. L., Hurt, E. C., Grosjean, H., and Branlant, C. (1999). Pseudouridine mapping in the *Saccharomyces cerevisiae* spliceosomal U small nuclear RNAs (snRNAs) reveals that pseudouridine synthase pus1p exhibits a dual substrate specificity for U2 snRNA and tRNA. *Mol. Cell Biol.* **19,** 2142–2154.

Meier, U. T., and Blobel, G. (1994). NAP57, a mammalian nucleolar protein with a putative homolog in yeast and bacteria. *J. Cell Biol.* **127,** 1505–1514.

Ni, J., Tien, A. L., and Fournier, M. J. (1997). Small nucleolar RNAs direct site-specific synthesis of pseudouridine in ribosomal RNA. *Cell* **89,** 565–573.

Ofengand, J. (2002). Ribosomal RNA pseudouridines and pseudouridine synthases. *FEBS Lett.* **514,** 17–25.

Ofengand, J., and Bakin, A. (1997). Mapping to nucleotide resolution of pseudouridine residues in large subunit ribosomal RNAs from representative eukaryotes, prokaryotes, archaebacteria, mitochondria and chloroplasts. *J. Mol. Biol.* **266,** 246–268.

Rashid, R., Liang, B., Baker, D. L., Youssef, O. A., He, Y., Phipps, K., Terns, R. M., Terns, M. P., and Li, H. (2006). Crystal structure of a Cbf5-Nop10-Gar1 complex and implications in RNA-guided pseudouridylation and dyskeratosis congenita. *Mol. Cell* **21,** 249–260.

Rozenski, J., Crain, P. F., and McCloskey, J. A. (1999). The RNA Modification Database: 1999 update. *Nucleic Acids Res.* **27,** 196–197.

Rozhdestvensky, T. S., Tang, T. H., Tchirkova, I. V., Brosius, J., Bachellerie, J. P., and Huttenhofer, A. (2003). Binding of L7Ae protein to the K-turn of archaeal snoRNAs: A shared RNA binding motif for C/D and H/ACA box snoRNAs in Archaea. *Nucleic Acids Res.* **31,** 869–877.

Schattner, P., Barberan-Soler, S., and Lowe, T. M. (2006). A computational screen for mammalian pseudouridylation guide H/ACA RNAs. *RNA* **12,** 15–25.

Schattner, P., Decatur, W. A., Davis, C. A., Ares, M., Jr., Fournier, M. J., and Lowe, T. M. (2004). Genome-wide searching for pseudouridylation guide snoRNAs: Analysis of the *Saccharomyces cerevisiae* genome. *Nucleic Acids Res.* **32,** 4281–4296.

Tang, T. H., Bachellerie, J. P., Rozhdestvensky, T., Bortolin, M. L., Huber, H., Drungowski, M., Elge, T., Brosius, J., and Huttenhofer, A. (2002a). Identification of 86 candidates for small non-messenger RNAs from the archaeon Archaeoglobus fulgidus. *Proc. Natl. Acad. Sci. USA* **99,** 7536–7541.

Thebault, P., de Givry, S., Schiex, T., and Gaspin, C. (2006). Searching RNA motifs and their intermolecular contacts with constraint networks. *Bioinformatics* **22,** 2074–2080.

Torchet, C., Badis, G., Devaux, F., Costanzo, G., Werner, M., and Jacquier, A. (2005). The complete set of H/ACA snoRNAs that guide rRNA pseudouridylations in *Saccharomyces cerevisiae. RNA* **11,** 928–938.

Vidovic, I., Nottrott, S., Hartmuth, K., Luhrmann, R., and Ficner, R. (2000). Crystal structure of the spliceosomal 15.5kD protein bound to a U4 snRNA fragment. *Mol. Cell* **6,** 1331–1342.

Wang, C., and Meier, U. T. (2004). Architecture and assembly of mammalian H/ACA small nucleolar and telomerase ribonucleoproteins. *EMBO J.* **23,** 1857–1867.

Watkins, N. J., Gottschalk, A., Neubauer, G., Kastner, B., Fabrizio, P., Mann, M., and Luhrmann, R. (1998). Cbf5p, a potential pseudouridine synthase, and Nhp2p, a putative RNA-binding protein, are present together with Gar1p in all H BOX/ACA-motif snoRNPs and constitute a common bipartite structure. *RNA* **4,** 1549–1568.

Yuan, G., Klambt, C., Bachellerie, J. P., Brosius, J., and Huttenhofer, A. (2003). RNomics in *Drosophila melanogaster.* Identification of 66 candidates for novel non-messenger RNAs. *Nucleic Acids Res.* **31,** 2495–2507.

CHAPTER SIXTEEN

RECONSTITUTION OF ARCHAEAL H/ACA SRNPS AND TEST OF THEIR ACTIVITY

Bruno Charpentier, Jean-Baptiste Fourmann, *and* Christiane Branlant

Contents

Laboratoire de Maturation des ARN et Enzymologie Moléculaire, Nancy Université, Faculté des Sciences et Techniques, Vandoeuvre-les-Nancy, France

Methods in Enzymology, Volume 425
ISSN 0076-6879, DOI: 10.1016/S0076-6879(07)25016-5

Abstract

Conditions for the reconstitution of archaeal sRNPs active in RNA-guided post-transcriptional modification of RNAs (2'-*O*-methylation or pseudouridylation) were recently developed. This has opened a vast field of research on structure–function relationships of the sRNP components. We present here an efficient method for *in vitro* reconstitution of H/ACA sRNPs with an active RNA-guided pseudouridylation activity. They are assembled from an *in vitro* transcribed H/ACA sRNA with recombinant L7Ae, aCBF5, aNOP10, and aGAR1 proteins. The protocol to test the activity of the assembled particles by the method of the nearest neighbor is also described. The combination of *in vitro* assembly of H/ACA sRNPs together with time course analysis of pseudouridine formation in the target RNA is useful for structure–function analysis of H/ACA sRNPs and also to identify the target sites of H/ACA sRNAs discovered by computer analysis of archaeal genomes. Furthermore, this efficient *in vitro* pseudouridylation system can be used for generation of pseudouridine residues at defined positions in RNAs.

1. INTRODUCTION

Isomerization of uridine into pseudouridine (5-ribosyluracil, Ψ) is one of the most frequent posttranscriptional modifications encountered in cellular RNAs (Rozenski *et al.*, 1999). The functional importance of Ψ in RNA folding (Davis, 1995; Newby and Greenbaum, 2002), ribosome synthesis (Badis *et al.*, 2003; King *et al.*, 2003; Mengel-Jorgensen *et al.*, 2006; Meroueh *et al.*, 2000; Sumita *et al.*, 2005), pre-mRNA splicing (Yang *et al.*, 2005; Yu *et al.*, 1998), and the specific recognition of anticodon by tRNAs (Grosjean *et al.*, 1998) is now illustrated by several experimental data.

In both Eukarya and Archaea, U to Ψ conversions in RNAs, which are termed pseudouridylations, can be catalyzed by a single protein with a specific RNA:Ψ-synthase activity or by an H/ACA ribonucleoprotein complex (RNP). These RNPs contain at least four proteins associated with a small RNA. The RNA component is named H/ACA snoRNA or H/ACA scaRNA in Eukarya, depending on its nuclear localization (nucleolus or Cajal bodies, respectively), and H/ACA sRNA in Archaea (Bachellerie *et al.*, 2002; Henras *et al.*, 2004; Kiss, 2002; Omer *et al.*, 2003). All these RNA components contain one or several elements folded into a characteristic stem-loop structure, and they carry conserved ANA trinucleotide sequences. The RNA component of the RNPs functions as a guide RNA that base pairs with the target RNA. Through this interaction, the guide RNA defines the position to be modified by the RNA:Ψ-synthase activity, that is carried by one of its protein partners (Fig. 16.1). A high degree of conservation exists between the eukaryal and the archaeal

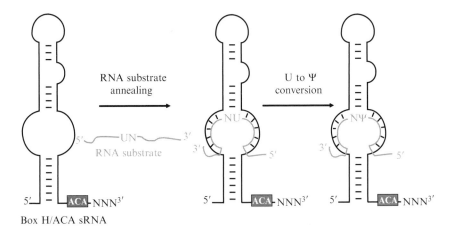

Box H/ACA sRNA

Figure 16.1   Generic scheme explaining the RNA-guided function of H/ACA sRNAs. Two segments (the antisense elements) of the internal loop present in the irregular stem-loop structure of the sRNA pseudouridylation pocket base pair with the sRNA substrate. A dinucleotide 5′-UN-3′ is single stranded, and the U residue is converted in a Ψ residue by the RNA:Ψ-synthase activity carried by the aCBF5 protein component of the sRNP. (See color insert.)

H/ACA RNPs. They all contain a protein that belongs to the TruB family of RNA:Ψ-synthases, NAP57/CBF5/Dyskerin in eukaryal H/ACA snoRNPs/scaRNPs, and its counterpart aCBF5 in Archaea. They contain three other proteins, NOP10, GAR1, and NHP2, in Eukarya and their homologs, aNOP10, aGAR1, and the L7Ae, in Archaea. Protein L7Ae is, in addition, a ribosomal protein. In Eukarya, as well as in Archaea, the characteristic stem-loop structures of the H/ACA guide RNAs are essential for their function. Their internal loop is complementary to the target RNA (Fig. 16.1) and is called the pseudouridylation pocket (Ganot *et al.*, 1997; for review, Bachellerie *et al.*, 2002). A pair of residues remains unpaired in the RNA duplex. The targeted residue is the 5′ residue in this pair. H/ACA snoRNAs most frequently contain two stem-loop structures that guide modification at two distinct positions in rRNAs that are linked by a hinge sequence carrying the ANANNA sequence (the H box). The ACA sequence is located downstream from the second stem-loop structure. It is always found three nucleotides upstream of the snoRNA 3′ end. In Archaea, the sRNA structure is more variable: it can include one, two, or even three stem-loop structures containing a pseudouridylation pocket. Here also, each of these stem-loop structures is flanked by an ANA trinucleotide (more generally ACA).

In addition to the importance of Ψ formation, the H/ACA RNP architecture is important by itself (Meier, 2005). Indeed, some of the box H/ACA snoRNPs are essential for the maturation of the pre-rRNA

(Bachellerie *et al.*, 2002). Furthermore, the vertebrate telomerase RNA, which is required for the synthesis of chromosome extremities, contains an H/ACA snoRNA domain (Mitchell *et al.*, 1999). This domain binds the H/ACA snoRNP proteins (Dez *et al.*, 2001; Pogacic *et al.*, 2000). Mutations in the gene encoding the telomerase RNA or the NAP57/CBF5 enzyme lead to bone marrow failure dyskeratosis congenita (Armanios *et al.*, 2005; Heiss *et al.*, 1998; Vulliamy *et al.*, 2001). This stresses the necessity to investigate the structure of the H/ACA RNP domains. Unfortunately, reconstitution of eukaryal snoRNPs is difficult to achieve mainly because of the low solubility of the eukaryal snoRNP proteins produced as recombinant proteins (Wang and Meier, 2004).

However, active archaeal H/ACA sRNPs can be reconstituted by use of purified recombinant proteins and an *in vitro* transcribed sRNAs (Baker *et al.*, 2005; Charpentier *et al.*, 2005). This opens the possibility of studying structure–function relationships for the various components of these RNPs. Information was already gained by use of this reconstitution system. It was shown that the ACA trinucleotide of archaea H/ACA sRNAs is required for H/ACA sRNPs assembly (Baker *et al.*, 2005; Charpentier *et al.*, 2005). In particular, it is needed for association of protein aCBF5 with the guide sRNA. Protein aCBF5 is required for protein aNOP10 recruitment in the sRNP (Baker *et al.*, 2005; Charpentier *et al.*, 2005). This minimal particle containing only two of the sRNP proteins has some RNA-guided RNA:Ψ-synthase activity (Charpentier *et al.*, 2005). However, the kinetics of the pseudouridylation is markedly increased by addition of the L7Ae and aGAR1 recombinant proteins (Charpentier *et al.*, 2005). When the full set of recombinant proteins is used, the targeted U residue can be totally converted into a Ψ residue in the substrate RNA within 10 min.

The 3D structures of components and, more recently, of a fully assembled sRNP (a H/ACA sRNA and the four proteins) were very recently established (Hamma *et al.*, 2005; Khanna *et al.*, 2006; Li and Ye, 2006; Manival *et al.*, 2006; Rashid *et al.*, 2006); several hypotheses formulated on the basis of these 3D structures can now be tested by use of the *in vitro* reconstitution system. Testing the effect of mutations in the H/ACA sRNA and/or the proteins will provide a deep understanding of the catalytic mechanism of H/ACA sRNPs and also of the relative roles of the various components of the H/ACA sRNPs in the assembly of catalytically active particles. Information will also be gained on the molecular basis for target specificity of H/ACA sRNPs. Noticeably also, the *in vitro* assembly system offers the possibility of defining the specificity of H/ACA sRNAs identified by computational analyses of archaeal genomes (see Chapter 15 by Muller *et al.*). Indeed, computer analyses now allow the identification of putative H/ACA sRNA genes in archaeal genomes with a high degree of reliability (see Chapter 15 by Muller *et al.*). However, it is often more difficult to predict their target site with accuracy. The *in vitro* assembly system described in this chapter can be used to test the activity of the H/ACA sRNA on its putative targets after production of the putative H/ACA sRNA and its putative target RNA sequences by *in vitro* transcription and of the H/ACA sRNP proteins as soluble recombinant proteins in *E. coli*.

2. OVERVIEW OF THE ISSUES THAT CAN BE ADDRESSED BY THE METHOD

- Structure–function analysis of the protein and RNA components of H/ACA sRNPs
- Investigation on the binding of a target RNA (designated as the RNA substrate) to the H/ACA sRNPs
- Investigation of the catalytic mechanism of the H/ACA sRNPs
- Identification of the RNA substrate for H/ACA sRNAs isolated by biochemical or by computational approaches

In addition, because RNA-directed Ψ formation is very efficient in this *in vitro* system, an important use of this method can be the production of RNAs with Ψ residues at defined positions. This application will require the design of sRNA guides adapted to the RNA substrate. This will be facilitated by the data accumulated in the course of the structure–function analysis of H/ACA sRNPs.

3. THE BASIC PRINCIPLES OF THE EXPERIMENTS

RNPs formed on incubation of an sRNA with the proteins are characterized by electrophoresis mobility shift assays (EMSA). For instance, we characterized various sRNPs assembled on incubation of an H/ACA sRNA (Pab91) and different sets of proteins (aCBF5, L7Ae, aNOP10). Their coding sequences were PCR amplified from the archaeon *P. abyssi* genomic DNA (Charpentier *et al.*, 2005). Examples of EMSA experiments performed with this experimental system are shown in Fig. 16.2B and C, as a reference for scientists who want to compare their data with the ones of these reconstitution experiments.

The RNA:Ψ-synthase activity of the assembled sRNPs is measured by the nearest neighbor approach by use of thin-layer chromatography (TLC) for fractionation of the 3′-labeled mononucleotides. Residue Ψp has a distinct mobility compared with Up in the TLC system used (Keith, 1995) (examples are shown in Fig.16.5).

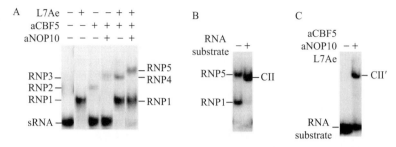

Figure 16.2 Examples of sRNP assembly experiments and RNA substrate binding by use of components of the Pab91 H/ACA sRNP from *P. abyssi*. (A) sRNPs were assembled by incubation of components of the *P. abyssi* archaeon: the radiolabeled guide sRNA Pab91 was incubated with either L7Ae, aCBF5, L7Ae-aCBF5, aCBF5-aNOP10, or L7Ae-aCBF5-aNOP10 proteins to obtain the RNP1, RNP2, RNP3, RNP4, and RNP5 complexes, respectively. The complexes are fractionated by electrophoresis mobility shift analysis (EMSA). (B) In the presence of an unlabeled RNA fragment which sequence is complementary to the pseudouridylation pocket of the sRNA Pab91, a unique CII complex is formed. (C) A unique CII′ complex is also obtained on incubation of a radiolabeled RNA substrate with the unlabeled sRNA and the three proteins L7Ae, aCBF5, and aNOP10.

 ## 4. Materials and Reagents

4.1. Chemicals

The four ribonucleotides triphosphate ATP, GTP, CTP, UTP (100 mM each, Fermentas).
[γ-^{32}P]ATP (10 mCi/ml, 3000 Ci/mmol; Amersham Biosciences) and [α-^{32}P]NTP (10 or 20 mCi/ml, 800 Ci/mmol; Amersham Biosciences)

4.2. General reagents

RNase-free water obtained from an ionization system (e.g., Milli-Q), sterilized and filtered through a 0.22-μm membrane filter (Millex, Millipore)
Glycerol 87% (Merck)
A phenol/chloroform solution
Na-Acetate (0.3 M)
RNasin (Amersham Biosciences)
Total yeast tRNAs (0.5 μg/μl, Roche)
A 40% Acrylamide/Bis-acrylamide solution (38:2)

4.3. Equipment

16 × 15 cm plates for vertical electrophoresis and 1- and 1.5-mm thick spacers

Plastic sheets precoated by a thin layer (0.1 mm) of cellulose (Macherey-Nagel)

The acrylamide gels used for electrophoresis mobility shift assays (EMSA) and the cellulose sheets used for 1D and 2D TLC were visualized and quantified on a PhosphorImager (Typhoon 9410, Amersham Biosciences) with the ImageQuant software.

4.4. Enzymes

Commercial T7 RNA polymerase (20 U/μl, Ambion) or purified from *E. coli* BL21 strain (Zawadzki and Gross, 1991)

Calf intestinal alkaline phosphatase (CIP, MBI-Fermentas, 0.1 U/μl) and its 10× reaction buffer (100 mM MgCl$_2$, 100 mM Tris-HCl, pH 7.5)

T4 Polynucleotide kinase (PNK, MBI-Fermentas 10 U/μl) and its 10× reaction buffer (100 mM MgCl$_2$, 50 mM DTT, 1 mM spermidine, 1 mM EDTA, 500 mM Tris-HCl, pH 7.6)

RNAse-free DNase I (10 U/μl, Amersham Biosciences)

RNase T2 (Invitrogen, 0.1 U/μl)

4.5. Buffers

20× TBE: 1780 mM borate, 40 mM EDTA, 1780 mM Tris-borate, pH 8.3

3× transcription buffer (TrB): 60 mM MgCl$_2$, 400 mM HEPES-KOH, pH 7.6

10× Ambion transcription buffer: 100 mM DTT, 20 mM spermidine, 60 mM MgCl$_2$, 400 mM Tris-HCl, pH 7.9

10× buffer D: 150 mM KCl, 1.5 mM MgCl$_2$, 0.2 mM EDTA, 20 mM HEPES, pH 7.9

Denaturing gel loading solution (i.e., formamide dye): 20 mM EDTA, 0.05% bromophenol blue, 0.05% xylene cyanol in deionized formamide (Merck)

Elution buffer (EB): 1 mM EDTA, 300 mM NaCl, 1% SDS, 10 mM Tris-HCl, pH 7.5

Complex separation loading dye (Cs): 40% glycerol, 0.03% bromophenol blue, 0.03% xylene cyanol, 20 mM HEPES-KOH, pH 7.9

EMSA running buffer: 115 ml glycerol 87%, 50 ml 20× TBE, total volume 2 liters

2× RNase T2 buffer: 100 mM NH$_4$-Ac, pH 4.6

Solvent N1 for 1D TLC: 50 ml isobutyrate, 1.1 ml ammoniac 25%, H$_2$O to 80 ml

Solvent N2 for the second dimension in 2D TLC: 70 ml isopropanol, 15 ml HCl 37%, H$_2$O to 100 ml

5. METHODS

5.1. Production/purification of the archaeal proteins aCBF5, aNOP10, L7Ae, and aGAR1

The ORFs encoding the archaeal proteins aCBF5, aNOP10, L7Ae, and aGAR1 are PCR amplified from the genomic DNA. As an example, we will describe the application of the methods to the *P. abyssi* archaeon. However, they are of general use and can be applied to other archaeons. The ORFs with GenBank accession nos. CAB49444 (aCBF5), CAB49761 (aNOP10), C75109 (L7Ae), and CAB49230 (aGAR1) were PCR amplified from the *P. abyssi* GE5 strain genomic DNA (Charpentier *et al.*, 2005).

The PCR-amplified DNA fragment can be cloned into an *E. coli* expression vector. We used plasmid pGEX-6P-1 (Pharmacia) to produce GST fusion proteins. M. Terns and colleagues produced the proteins from *P. furiosus* as His-tagged proteins (Baker *et al.*, 2005). The recombinant GST-aCBF5, GST-aNOP10, GST-L7Ae, and GST-aGAR1 proteins are routinely produced in *E. coli* BL21 CodonPlus cells (Novagen) and purified under native conditions in the PBS buffer (140 mM NaCl, 2.7 mM KCl, 10 mM Na$_2$HPO$_4$, 1.8 mM KH$_2$PO$_4$, pH 7.3) with glutathione-Sepharose 4B (Pharmacia) as previously described (Charron *et al.*, 2004). The GST part of the fusion proteins are directly cleaved on beads with 80 U of PreScission protease (Pharmacia) per ml of glutathione-Sepharose bead suspension. Cleavage is performed overnight at 4° followed by a 15 min incubation at 65°. The contaminant *E. coli* proteins are precipitated in these conditions, and they are eliminated by a 20-min centrifugation at 16,000g. The quality of the protein preparation is assessed by SDS–PAGE and Coomassie staining. Absorbance at 280 nm (A$_{280\,nm}$) is measured for each sample, and the protein concentration is calculated by use of their theoretical ε_{280nm}. The proteins are stored at −80° at a 4.5 μM concentration in 150 mM NaCl, 1 mM EDTA, 1 mM DTT, glycerol 10%, 50 mM Tris-HCl, pH 7.

5.2. Production of the box H/ACA guide sRNAs

The guide sRNAs are synthesized by T7 RNA polymerase with, as the template, PCR-amplified DNA fragments containing the H/ACA sRNA coding sequence under the control of a T7 RNA polymerase promoter (Fig. 16.3). These fragments are obtained by PCR amplification by use of an archaeal genomic DNA as the template and as the forward primer, an oligonucleotide carrying at its 5′ extremity a T7 RNA polymerase promoter sequence (5′ TAATACGACTCACTATAGGG 3′). Alternately, the DNA sequence of an sRNA can be obtained by use of two complementary oligonucleotides as the template for PCR amplification. The sequence of

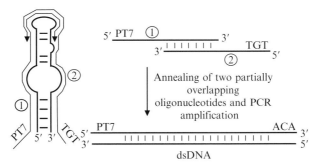

Figure 16.3 Strategy used to obtain the DNA fragment for T7 *in vitro* transcription of a H/ACA sRNA with a single hairpin structure (see text for details).

the T7 RNA polymerase promoter is included in the left template oligonucleotide (named 1 in Fig. 16.3), or it is introduced by a forward primer in a second PCR step. This second approach for production of the template RNA sequence opens the possibility to introduce mutations within the RNA sequence. Classical protocols are used for the PCR amplifications. Purification of the amplified fragments is not necessary for subsequent transcription with the T7 RNA polymerase.

The transcription reaction is performed in a 30-μl solution obtained by mixing 9 μl of H_2O, 3 μl of 10× transcription buffer, 10 μl of a mix of the four NTPs (12.5 mM each), 6 μl of the PCR reaction, 1 μl of Rnasin, and 1 μl of T7 RNA polymerase (20 U). When T7 RNA polymerase purified from *E. coli* is used, the reaction is performed in 1× TrB buffer. Incubation is performed at 37° for at least 4 h; routinely incubation is done overnight. Then, the DNA template is hydrolyzed by 1 μl of DNase I (10 U) at 37° for 1 h. After addition of 15 μl of the denaturing gel loading solution and heating at 95° for 2 min, the products of the transcription reaction are fractionated on a 1-mm thick, 8% denaturing polyacrylamide-8M urea gel prepared in 1× TBE buffer. RNAs are visualized by UV shadowing, the gel slices are cut with a razor blade, and the RNA molecules are eluted from the gel piece by an overnight incubation in 300 μl of elution buffer EB at 4°. The elution solution is collected and transferred into a microcentrifuge tube. The RNA transcripts are precipitated at −80° after addition of 1 ml of ethanol. The precipitated RNAs are collected by microcentrifugation at top speed for 10 min at 4°, washed with ethanol 70%, and dried by leaving the tube open on the bench. The dry pellet is dissolved in 20 μl of H_2O. The RNA concentration is quantified by measurement of the $A_{260\,nm}$. Milli-Q H_2O is added to get a 200 ng/μl concentration and RNA is stored at −80°.

Binding of the proteins on the sRNAs is analyzed by EMSA (see later) by use of a radiolabeled sRNA. To this end, the unlabeled sRNA transcripts

prepared as previously described are labeled at their 5′ end after dephosphorylation of their 5′ triphosphate extremities by calf intestinal phosphatase
(CIP). Dephosphorylated sRNAs are prepared by the following procedure:
5 μl (1 μg) of sRNA is treated by 1 μl of CIP enzyme (0.1 U) in 10 μl of the
1× CIP buffer for 1 h at 37°. The reaction is stopped by addition of 150 μl
of Na-Ac (0.3 *M*), followed by a phenol-chloroform extraction, and the
dephosphorylated RNAs are ethanol precipitated. The dried pellets are
dissolved in 10 μl of Milli-Q H_2O and stored at −80°.

For 5′-end labeling, 2.5 μl (250 ng) of the dephosphorylated RNA
preparation are mixed with 1 μl of 10× PNK buffer, 1 μl of [γ-^{32}P]ATP
(10 mCi/ml, 3000 Ci/mmol), 1 μl of T4 PNK (10 U/μl), and 4.5 μl H_2O
(total volume 10 μl), and incubated for 45 min at 37°. The reaction is
stopped by addition of 10 μl of denaturing gel loading solution. After heating
at 96° for 2 min, the sample is loaded on a 1-mm thick 8% polyacrylamide-
8 *M* urea denaturing gel. The radiolabeled band is cut, the 5′-end labeled
sRNA is eluted overnight at 4° with the procedure described previously,
followed by ethanol precipitation. The pellet is dissolved in Milli-Q H_2O to
get 10,000 cpm/μl, and the labeled RNA is stored at −20°.

5.3. Production of the RNA substrate

The unlabeled RNA substrate, complementary to the sRNA pseudouridylation pocket (Fig. 16.1), is synthesized by T7 RNA polymerase with the
same protocol as for sRNA production. For RNA:Ψ-synthase activity
measurements, the RNA substrate needs to be labeled. This labeling is
obtained by incorporation of a radiolabeled NTP during the transcription
reaction. The choice of the labeled nucleotide depends on the sequence of
the RNA substrate. To demonstrate the occurrence of the U to Ψ conversion in the substrate by the nearest-neighbor method on the basis of RNase
T2 digestion, a labeled phosphorus atom should be present 3′ to the targeted
U residue (Fig. 16.4). Thus, the identity of the residue 3′ to the targeted U
residue defines the identity of the NTP to be used in the labeling reaction.
For instance, in the case of the *P. abyssi* Pab91 sRNA, the needed labeling
was obtained by incorporation of [α-^{32}P]CTP in the course of transcription
(Fig. 16.4). The following conditions are used for transcription: the reaction
is performed by mixing 5 μl of 3× TrB buffer, 0.5 μl of RNasin, 3 μl of
a mix of the ATP, UTP, GTP nucleotides (20 m*M* each), 1 μl of CTP
(2 m*M*), 1 μl of [α-^{32}P]CTP (800 Ci/mmol), 1.5 μl of BSA (2 μg/μl), 1 μl
of the PCR reaction, and 1.5 μl of the T7 RNA polymerase. Incubation is
performed at 37° for 2 h, and the DNA is digested with 10 U of DNase I for
1 h. The labeled transcripts are fractionated by electrophoresis on the same
type of polyacrylamide denaturing gel as described previously and visualized
by autoradiography. The radiolabeled bands are cut, and the labeled RNAs

RNA substrate labeled with [α-^{32}P]CTP in the course
of *in vitro* transcription

5′ pppGpAp*Cp*CpGp^{U}p*CpGpUpGpApAp*Cp*CpGp*C 3′

| Incubation with proteins
▼ and the H/ACA sRNA

5′ pppGpAp*Cp*CpGp$^\Psi$p*CpGpUpGpApAp*Cp*CpGp*C 3′

| RNase T2 digestion

3′-labeled mononucleotides   2 Ap*   1 Cp*   1 Gp*   1 Up* 1 Ψp*
unlabeled mononucleotides
                1 Ap   3 Cp   4 Gp   1 Up
                     1 C   1 pppGp

| 2D Thin Layer Chromatography (TLC)

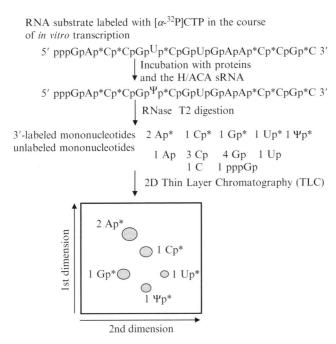

Figure 16.4   Detection of Ψ residues on a 2D TLC by the technique of the nearest neighbors. The radiolabeled phosphates of the RNA substrate sequence are marked by an asterisk.

are eluted overnight as described previously. After ethanol precipitation, RNAs are diluted in Milli-Q H$_2$O to get 10,000 cpm/μl.

5.4. Assembly of box H/ACA sRNPs and their characterization by EMSA

To avoid nonspecific binding, a mixture of yeast tRNAs is used as competitor. Thus, the ^{32}P-labeled sRNA (10,000 cpm, approximately 50 fmoles) together with 500 ng of yeast tRNA mixture are added in 1× D buffer in a final volume of 3.5 μl. The RNAs are denatured by a 5-min incubation at 65°. After a brief centrifugation, they are renatured by slow cooling at room temperature. The mixture of the recombinant proteins aCBF5, aNOP10, L7Ae, and aGAR1 at a 900 n*M* concentration each is prepared in 1× D buffer. RNPs are formed by adding 1 μl of the protein mixture to the 3.5-μl RNA solution (each recombinant protein is at a 200 n*M* final concentration). After incubation for 10 min at 65°, 4 μl of complex separation loading dye (Cs) is added. The formation of RNA–protein complexes is tested by electrophoresis at room temperature in 1.5-mm thick 6% nondenaturing polyacrylamide gels. For 40 ml of gel solution, 1 ml of 20× TBE is mixed with 2.3 ml of 87% glycerol, 6 ml of 40% acrylamide/bisacrylamide (38/2), and Milli-Q H_2O is added to 40 ml. Electrophoresis is carried out at 6.5 V/cm for 2 h.

5.5. Analysis of the formation of an RNA duplex between the guide sRNA and the RNA substrate

The association of the sRNP with the target RNA can be tested by two alternative procedures. In a first procedure, the guide sRNA is labeled, and the RNA substrate is unlabeled. This unlabeled substrate is added to the RNP assembly reaction described in the preceding section. The quantity of RNA substrate required to observe a complete shift of the electrophoretic mobility of the assembled RNP is variable, depending on the RNA guide and the substrate used. For the Pab91 sRNP, 15 ng ($\sim$2 pmol) of target RNA is used. Incubation and analysis by EMSA are carried out as described in the preceding section (Fig. 16.2B).

In the second procedure, the labeled RNA is the RNA substrate. The RNP assembly is performed with an unlabeled sRNA, and thus binding of its labeled target RNA is monitored (Fig. 16.2C). In that case, the RNP is assembled by combining in 1$\times$ D buffer, 200 ng of unlabeled sRNA ($\sim$10 pmol), 500 ng of tRNAs, and 200 nM of each protein in the presence of the labeled RNA substrate (10,000 cpm, $\sim$150 fmoles). The RNP/RNA substrate complex is resolved as described previously by gel electrophoresis in nondenaturing conditions.

5.6. Measurement of the RNA:Ψ-synthase activity of the reconstituted sRNPs

The RNA:Ψ-synthase activities of the reconstituted sRNPs are tested by the nearest-neighbor approach (Fig. 16.4). The substrate RNA, uniformly labeled by the strategy explained in the section "Production of the RNA Substrate," is incubated with the modification machinery (the guide sRNA and the proteins). The incubation is performed at 65°, which corresponds to the optimal temperature for the Ψ-synthase activity of reconstituted archaeal H/ACA sRNPs (data not shown). The RNA substrate is next totally digested by an RNase to get labeled 3'-mononucleotides that can be fractionated by 2D or 1D TLC.

The unlabeled sRNA (10 pmoles, $\sim$2.2 mM) and ^{32}P-labeled target RNA (150 fmoles, $\sim$30 nM) are mixed at room temperature in buffer D and treated as described for EMSA assays. The proteins (200 nM each) are added at room temperature. The final 4.5-μl solution is incubated at 65°. For time course experiments, the reaction volume is scaled up to 45 μl, by use of the same RNA and protein concentrations, and the reaction is started

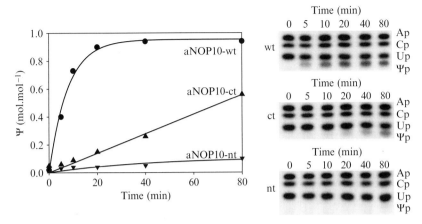

Figure 16.5 Time course analysis of Ψ formation in the RNA substrate of the Pab91 sRNA. The unlabeled Pab91 sRNA was mixed with the labeled RNA substrate and the four proteins L7Ae, aCBF5, aNOP10, and aGAR1. The amounts of Ψ residues formed were estimated by 1D TLC. They are expressed in moles of Ψ residue formed per moles of RNA substrate. Variants of the aNOP10 protein were tested: either a fragment carrying the N-terminal residues (aNOP10-nt) or the C-terminal residues (aNOP10-ct) was tested.

by addition of the proteins after incubation of the RNA solution at 65° for 10 min. At time intervals, 4.5-μl aliquots are collected. For each aliquot, the reaction is stopped by the addition of 150 μl of Na-Ac (0.3 M), 2 μg tRNAs, which is immediately followed by extraction with 150 μl of phenol/chloroform (1:1). The extracted RNAs are ethanol precipitated.

The dried pellet of the modified RNA targets is dissolved in 8 μl of $1\times$ T2 buffer containing 0.4 U of T2 RNase (0.1 U/μl) for 12 h at 37°. The resulting 3'-mononucleotides (3 μl) are chromatographed on thin-layer cellulose plates as described elsewhere (Keith, 1995). For reactions performed on RNA substrates with no 3'-labeled guanosine, 1D chromatography in solvent N1 is sufficient to fractionate the other labeled 3'-mononucleotides (Fig. 16.5). If 3'-labeled G residues are obtained after RNase T2 digestion, a 2D TLC fractionation is required, and migration in the second dimension is performed in solvent N2.

5.7. Quantification of the data

The dried gels of the EMSA experiments and the plates used for TLC are analyzed with a PhosphorImager (Typhoon 9410, Amersham Biosciences). The amounts of radioactivity in the bands of gels and in the spots are estimated with the ImageQuant software.

5.7.1. Quantification of the amount of assembled sRNP

The percentage of RNA in each RNP is calculated, taking into account the percentage of radioactivity in each band on the gel relative to the total amount of radioactivity in the lane. As can be seen in Fig. 16.2A, >90% of the Pab91 sRNA is shifted into a unique RNP1 complex in the presence of protein L7Ae. Incubation of this sRNA with protein aCBF5 leads to formation at a low yield of the RNP2 complex (~7%). The direct binding of aCBF5 is more efficient on other sRNAs with yields of RNP2 formation >50% (Fourmann *et al.*, unpublished results). The Pab91 sRNA is shifted partially into a complex of low electrophoretic mobility (RNP5) on its incubation with the three proteins aCBF5, aNOP10, and L7Ae. The residual RNP1 complex remains the more abundant (>50%). Nevertheless, complex RNP5 can be formed at higher yields (>80%) with other guide sRNAs (Fourmann *et al.*, unpublished results).

In the presence of the unlabeled RNA substrate, a unique complex we named CXII is formed (Fig. 16.2B). We named CII' the complex detected on incubation of the proteins, the unlabeled guide sRNA, and the radiolabeled RNA substrate (Fig. 16.2C). In various experiments, we found that the amount of this complex was variable with a maximal value of ~40%.

5.7.2. Quantification of the Ψ formed

A calculation table, prepared with the classical Excel software, can be used to determine the number of moles of Ψ formed per mole of RNA substrate. A table with 5 columns (A, B, C, D, and E) and 5 lanes (1, 2, 3, 4, and 5) is built (Fig. 16.6). This table is filled for each assay in the time course analysis of Ψ formation. The number of 3' end-labeled nucleotides (A, C, G, U, and Ψ), which are expected to be obtained after complete digestion of the RNA substrate by the T2 RNase, are used to fill the cells of the B column. For the example given in Fig. 16.4, the values are of 2 for cell B1, and 1 for the B2, B3, and B4 cells. A value of 1 is used for the expected number of Ψ nucleotide. The values experimentally obtained by quantification of the TLC spots by the ImageQuant software are used to fill the cells in column C: for example, the value of the quantification of the Ap spot is written in cell C2. The D column will give the ratios between the expected nucleotide numbers and the measured radioactivity (Cx/Dx). When incorporation of radioactive residues in the substrate has occurred uniformly, the C1/B1, C2/B2, C3/B3, and C4/B4 values should be very similar. Discrepancies between these values in column D may reflect a bias in the labeling reaction and/or some degradation of the RNA in the course of the experiment. Therefore, it is very important to take these values into consideration.

The number of moles of Ψ per mole of RNA substrate is calculated as the mean value determined by taking into account the B and C values for each nucleotide. To this end, for each determination, the C5 value is

	A	B	C	D	E
	Nucleotides	Number of 3′ labeled nucleotides (1)	Values of the Phosphor Imaging quantification (2)	Relative quantity	Yield of Ψ formation (3)
1	A	B1	C1	C1/B1	E1 = C5 × (B1/C1)
2	C	B2	C2	C2/B2	E2 = C5 × (B2/C2)
3	G	B3	C3	C3/B3	E3 = C5 × (B3/C3)
4	U	B4	C4	C4/B4	E4 = C5 × (B4/C4)
5	Ψ	1	C5		E5 = (E1 + E2 + E3 + E4)/4

(1) Number of 3′ end-labeled nucleotides calculated from the nucleotide sequence of the RNA substrate taking into account the identity of the labeled nucleotide used for the *in vitro* transcription. The number of Ψ written in this cell is 1, based on the assumption that only one U residue is specifically modified within the RNA substrate sequence.

(2) Values of the PhosphorImager quantification, corrected for background.
Values are given in Volume by the Molecular analyst software.

(3) The value in moles of Ψ formed per moles of RNA substrate is obtained in cell E5. If only one Ψ is present in the RNA substrate sequence, the maximal values in this column cannot be higher than ~1 for each labeled nucleotide.

Figure 16.6 Calculation table in Excel format used for the determination of the amount of Ψ formed.

multiplied by the theoretical B value for one given nucleotide and divided by the experimental C value measured for this residue. One E value is thus obtained (column E), and then the mean value of the E numbers is calculated in cell E5.

In the conditions used in the experiment displayed in Fig. 16.5, ~90% of the RNA substrate molecules are modified after 20 min of incubation.

6. GENERAL COMMENTS

6.1. Choice of the size of the RNA substrate

We tested the effect of the RNA substrate length on the activity of the reconstituted Pab91 H/ACA sRNP. We found that addition of 19nt at the 3′ extremities of the sequences complementary to the antisense element of the guide RNA does not change the amount of Ψ residue formed after a 1-h incubation. However the kinetics of the reaction is slower in the presence of the extensions (Charpentier *et al.*, 2005).

6.2. No activity is detected

The binding of each protein should be tested by EMSA, in particular the binding of proteins L7Ae and aCBF5 that bind the sRNA directly (Fig. 16.2A).

ACKNOWLEDGMENTS

J.-B. F. is a fellow from the French Ministère Délégué à l'Enseignement Supérieur et à la Recherche. The work was supported by the Centre National de la Recherche Scientifique (CNRS), the French Ministère de la Recherche et des Nouvelles Technologies (MRNT), the Actions Concertées Incitatives (ACI) Microbiologie and BCMS of the French MRNT, and the Bioingénierie PRST of the Conseil Régional Lorrain.

REFERENCES

Armanios, M., Chen, J. L., Chang, Y. P., Brodsky, R. A., Hawkins, A., Griffin, C. A., Eshleman, J. R., Cohen, A. R., Chakravarti, A., Hamosh, A., and Greider, C. W. (2005). Haploinsufficiency of telomerase reverse transcriptase leads to anticipation in autosomal dominant dyskeratosis congenita. *Proc. Natl. Acad. Sci. USA* **102,** 15960–15964.

Bachellerie, J. P., Cavaille, J., and Huttenhofer, A. (2002). The expanding snoRNA world. *Biochimie* **84,** 775–790.

Badis, G., Fromont-Racine, M., and Jacquier, A. (2003). A snoRNA that guides the two most conserved pseudouridine modifications within rRNA confers a growth advantage in yeast. *RNA* **9,** 771–779.

Baker, D. L., Youssef, O. A., Chastkofsky, M. I., Dy, D. A., Terns, R. M., and Terns, M. P. (2005). RNA-guided RNA modification: Functional organization of the archaeal H/ACA RNP. *Genes Dev.* **19,** 1238–1248.

Charpentier, B., Muller, S., and Branlant, C. (2005). Reconstitution of archaeal H/ACA small ribonucleoprotein complexes active in pseudouridylation. *Nucleic Acids Res.* **33,** 3133–3144.

Charron, C., Manival, X., Charpentier, B., Branlant, C., and Aubry, A. (2004). Purification, crystallization preliminary X-ray diffraction data of L7Ae sRNP core protein from *Pyrococcus abyssi. Acta Crystallogr. D Biol. Crystallogr.* **60,** 122–124.

Davis, D. R. (1995). Stabilization of RNA stacking by pseudouridine. *Nucleic Acids Res.* **23,** 5020–5026.

Dez, C., Henras, A., Faucon, B., Lafontaine, D., Caizergues-Ferrer, M., and Henry, Y. (2001). Stable expression in yeast of the mature form of human telomerase RNA depends on its association with the box H/ACA small nucleolar RNP proteins Cbf5p, Nhp2p and Nop10p. *Nucleic Acids Res.* **29,** 598–603.

Ganot, P., Bortolin, M. L., and Kiss, T. (1997). Site-specific pseudouridine formation in preribosomal RNA is guided by small nucleolar RNAs. *Cell* **89,** 799–809.

Hamma, T., Reichow, S. L., Varani, G., and Ferre-D'Amare, A. R. (2005). The Cbf5-Nop10 complex is a molecular bracket that organizes box H/ACA RNPs. *Nat. Struct. Mol. Biol.* **12,** 1101–1107.

Heiss, N. S., Knight, S. W., Vulliamy, T. J., Klauck, S. M., Wiemann, S., Mason, P. J., Poustka, A., and Dokal, I. (1998). X-linked dyskeratosis congenita is caused by mutations in a highly conserved gene with putative nucleolar functions. *Nat. Genet.* **19,** 32–38.

Henras, A. K., Dez, C., and Henry, Y. (2004). RNA structure and function in C/D and H/ACA s(no)RNPs. *Curr. Opin. Struct. Biol.* **14,** 335–343.

Keith, G. (1995). Mobilities of modified ribonucleotides on two-dimensional cellulose thin-layer chromatography. *Biochimie* **77,** 142–144.

Khanna, M., Wu, H., Johansson, C., Caizergues-Ferrer, M., and Feigon, J. (2006). Structural study of the H/ACA snoRNP components Nop10p and the 3' hairpin of U65 snoRNA. *RNA* **12,** 40–52.

King, T. H., Liu, B., McCully, R. R., and Fournier, M. J. (2003). Ribosome structure and activity are altered in cells lacking snoRNPs that form pseudouridines in the peptidyl transferase center. *Mol. Cell* **11,** 425–435.

Kiss, T. (2002). Small nucleolar RNAs: An abundant group of noncoding RNAs with diverse cellular functions. *Cell* **109,** 145–148.

Li, L., and Ye, K. (2006). Crystal structure of an H/ACA box ribonucleoprotein particle. *Nature* **443,** 302–307.

Manival, X., Charron, C., Fourmann, J. B., Godard, F., Charpentier, B., and Branlant, C. (2006). Crystal structure determination and site-directed mutagenesis of the *Pyrococcus abyssi* aCBF5-aNOP10 complex reveal crucial roles of the C-terminal domains of both proteins in H/ACA sRNP activity. *Nucleic Acids Res.* **34,** 826–839.

Meier, U. T. (2005). The many facets of H/ACA ribonucleoproteins. *Chromosoma* **114,** 1–14.

Mengel-Jorgensen, J., Jensen, S. S., Rasmussen, A., Poehlsgaard, J., Iversen, J. J., and Kirpekar, F. (2006). Modifications in *Thermus thermophilus* 23 S ribosomal RNA are centered in regions of RNA-RNA contact. *J. Biol. Chem.* **281,** 22108–22117.

Meroueh, M., Grohar, P. J., Qiu, J., SantaLucia, J., Jr., Scaringe, S. A., and Chow, C. S. (2000). Unique structural and stabilizing roles for the individual pseudouridine residues in the 1920 region of *Escherichia coli* 23S rRNA. *Nucleic Acids Res.* **28,** 2075–2083.

Mitchell, J. R., Cheng, J., and Collins, K. (1999). A box H/ACA small nucleolar RNA-like domain at the human telomerase RNA 3′ end. *Mol. Cell. Biol.* **19,** 567–576.

Newby, M. I., and Greenbaum, N. L. (2002). Investigation of Overhauser effects between pseudouridine and water protons in RNA helices. *Proc. Natl. Acad. Sci. USA* **99,** 12697–12702.

Omer, A. D., Ziesche, S., Decatur, W. A., Fournier, M. J., and Dennis, P. P. (2003). RNA-modifying machines in archaea. *Mol. Microbiol.* **48,** 617–629.

Pogacic, V., Dragon, F., and Filipowicz, W. (2000). Human H/ACA small nucleolar RNPs and telomerase share evolutionarily conserved proteins NHP2 and NOP10. *Mol. Cell Biol.* **20,** 9028–9040.

Rashid, R., Liang, B., Baker, D. L., Youssef, O. A., He, Y., Phipps, K., Terns, R. M., Terns, M. P., and Li, H. (2006). Crystal structure of a Cbf5-Nop10-Gar1 complex and implications in RNA-guided pseudouridylation and dyskeratosis congenita. *Mol. Cell* **21,** 249–260.

Rozenski, J., Crain, P. F., and McCloskey, J. A. (1999). The RNA Modification Database: 1999 update. *Nucleic Acids Res.* **27,** 196–197.

Sumita, M., Desaulniers, J. P., Chang, Y. C., Chui, H. M., Clos, L., 2nd, and Chow, C. S. (2005). Effects of nucleotide substitution and modification on the stability and structure of helix 69 from 28S rRNA. *RNA* **11,** 1420–1429.

Vulliamy, T., Marrone, A., Goldman, F., Dearlove, A., Bessler, M., Mason, P. J., and Dokal, I. (2001). The RNA component of telomerase is mutated in autosomal dominant dyskeratosis congenita. *Nature* **413,** 432–435.

Wang, C., and Meier, U. T. (2004). Architecture and assembly of mammalian H/ACA small nucleolar and telomerase ribonucleoproteins. *EMBO J.* **23,** 1857–1867.

Yang, C., McPheeters, D. S., and Yu, Y. T. (2005). Psi35 in the branch site recognition region of U2 small nuclear RNA is important for pre-mRNA splicing in *Saccharomyces cerevisiae*. *J. Biol. Chem.* **280,** 6655–6662.

Yu, Y. T., Shu, M. D., and Steitz, J. A. (1998). Modifications of U2 snRNA are required for snRNP assembly and pre-mRNA splicing. *EMBO J.* **17,** 5783–5795.

Zawadzki, V., and Gross, H. J. (1991). Rapid and simple purification of T7 RNA polymerase. *Nucleic Acids Res.* **19,** 1948.

Author Index

Subject Index

427